Crucial Coalition

Anglo-Danish Military Collaboration and the Message of History

Kjeld Hald Galster

Legacy Books Press

Published by Legacy Books Press
RPO Princess, Box 21031
445 Princess Street
Kingston, Ontario, K7L 5P5
Canada

www.legacybookspress.com

The scanning, uploading, and/or distribution of this book via the Internet or any other means without the permission of the publisher is illegal and punishable by law.

This edition first published in 2012 by Legacy Books Press
1

© 2012 Kjeld Hald Galster, all rights reserved.

Printed and bound in the United States of America and Great Britain.

This book is typeset in a Times New Roman 11-point font.

Library and Archives Canada Cataloguing in Publication

Galster, Kjeld Hald, 1952-
    Crucial coalition : Anglo-Danish military collaboration and the message of history / Kjeld Hald Galster.

Includes bibliographical references and index.
ISBN 978-0-9880192-8-7

    1. Combined operations (Military science)--History. 2. Military history. 3. Denmark--Military relations--Great Britain. 4. Great Britain--Military relations--Denmark. 5. Afghan War, 2001- --Participation, Danish. 6. Afghan War, 2001- --Campaigns--Case studies. I. Title.

U260.G34 2012   355.4'609   C2012-904317-6

This book categorically denies any involvement in putting laxatives in Stalin's coffee during the Tehran conference.

To the memory of my friend and mentor
Professor and Brigadier Richard Holmes

Table of Contents

Acknowledgements. vii

Foreword. ix

Preface. xii

Introduction. 1
    The General Context
    The Political Context: Denmark in the Twenty-First Century
    The Military Context
    Constraints
    Aim

Chapter I – Afghanistan and Foreign Power Projection. 23
    Post-British Afghanistan

Chapter II – ISAF Team 10. 49
    Fiction and Reality
    Co-operation within the Coalition
    Pre-Deployment Issues
    Intelligence
    Operations
    Afghan National Army Build-Up
    Afghan National Police Build-Up
    Public Debate
    The Downsides
    Operational Matters
    Communication
    Cultural Differences
    Discipline, Spirituality and Mental Health
    Partners and Dominance
    Disunity and Dissolution
    Experiences
    Post-Redeployment

Chapter III – Vital Partnership. 135
    Ancient Greece
    Medieval Italy

Early Modern Europe
Late Modern Era
World War I
World War II
Korea

Chapter IV – The Crucible............................... 220
    Military Aspects
    Constraints: Culture
    Coalition Partners and Dominance
    The Downsides of Coalition Warfare
    Coalition Disunity and Dissolution, Defection of Coalition Partners
    Coalition Interoperability

Chapter V – The Message of History...................... 247
    Interests
    Political Purpose
    War Aims
    Strategy
    Doctrine
    Planning
    Logistics
    Disparities
    Disagreement
    Lead-Nation
    Leadership
    Personal Understandings
    Languages
    Asymmetry
    Terrorism
    The End of Coalitions

Epilogue... 271

Biographies.. 274

Bibliography... 318

Index.. 329

About the Author....................................... 338

Acknowledgements

THIS BOOK WOULD have never gone to print had it not been for the help and suggestions by many friends and colleagues to whom I owe much of the insight laid out in the chapters.

I would like to thank Professor Doug Delaney of the Royal Military College of Canada for valuable support on global and, especially, Canadian experiences of coalitions in the past as well as those of recent years. Much credit is due to my erstwhile employer, the Royal Danish Defence College, for having generously supported my endeavours with research time and travel grants and to Brigadier Michael Clemmesen, MA and Dr Niels Bo Poulsen at the Centre for Military History of that college for having volunteered much constructive comment as the work proceeded and I needed my ideas to be vetted by competent and critical associates. Moreover, thanks are due to friends and colleagues amongst military historians at home and abroad.

I also like to thank Professor Andrew Lambert, Professor Doug Delaney, Professor Paul Dover, Dr Thomas Vogel, Dr Adam Schwartz, Dr Thomas Heine Nielsen and Captain Ivan Cadeau for their permission to quote from their conference papers presented at the Conference on Coalition Warfare held in Copenhagen in May 2011.

My sincere gratitude goes to Brigadier James Chiswell, the Commander of Task Force Helmand October 2010 to March 2011, to Mr Edward

Ferguson and Brigadier, since December 2011 Major-General, Tim Radford of the British Ministry of Defence, to Colonel Lennie Fredskov, Commanding Officer of the Danish ISAF Team 10, the chief of the operations branch Major Christian Bach Byrholt, the chief of intelligence Captain Thomas Larsen, the deputy commander Lieutenant-Colonel Thomas Funch Pedersen, the officer commanding B Company, Major Michael C. Toft, Major Kaj Vincent Frederick Ahlefeldt-Laurvig and Major Kim Kristensen for most kindly giving me time and candidly expressing their views in interviews and correspondence during the months from summer 2010 to spring 2011.

Finally, I wish to thank Lieutenant-Colonel Crispin Lockhart and Ms Sonja Hall of the British Ministry of Defence for valuable support and kind co-operation as well as Major Peter Dahl and Captain Lars Kristensen of the Danish Army Operational Command for their generous provision of information and extensive assistance en route through the massive material as well as for vetting the chapter on operations in Afghanistan.

Foreword

COALITION WARFARE IS not only a phenomenon with deep historical roots dating back to antiquity; it is also a "must" in contemporary strategies of most states today. Save a handful of bilateral conflicts, virtually all military action in the current global security environment takes place within the framework of a coalition or an alliance. Recent examples are the ongoing anti-piracy endeavour off the Horn of Africa, the enforcement of a no-fly zone over Libya in spring 2011 and ISAF's support of the Afghan government. All these examples involve/have involved Danish force contributions, and it is most likely that also in the future the Danish Defence will be part of such coalitions. This is true also for the United States, Canada, the UK and many others. In order to prepare for this, it is imperative to examine the historical record of coalition warfare and identify the lessons of the past. This is exactly what Dr Kjeld Galster – a now retired senior researcher at the Royal Danish Defence College and a former Royal Life Guards officer – does in this book.

Politically, coalitions (especially when acting under a UN mandate) are important as a way of securing international support and legitimacy. Militarily, coalitions are seen by many as especially attractive to small states. However, large, technologically advanced states may also gain from entering into a coalition, and not only due to political expediency. Specialised assets, intelligence, bases or additional troops may be offered by the junior partner(s).

Despite their many advantages, military coalitions are also fraught with problems: Coalition partners may have differing objectives, the rules of engagement, doctrines and equipment may be so diverse that they hamper the operational and tactical efficiency, and while coalitions may serve as a bridge between different cultures and societies, they may under certain circumstances also serve to amplify existing prejudices and disagreements.

Dr Galster's book combines a historical enquiry and a strategic perspective on the subject of coalition warfare with a deep scrutiny of a specific case of current coalition warfare viewed "bottom-up," namely Anglo-Danish co-operation within the framework of ISAF – The International Security Assistance Force in Afghanistan. His case study of the Danish ISAF Team 10 and its deployment to Afghanistan as part of the British led 'Task Force Helmand' 2010-11 offers a detailed insight into a contemporary case of coalition warfare. While coalitions are forged by politicians and the military top-brass at the strategic level, they are implemented by servicemen and women at the sharp end. One of the strengths of this book is that it takes us to both levels and thus presents a multi-layered analysis of what may challenge and strengthen coalitions both in the capitals and on the battlefield.

Dr Galster's book is not only rich on historical material but – as its title indicates – also serves as an important reminder that while history may offer specific lessons, it is also a most forceful provider of food-for thought for politicians and civilian scholars as well as military practitioners.

Thus, the realisation of the need for strategic agreement has come a long way since World War I, when distrust of allies, absence of common doctrines and lack of a common supreme leadership made co-ordinated action extremely difficult. Moreover, as we have seen in the recent case of Afghanistan, possibly for the first time ever, a coalition has been deployed to a war torn country with development of local military forces and hand-over to indigenous authorities as some of their primary 'war aims.' And we may indeed find these experiences coming in usefully in future.

In my capacity of commandant of the Royal Danish Defence College, I highly recommend this book not only due to its richness on historical details and vivid analysis of past events and problems, but also because several of its parts, including its chapter on the Danish Defence Forces' participation in coalition warfare in Afghanistan over the last decade, provides reflections and insights which point towards to the future. Thus, D. Galster has brilliantly managed to analyse past and present with an eye to the future. This is – in my opinion – the true value of this work.

The current book came into being as part of a larger project on coalition warfare undertaken by Dr Galster at the Royal Danish Defence College's

Centre for Military History. As a matter of formality, I would like to stress, that Dr Galster's work does not necessarily reflect the official views of the Danish Defence. This is only natural as this book should be seen a scholarly-analytic contribution and not an official document. And in contrast with what often happens to official documents, I hope that this insightful and thought-provoking book will generate discussion and spur further research of this hitherto somewhat neglected field of study.

Nils Wang
Rear Admiral
Commandant of the Royal Danish Defence College
Copenhagen, March 2012

Preface

IN THE TWO worldwide conflicts of the first half of the 20th century opposing coalitions fought each other for six years until the complete break down and unconditional surrender of one of them. During the Cold War two daunting alliances lay ready to commence within minutes' notice the destruction of civilisation as we know it. While since the end of the Cold War wars have been on a limited scale, they have also been numerous and highly intensive, and the world has seen successive coalitions-of-the-willing fighting in the Balkans, in Iraq, in Afghanistan and in various other places of unrest to contain them. However, although the opposing Cold War coalitions never actually went as far as engaging each other militarily the concept of coalition warfare is no novelty – indeed it has existed for millennia.

The book you are holding in your hand now is about the distinctive features of wars waged by coalitions, but it is not a comprehensive, chronological survey of this kind of martial activity. My aim is to demonstrate key characteristics and their appearance across history, highlighting changes and continuity over time rather than an all-inclusive examination. Since for many years before becoming a military historian I served with the Danish armed forces, which for centuries have had close professional ties with the British, I have chosen – to a wide extent – to illustrate features of coalition wars through examples from the Anglo-

Danish military co-operation. In many instances of such wars, Anglo-Danish partnership has been extant and equally as successfully performed in the past as it is happening today in Helmand Province.

Although this book does indeed take previous scholarly research into account, my key sources of inspiration have been the correspondence between partners of historical coalitions and memoirs of key decision-makers, as well as eye witness reports by some of the British and Danish officers fighting in Afghanistan 2010-11. Field-Marshals Lords Alanbrooke's and Montgomery's are impressive sources not only presenting successful instances of collaboration, but equally as much pointing at the difficulties arising when working with allies and when military men must work in tandem with politicians. Malte König's *Kooperation als Machtkampf* [Co-operation as a power struggle] has given me a thorough insight in the dealings and envies between the coalition partners of the Axis powers Germany and Italy during the Second World War, and Gal Luft's incisive work on cultural clashes amongst coalition partners *Beef, Bacon and Bullets* has provided me with a glut of examples of religious, linguistic and behavioural incompatibilities which have endangered the smooth operation of historic coalitions. Modern coalition warfare, which I have chosen to illustrate by examples from the Anglo-Danish Task Force Helmand 2010-11, is described in an overabundance of literature, much of it written by journalists. While some of this material is indeed useful as reference, my interviews with key players in the actual drama have been placed in centre stage and merged with the general picture most kindly provided by the Ministry of Defence in London and the Army Operational Command in Karup, Denmark.

For obvious reasons, much of this cannot be substantiated by references or detailed descriptions of the minutiae of operational dealings as this might jeopardise personnel as well as operational security. Thus, in the chapter concerning the operations in the Helmand Province of Afghanistan 2010-11 you will find only a limited number of footnotes. However, all those interviewed as well as their military employers have agreed to publication and have had the opportunity to comment on my text. This has happened without in any way compromising the historical reliability and independence of the work, and any inconsistency or omission you might notice remains entirely my responsibility.

It has been my intention to write a book which can be read equally as well by the lay person as by the military professional. For this reason some pieces of practical information may be useful to the majority of readers.

While the languages used by the sources of my investigation are primarily Danish and English, with a clear dominance by the latter, French and German material has been exploited too. However, all Danish, French and German quotations have been translated into English by me, and any misunderstanding that might be observed, consequently will be a fault of mine.

The vocabulary may confuse readers of military history of earlier periods, as some expressions have assumed new or slightly tweaked meanings. Thus for instance, while in a Second World War context the terms 'combined' and 'joint' meant multiple service co-operation and multi-national operations respectively, today they have swapped meanings and we will stick to the modern use of these terms throughout. Moreover, in the parlance associated with the coalitions-of-the-willing of the 2000s, traditional military organisational terminology comprising terms like corps, division, brigade, battalion etc. is frequently replaced by designations better fitting a theatre-specific ad hoc organisation, which is not necessarily composed of the traditional building blocks. Thus, for instance, the coalition-of-the-willing in Afghanistan, the 'International Security Assistance Force,' or ISAF in the vernacular of the day, is made up of a number of 'regional commands' – which are neither divisions nor brigades although they are commanded by major-generals and brigadiers. Similarly, beneath this level we saw, among others, the 'Task Force Helmand' which comprised c. 5,500 troops and was commanded by a brigadier, who had under his command the 'Danish Battle Group', which was actually a considerably reinforced battalion, which was led by a full colonel. All measurements are stated in the metric system.

Where modern 'coalitions-of-the-willing' are concerned, the point of departure has been the small-state participant's aspect exemplified by Denmark. Danish contingents joining such coalitions have been designated in a variety of manners. As an overall generic term DANCON [Danish Contingent] remains the most commonly used expression, but task specific designations such as the DABG [Danish Battle Group], momentarily serving in Afghanistan, are seen as well. Moreover, since the Danish Battle Group in the Helmand Province 2010-11 has had British sub-units attached as well as *tolays* [companies] of the Afghan 3rd *Kandak* [brigade], whom they partnered and mentored, it was mostly designated the Combined Force Nahr-e Saraj North – or CF NES (N) – referring to its multi-national character and its geographical area of operations. For reasons of variety both designations has been used throughout Chapter III of this work.

As briefly intimated above, this book is intended for the general reader of military history as well as scholars and military officers. I have therefore endeavoured to write in a language not too heavily infested by staff course jargon. Thus, by and large military acronyms – abounding in modern armed forces documents and colloquial military Newspeak – have been avoided, but some obviously well known and commonly used abbreviations have been retained. These include unit designations such as "coy" for company (when it appears as the name of a given sub-unit like e.g. A Coy), HQ for headquarters, certain weapons systems like the RPG (rocket propelled grenade) etc. However, all abbreviations and acronyms used in this book, and quite a few commonly found in military texts on the subject, though avoided by this author, are explained in the List of Abbreviations.

In addition to the military terminology, which will be explained throughout the narrative, there is a need to touch on the vocabulary used for describing the troops involved. While generally avoiding military technicalities, some key professional terms are unavoidable. These, however, have been used sparingly.

Many books have been published in recent years, most of them describing particular events, operations or commanders and their deliberations as well as the political situations under which coalitions-of-the-willing have been formed. This work endeavours to take a slightly different approach, and it is my hope that it may contribute to the understanding of the intricacies of coalition warfare in general and, in particular, to putting the Anglo-Danish military co-operation in the Helmand Province 2010-11 into an historical perspective.

Kjeld Hald Galster
Royal Danish Defence College, December 2011

Introduction

WE ALL NEED friends whom we can trust and rely on to support us in times of need. This is a simple fact of life. It is true in our every-day relationships and it is crucial when we come under attack – personally or as a larger society. Our forebears looked to likeminded groups or nations when greedy and cantankerous neighbours tried to rob them of their land, women and resources or to suppress their freedom of decision to choose a way of life, creed or political system. So do we today, and we also do our best to place ourselves at the disposal of those needing our assistance for similar reasons.

While, over recent years, the warfare of various post-Cold War 'coalitions-of-the-willing' have quite justly drawn much public attention, associations of nations fighting, or preparing to fight, for common causes are not inventions of the twenty-first century. From times immemorial, nations have combined to achieve common goals – or united militarily just in order to survive. They have done so by pooling their resources in either one of two slightly different types of organisation honed for multi-national warfare, namely alliances and coalitions. Although first and foremost this book will address the broader perspective of coalition warfare it will, invariably, also touch upon the alliance niche of organising communal martial struggle. The characteristic reality distinguishing between these two sides of the same coin concerns the extent of the shared interests that causes their genesis. Canadian scholar Paul T. Mitchell has aptly pointed out that while coalition partners may share some interests, they do not necessarily do so to the same depth or for the same length of time as do true allies. Indeed, they may agree in some vital fields but be competitors in other areas, or they may choose to oppose each other even on related issues. A coalition organised around a single issue thus tend to be driven by the state that possesses the greatest interest in that matter. The level of co-operation that this state can expect will vary depending on how closely its interests match those of its partners.[1]

Not surprisingly, nations feeling themselves menaced by the same threat or sharing a number of strategic interests will realise the advantages of combining their military forces and efforts; but multi-national co-operation in fields as costly and as fateful as war will always depend on considerations and caveats concerning political purpose, risks, mutual trust, national wealth and pride, compatibility of military forces and a glut of intangible forces and effects characterising human interaction in combat. Therefore, like war itself, the forming of a coalition is a highly political matter. Thus, in this book we shall try to make a close inspection of the workings of military coalitions past and present highlighting some perspectives of the commonalities as well as the changes which can be observed taking place across history. We shall look into such issues, which are extant regardless of time, but we shall see also that, while some remain unchanged, others vary depending on the specific situation.

The General Context

The United Nations was founded in 1945 after the end of World War II to replace the defunct League of Nations and to prevent war between countries as well as to provide a platform for dialogue. It still is an international organisation whose aims were and remain facilitating co-operation in international law, international security, economic development, social progress, human rights, and the achievement of world peace. The original intention of providing this organisation with an international military staff and executive power failed, however, but over the years its Security Council has lent its authority to many a peacekeeping or peace enforcing effort, whose execution has been laid in the hands of collaborating member nations.

---

[1] Paul T. Mitchell, *Network Centric Warfare and Coalition Operations: The New Military Operating System* (Oxon: Routledge Global Security Studies, 2009) pp. 47-48.

It is a matter which merits our attention that, since the fighting ceased after World War II, western and westernised liberal democracies – apart from Israel – have not fought major wars for reasons of national survival, and they have rarely waged their wars alone. With a few exceptions – such as the Suez Campaign in 1956 (Operation Musketeer) and the US led invasion of Iraq in 2003 (Operation Iraqi Freedom) – wars of the late twentieth and early twenty-first centuries have been mandated by the United Nations' Security Council and conducted as coalition enterprises led by an alliance or a by a great power. This has enormously boosted the legitimacy of these actions and eased the burden on the individual nation's treasury, but at the same time it has brought about a plethora of novel hitches and challenges.

Abiding issues of contention are those of the sharing of costs, of influence and of glory, but there are other and much more subtle snags. Amongst these rips we find the hurdles of multi-national military collaboration in general – such as incompatibility of weapons and equipment and disparity of doctrines – as well as those stemming from cultural clashes and dissimilarities in particular. One striking example of the cultural complexity of coalition warfare we may observe when looking into the Turco-German military co-operation during World War I. Before the outbreak of war, frequent reports with vivid descriptions of what the Germans perceived as Turkish barbarism had been sent home by German military advisors working with the Ottoman forces. These negative – or sometimes even hostile – sentiments were conveyed by a handful of officers, but they pervaded German military and civic opinion of the early 1910s and as, when war came, German troops were massively deployed in Turkey, many of them had already been infected with a strong cultural bias.[2] The lack of cultural correspondence that had been generating for some years endangered the collaborative achievements of the Central Powers: the coalition's war aims, because it was not properly addressed and it increased the potential for misunderstandings and mutual antagonism. Central Powers coalition warfare, therefore, was fraught with inherent pitfalls of this nature, and we will try to find out whether this can be observed more generally and if it remains so today.

Like many other UN member nations, Denmark of this day is found among the small players on the international stage and, though she is an active member of NATO and various other international security related fora, so far she does not take any part in military actions launched by the

---

[2] Luft, Gal, Beef, *Bacon and Bullets: Culture in Coalition Warfare from Gallipoli to Iraq* (no place: Gal Luft, 2009), p. 39.

European Union. Denmark was a neutral country from 1864 to 1949[3], and during the following period up until 1991 she was a loyal NATO member but took part in no active combat. However, after the end of the Cold War Denmark shed her centuries-old tradition of leaving international power politics exclusively to the major powers and embarked upon an activist foreign policy resting on the concept of indirect security and a shared responsibility for the maintenance of peace, prosperity, law and order and human rights. Consequently, from then onwards the country engaged not only in peacekeeping operations under the aegis of the United Nations, but also in coalitions-of-the-willing employed in endeavours to topple dictatorships and combating terrorism worldwide.

Thus, unlike during the years of neutrality stretching from Denmark's military defeat by Austria and Prussia in 1864 until the joining of NATO in 1949, and the Cold War interlude of deep alliance dependence 1949–91, today Danish troops, ships and aircraft are now operating within ad hoc coalitions as far afield as Libya, Afghanistan, the Sudan and the Arabian Sea off the Horn of Africa. This change, of course, has not come about overnight, but has happened as a gradual, though apparently irreversible, process over around twenty years. Since the early 1990s, Denmark has pursued an increasingly active foreign and defence policy and Danish armed forces have been engaged in the Balkans, in the Middle East and in Africa in United Nations peacekeeping missions as well as in peace enforcement operations run by NATO on a United Nations mandate, and this remains so today.

In Denmark, the western invasion of Iraq never received the same popular backing as has the intervention in Afghanistan. However in 2003, backed by a majority in parliament, the Danish government decided that Denmark ought to do her bit as part of the coalition-of-the-willing being assembled in order to rid the world of what was then believed to be Saddam Hussein's weapons of mass destruction production facilities. Therefore, from 2003-2007 Danish troops fought within the multi-national coalition under the leadership of the British formation in the Basra Region of Iraq, as they do now in the Helmand Province in Afghanistan. Thus, they provide a contribution to modern coalition warfare combating what is perceived, rightly or wrongly, as the looming dangers of the twenty-first century.

If a small country of limited resources wants to retain its armed forces

---

[3] Although a number of Danes joined the Waffen-SS, this happened on an individual basis, not as a co-operation by the Kingdom of Denmark. Similarly, this happened as far as Danes in British or American service were concerned. Thus, Denmark was officially non-aligned during WW II.

at a level adequate for territorial defence as well as overseas deployment it must make sure that the forces are seen by tax payers to be of some use as a tool in the hands of the political leadership. Thus politicians, whose ambition it is to let their country play an active rôle internationally, must be willing to send their armed forces to the hot spots of the world, making sure that they possess the weapons and the equipment necessary for fulfilling the tasks assigned to them with a minimum of casualties. Although today Danish forces are trained and equipped to be fully compatible with those of partner nations, in comparison with most alliance or coalition contributors Danish contingents are always of modest size and limited punch – though of excellent training and proficiency. For these reasons, they are employed as component units within larger coalition contingents subordinated to a lead-nation and, as in the case of Helmand, to a military lead-formation. The leadership may be provided by the French, as was the case during some of the operations in the Balkans, or by the British, as has been seen in Iraq and in Afghanistan today.

Observed against the backdrop of the Cold War, small state participation in military operations in distant countries is a novelty, but over the centuries there has been noticeable evidence that such endeavours were by no means uncommon.

Wars are fought for a reason, and any government will do its utmost to make this a convincing one likely to persuade the population that the war is just and the aims worth fighting for. The reasons given by governments for entering war-time coalitions or alliances have differed from time to time – and, more often than not, the true motivation is at odds with the one that is publicised – but the crux of the matter has always been need for security within the country's own borders. The notion of 'state security' must not be defined too narrowly. It should rather be understood in its widest possible sense including such elements as dynastic interests, territorial assertion, pecuniary needs, deterrence, freedom of expression, freedom of economic and political initiatives and many other considerations pertinent to the country's and its inhabitants' liberty, prosperity, integrity and security at any specific moment. Thus, while the dangers from terrorism and other trans-national crimes are abiding key concerns in the first decades of the third millennium, throughout centuries past various other phenomena have been the prompters of nations' decisions to rally to multinational coalitions led by, presumably friendly, great powers. During the Thirty Years War (1618–48), prestige and hopes of territorial gains were the justifications behind the four year-long Danish participation (1625-29); and a national policy including security considerations, geo-strategic ambitions, economy and international standing was the basis for the Danish contribution to the Williamite coalition in the Nine Years War fought in Ireland 1689–91 and in Flanders from 1692-7. On the other hand, while it was external pressure that compelled Denmark to ally herself with Napoleon in 1807, coalition participation in the twenty-first century seems to be primarily a matter of showing solidarity with kindred democracies whose support Denmark might eventually want to solicit, if things should go awry in Europe.

The subsequent chapters of this book will address the intricacies of coalition warfare giving pride of place to the British and Danish contributions to the coalition endeavours of stabilising and developing war torn Afghanistan in recent years. The modern day struggle in Afghanistan has gone on now for more than a decade and has been performed by a coalition called the International Security Assistance Force (ISAF) led by NATO, though mandated by the United Nations and strongly influenced by the perceived security needs of the United States. However, we shall deal neither with the coalition as a whole nor with the entire length of NATO's engagement in Afghanistan, but, for reasons of simplicity and clarity, rather focus the central narrative on the Danish contingent fighting under a British lead-formation in the Helmand Province from August 2010 through February 2011.

By and large, so-called 'civilised warfare' is conducted in accordance with treaties approved by the international community over more than 350 years. In 1648, the Peace of Westphalia ending the Thirty Years War settled relations among modern, sovereign, European countries and their rulers, drawing up the map, and making rules for civilised interaction among states including regulations for armed struggle. This martial 'code of conduct' made war an accepted tool in power politics reserving its use, though, for sovereign and legitimate rulers; and it imposed a number of restrictions on how this instrument might be wielded. In many respects post-Thirty-Years-War warfare was civilised, dignified and not particularly bloody. Manœuvres in order to split or isolate enemy forces and sieges aimed at starving stubborn garrisons, towns or cities into submission were key features. Opposing armies were comparatively small compared with those of the Napoleonic and later eras, and fighting normally took place only in the summer season when fodder for the huge number of riding horses as well as draught animals was to be found in the fields. During the nineteenth century, the Industrial Revolution allowed wars to spread in terms of participation and time as well as in space, and as weapons and communications technology grew increasingly more sophisticated and as armed combat tended towards becoming all-encompassing new rules were needed – rules that were more comprehensive and more explicit than the time-honoured chivalric ideals. Regulating interaction amongst warring parties as well as the rights and obligations of neutral countries, the handling of prisoners-of-war, the humanitarian treatment of the victims of war etc. the Hague Conventions of 1899 and 1907 and the Geneva Conventions of 1949 saw the light of day. These codes were the first formal statements of the laws of war and war crimes appearing in the nascent body of comprehensive, secular, international law. However, such laws on the obligations and rules connected with armed conflict presupposed organised and uniformed armed forces under the responsibility of the governments of sovereign states. No mention was made of international crime, terrorists, pirates or so-called holy warriors, who have since appeared as the primary nuisances of the post-Cold War Era.

The king's coffers used to be the restrictive factor in warfare, and the state economy has always been part and parcel of the true – if not openly admitted – reason for war. In 1689, Christian V of Denmark, having for many years retained an army much too big for being paid to stand idle, seized the moment and initiated negotiations to hire out a large portion of his military – *c.* one fifth of its total strength of 34,000. This was an option open to him, because King William III, who had just recently been crowned king of England and Scotland, was faced with invasion by the French, who were backing his predecessor King James II. Since William could not send his entire army to Ireland, whence the primary threat of invasion appeared, laying Britain open to any possible invader, he needed allies to combat King James' French supported troops wishing to re-conquer Ireland as a preliminary step before taking Scotland and England back. Hiring out his best troops would give King Christian's officers combat experience, hence keeping the military instrument sharp until the day might come when he would need it himself; and it would save his treasury the burden of paying the troops, since this obligation would fall upon Britain – the lead-nation of the proposed coalition.

Now, as in the 17th century, affluent lead-nations do have the option of dangling tempting remunerations before their potential partners and history presents a succession of cases in which this has been done. In June 1803, wishing to create a third coalition against Napoleonic France the British Prime Minister William Pitt suggested that Austria be offered £ 2 million, Prussia £ 1 million, and Saxony and Sweden each £ ½ million for entering the war on the British side.[4] Similarly, following the 11 September 2001 atrocities, the United States' administration under President George W. Bush lifted all sanctions imposed upon the military quasi-dictatorship of Pakistan as well as promised huge loans and grants in order to bring this country in on the anti-terrorist side. Moreover, the Americans persuaded

---

[4] Frederick C. Schneid, *Napoleon's Conquest of Europe: The War of the Third Coalition* (Westport, Connecticut, London: Praeger, 2005) pp. 77, 80-81.

Pakistan to open her bases, harbours and air space to American and allied armed forces combating al Qaeda in Afghanistan.[5]

Today, an 'ordinary' western partner participating in a coalition-of-the-willing saves none of its treasury's money, but it does create goodwill amongst great power coalition partners, which may indeed make it worth the effort. Moreover, the value of keeping the military instrument sharp, training officers and soldiers and providing valuable lessons on modern combat and military co-operation remain undiminished.

However, there are considerations to be made other than those concerned with military needs – political advisability, bi- or multi-national agreements, matters fiscal as well as calculations on the possible benefits and risks.

The legitimacy and justification of deploying a state's national military assets in far away countries is frequently being challenged by today's politicians, media and non-governmental organisations. Is this at all a legitimate cause of action? Is it in accordance with international law, and are the human rights of all concerned considered when an army ventures abroad for purposes not directly relevant to the state's territorial defence? Regrettably, there are no simple answers to these queries, but from a liberal, democratic point of view it seems acceptable as long as no conscript servicemen and women are forced to go, as long as such a deployment rests on a resolution by the United Nations' Security Council, and the expeditionary forces act in conformity with the Geneva conventions and other international accords acceded to by the nations forming the coalition as well as the receiving state. While these restrictions apply today and are being keenly monitored by Amnesty International and similar organisations, considerations of that kind have hardly been seen as relevant by decision-makers of coalitions up until the end of the nineteenth century, to whom the needs of their respective monarchs and the long term interests of their states were probably the sole *raisons d'être* of their endeavours. Today's terrorists and so-called "holy warriors" are equally dismissive.

The modern notion of indirect security in the sense of combating the perceived evils wherever on the Earth's surface these might be found may be a link between the past and the present. Firstly, restraining a despotic ruler's appetite for conquest on the continent of Europe by sending troops to Ireland in 1689 was as reasonable an indirect security measure as was combating Hitler's Germany 250 years later by giving financial support to the Soviet Union, or employing NATO forces to prop up Hamid Karzai in

---

[5] Ahmed Rashid, *Descend into Chaos: Pakistan, Afghanistan and the Threat to Global Security* (London: Penguin Books, 2008), p. 89.

the early 2000s to topple the Taliban regime in Afghanistan and eradicate the Al Qaeda terrorist network. Secondly, since modern war is expensive the need for burden-sharing must not be forgotten, and the recent worldwide financial crisis has again buttressed this necessity. Today, the United States wants foreign support for her military adventures not only to lend them a veil of legitimacy, but just as much in order to make it clear to American tax-payers that they are not alone in funding the upkeep or introduction of law, order and democracy around the world. And, thirdly, although this might be less urgent in the post-Cold War Era than before, negotiators must always bear in mind that troop contributors might need their forces back for domestic purposes at an earlier moment than initially expected. While during the Cold War the Warsaw Pact loomed large on the horizon, today it will most probably be the terror menace to the homeland that must be considered. Finally, appreciating the requirements to coalitions of indirect security we must not forget that it is the duty of military commanders to fulfil the political purpose intended by their own governments simultaneously with co-operating constructively with their foreign fellow officers in the *ad hoc* assembled coalition military hierarchies.

The late British historian Richard Holmes fittingly described these frequently conflicting interests in the context of Wellington's leadership.[6] In the case of the Iron Duke, as he was later to become known, leadership was a matter of overall command by the lead-nation in the large European coalition against Napoleon, thus being a top-down observation revealing the commander's trouble de-conflicting the individual interests of subordinate generals to ensure fulfilment of the common goal. In this book, dealing in particular depth with the micro-cosmos of the Helmand Province 2010-11, we shall endeavour to make a bottom-up assessment of the issue seen with the eyes of the lesser coalition partner, Denmark, whence the need is rather to make sure that one's own formation is given tasks according to its particular capabilities and strengths. In both of these aspects it can be observed that agreement amongst coalition partners is not a given, and that demands by the lead-nation commander may not always be acceptable to the subordinate commanders of individual troop contingents. However, it is also worthwhile noticing that their possible refusal to carry out orders issued by the lead nation commander is not only a simple matter of discipline. It raises questions concerning the cohesion within and the common purpose of the coalition. American political scientist Patricia Weitsman takes the 'from above' position when she argues that when in

---

[6] Richard Holmes, *Wellington: The Iron Duke* (London: HarperCollinsPublishers, 2002).

March 2003 Australian fighter pilots in Iraq refused to obey their American commanding officer's orders, which they regarded as illegal, and were backed by their government, this was an illustration of one of the many shortcomings of coalition warfare.[7] Similarly, but observed from below, Canadians were justly infuriated that American authorities did not court-martial two US pilots who, in April 2002, contrary to instructions from their air controller bombed and killed four Canadians exercising with their infantry platoon near Kandahar in Afghanistan.[8] Such incidents are all too common occurrences and they apply severe strain on relations between commanders as well as between contributing nations, but they are of no recent origin. The tension caused by 'friendly fire', such as that suffered by the Canadians and by so many others in recent years, did indeed occur in the seventeenth century as well. On 1 July 1690, when King William III led his cavalry across the river Boyne and charged up Donore Hill at the head of the Enniskillen Regiment, the troopers and the Danish infantry mistook each other for the enemy. They got involved in a *mêlée*, and William was lucky not being hit by their bullets.[9]

As war is a political tool for achieving national goals, a government involving its county in a conflict should have a sound and coherent national policy from which its strategy can flow, providing the military machinery with clear and obtainable war aims. Coalition warfare is in a different sphere of politics, but the need for realistic strategies is no less pressing. Coalition war aims must merge many interests and objectives so as to satisfy the security needs of all participating partners.

This book is not singular in its attempt to shed light on coalition warfare in general. Keith Neilson and Roy A. Prete have edited an excellent compilation of contributions, *Coalition Warfare: an Uneasy Accord*, originally presented in 1981 as scholarly papers at the Royal Military College of Canada's 8th Military History Symposium.[10] This anthology presented a *tour d'horizon* of coalition warfare including such diverse

---

[7] The Australian pilots aborted 40 bombing missions at the last minute, because they were likely to have led to excessive civilian casualties. Patricia Weitsman, 'With a Little Help from Our Friends?: The Costs of Coalition Warfare' in *Origins* accessed on 6 January 2009 through URL http://ehistory.osu.edu/osu/origins/article.cfm?

[8] http://news.bbc.co.uk/2/hi/south_asia/1936589.stm

[9] Harman Murtagh, *The Battle of the Boyne 1690: A Guide to the Battlefield* (Mell: The Boyne Valley Honey Company, 2006) p. 55.

[10] Keith Neilson and Roy A. Prete, Eds. *Coalition Warfare: an Uneasy Accord* (Waterloo, Ontario, Canada: Wilfrid Laurier University Press, 1983).

themes as the Austro-German World War I alliance, Soviet coalition theory, planning and performance, the so-called greater East Asian Co-Prosperity Sphere and the Germano-Ottoman coalition of 1914-18. Moreover, some outstanding works have delved into specific niches of coalition, such as Gal Luft's, *Beef, Bacon and Bullets: Culture in Coalition Warfare from Gallipoli to Iraq* dealing with the cultural aspects and the difficulties in holding coalitions together when language, world view, religion and various ethno-centric prejudices surface.[11] This is indeed a seminal work which sheds important light on many inter-coalition tensions arising from lack of personal empathy, ethnic arrogance and simple narrow-mindedness. *Napoleon's Conquest of Europe: the War of the Third Coalition* by Frederick C. Schneid nicely explains how commercial interests may help forge some coalitions while contributing to the ruin of others.[12] Especially pertinent to the coalition warfare of the 21st century are the incisive analyses in Paul T. Mitchell's *Network Centric Warfare and Coalition Operations: the New Military Operating System*.[13] Last but not least, not only historical research, but indeed political science, is relevant with respect to small states participating in this kind of war. Quincy Wright *A Study of War* (1942), Kenneth Waltz, *Theory of International Politics* (1979) and Birthe Hansen's work on the so-called bandwagoning phenomenon have contributed to the understanding of small state behaviour as members of coalitions.[14]

The Political Context: Denmark in the Twenty-First Century

Wartime coalitions are formed for a variety of reasons. There may be a need to strengthen hegemonial – or imperial – defence in order to bind dependencies together in a closely knit association, a notion cherished by Napoleon, who made scores of minor German states part of his Confederation of the Rhine (*États confédérés du Rhin*), and by Britain and her self-governing dominions, although in this case the notion of a relationship based on the Dominions' dependence on Britain and her

---

[11] Gal Luft's, *Beef, Bacon and Bullets*.

[12] Frederick C. Schneid, *Napoleon's Conquest of Europe: The War of the Third Coalition* (Westport, Connecticut, London: Praeger, 2005).

[13] Paul T. Mitchell, *Network Centric Warfare*.

[14] Hansen, Birthe "The Unipolar World Order," in Birthe Hansen and Bertel Heurlin, Eds. *The New World Order* (London: Macmillan Press Ltd.), pp. 112-133.

hegemonial interests was met with some scepticism by Canada as well as by South Africa. A second form of coalition materialises when a great power contemplates alliance with a peer nation – like Britain did with Japan in 1902 and, less formally, France with Russia a few years later. In this way nations seek the dual security which a coalition offers: aid from partners in the event of hostilities; and a pledge to assist partners, whose defeat by a common enemy might be disastrous to one's own interests.[15]

While in the case of the former the great, hegemonic, power taking the initiative will obviously be the lead-nation of the coalition, in the latter one coalition partner would rarely be allowed to take the general lead – although in specific theatres of war the nation with most interests at stake and the largest military formations deployed might frequently play a dominant rôle. This was the case in both world wars as. Towards the end of World War I, it was agreed that France, being the largest contributor of troops, would provide the strategic leadership of the Entente's forces including those of the United States. Thus, in 1918 Marshal Ferdinand Foch was appointed Allied Supreme Commander. Similarly, during the Second World War the United States became the *primus inter pares*, the first among equals, as their troop contributions grew out of all proportion of those of Great Britain.

While previously there have been at least two patterns of coalition leadership, in the twenty-first century coalitions usually unite under the aegis of a great power lead-nation. This leader is assumed to be one with great military and economic potentials and with ambitions of deciding the end state of the conflict, which, for almost all practical purpose, means the United States.

Throughout the first decade of the twenty-first century, in international affairs many small and medium powers have sustained a political ambition of aligning themselves with like-minded and stronger nations – notably Britain, France and the United States. The reason for this, we may assume, is the perceived need for support if and when military threats should materialise close to their borders. To a certain extent this also happened in 1807 when Britain had commandeered the Danish fleet and shelled Copenhagen, while at the same time French forces had taken up positions on Denmark's southern borders. Consequently, having no other option Denmark chose to ally herself with France taking up arms against the opposing coalition. Similarly, towards the close of the Napoleonic Wars, as the obvious security interests of the Danish monarchy was no longer served

---

[15] Paul Kennedy, "Military Coalitions and Coalitions Warfare over the past Century" in Keith Neilson and Roy A. Prete, Eds., *Coalition Warfare: an Uneasy Accord* (Waterloo, Ontario, Canada: Wilfrid Laurier University Press, 1983) pp. 5-6.

by the alliance with Napoleon, the country changed sides, joining the anti-French coalition which eventually defeated and occupied France.

Moreover, there may be political as well as commercial reasons for a small state to join an alliance with a dominant great power. Having lost, in 1864, the two South-Jutland Duchies to the German great powers, there was a strong Danish inclination to ally with France. Retrieving the Duchies this way was seen by many as a logical option – the only one open at the time. Therefore, in 1866 when Prussia attacked Austria as well as in 1870 when the German states invaded France, it was seriously discussed in Copenhagen if the possible outcome of either of these conflicts might be of a kind vindicating Danish alignment with France. In the end it was decided that this was too dangerous a path to tread, and Denmark – probably fortunately – kept out of these wars. Conversely, in the wars in Iraq and Afghanistan in the early 2000s Denmark found it advisable to show solidarity with the United States and the liberal ideals of the Western World, and in spite of considerable losses, the gains in terms of doors opened in Washington and London were substantial.

The key personalities influencing the choice of partners and the conditions for entering the coalitions-of-the-willing fighting the twenty-first century's conflicts are all to be found in the inner circle of the de facto decision-makers – the government and those members of parliament who support the administration in office. Decision-makers wish to get the most out of the risky enterprise, and if their co-operation is sufficiently important they can demand either economic remuneration, as Denmark did joining William III for the Irish campaign in 1689, or geo-strategic compensation such as bases like the Russians wished to get in Malta in 1805 and the United States wanted, and got, from Britain during World War II.[16]

Similarly, the realities of the early twenty-first century placed the Pakistani General-President Perves Musharraf in a position allowing him to demand from the United States political endorsement of his quasi-dictatorial governance, deliveries of high-tech military equipment, and huge economic recompenses. Although in large part this was due to the American need for a launching pad into Afghanistan and a supply line for its troops deployed there, the American embrace of Musharraf and his tainted autocratic regime found little sympathy in countries like Denmark, one of the reasons for accession to the NATO-led coalition bound for Afghanistan was Denmark's need to ascertain a favourable standing with her major trading partners and political trend-setters of the world. As the Cold War had been reasonably inexpensive, from the early 1990s onwards the treasury

---

[16] Schneid, *Napoleon's Conquest of Europe,* pp. 85-86.

could now provide the means necessary for fulfilling the ambitions of the political élite of playing a rôle along with powers, which for a long time had been the favourite collaborators and cherished similar humanistic ideals and liberal economic values. To underpin these endeavours, the administration of the country's finances relied on the ongoing boom in world economics making activism on the international stage all the more realistic. However, the world economic crisis from 2008 onwards reanimated the worry about funding of ambitious foreign and defence policy goals.

In order to develop a useful strategic concept, coalition partners must agree on common war aims. Prominent among such political and military aims are those concerning the conquest of land and resources as well as the creation of political spheres of influence. While, during the anti-French coalition negotiations between Britain and Russia in 1805, the restoration of the Kingdom of Piedmont-Sardinia and the independence of the Netherlands and Switzerland constituted such geo-strategic war aims, the 1991 coalition against Saddam Hussein's Iraq set up the restoration of Kuwait as a sovereign state as the basic, official aim of the common war effort.[17] Less openly, western access to Kuwaiti oil-fields might have figured prominently amongst some coalition partners' priorities.

For centuries coalition strategies have included new designs of the political layout of the contested areas so as to achieve improved security for the states involved and to prohibit future usurpation of their acquired rights and positions. The Romans fought in the Trans-Alpine parts of Europe to create a buffer-zone against the Germanic hordes, the Soviet Union got Churchill's acceptance of 'moving Poland westwards' at the end of World War II in order to create a *cordon sanitaire* for their homeland, just like NATO and the United Nations have tried, with limited success though, to re-design the political map of the Balkans after the collapse of Yugoslavia.

Although, when entering existing coalitions or forming new ones, most participating states will decide on their contributions on the basis of the above mentioned grand strategic issues, other considerations, mostly of a practical nature, exert considerable influence as well. High on the list of priorities is the military, professional experience, which might result from active participation in war. As during Cold War years officers of the armed forces of most NATO countries apart from those of France, Britain and the United States saw little active service, peace enforcement in the Balkans in the 1990s and co-operation in the coalitions-of-the-willing in Iraq and Afghanistan in the early 2000s, in many respects, came as a welcome opportunity to obtain realistic notions of the conditions and challenges of

---

[17] Ibid., p. 83.

combat. Moreover, the rank and file of military forces benefit from being sent on consecutive stints in overseas warzones. While in the new millennium to some extent conscription is still practised, as far as Denmark is concerned all personnel being sent on active service abroad are now regulars and Danish nationals, whose experiences from practising their profession is of immense importance to the development and efficiency of the armed forces.

No less important is today's emphasis on language training and proficiency. This, however, is not as easily obtained as some might believe. In a globalised world there is a trend of taking language skills for granted, but the reality is that in many countries the ordinary citizen's command of foreign tongues is deplorably limited. Although for operational purposes English is the language used in written as well as in oral communication at battle group and formation levels, amongst the NCOs and private soldiers the working language remains Danish. Very few speak French, and the languages spoken in the current areas of operations, such as Arabic or Dari, are commanded only by the military linguists or interpreters. As for future operations, which may take the troops anywhere, it is more than likely that the agreed common language will remain English and that communication with the indigenous people will still require a staff of skilled interpreters. Last but not least, there is always some disparity of cultural background and social attitudes, which are mostly insignificant in collaboration with western partners, but of quite considerable consequence where mentoring and partnering with the Afghan National Security Forces are concerned.

Formally, in Denmark political decisions on entering coalitions are taken by parliament, but for all practical purposes it is the prime minister who, in close collaboration with the secretaries for foreign affairs, finances and defence, decides which tasks to take on. Nonetheless, for reasons of legitimacy any government would want to have the backing of as many parties as possible, thus prohibiting such decisions from becoming an embarrassing theme in future general elections. Most Danish prime ministers, who have been faced with the choice between war and complacence, have gone out of their way to persuade a parliamentary majority to come on board the coalition vessel. Apart from that, over the last two decades there has been a trend that, increasingly often, it is in the prime minister's office that foreign policy issues are decided. Negotiations are being conducted by the subordinate ministries in their specific fields, but the initiative is mainly with the government leader.

To a large extent agreements on purely military matters are made by officials of the Ministry of Defence, the Defence Staff and the various services' operational commands with their foreign opposite numbers, and when actually deployed most decisions of a tactical or operational nature are taken by the local Danish commander after consultation with the lead-formation headquarters.

The Military Context

In military theory time and space are frequently seen as each other's complement. One fighting entity may yield space to gain time, allowing itself to organise a counter-attack, build up forces or adjust the nation to the needs of war such as introducing conscription, putting constraints on consumption and turn on full war production. Thus carving up Poland to share her with Nazi-Germany, in 1939 Stalin wished to gain some space, which might become useful later when time would be needed for getting ready before the German onslaught, which he was sure was in the offing. From a Russian point of view this was a wise precaution, because the Soviet Union was ill prepared for war and would have to make time-consuming preparations to be able to withstand the well-rehearsed German attack, looming on the horizon.[18]

Time may be essential to some coalition partners, while less of a problem to others. In 1805, Austria and Russia knew that the French war machine was being geared up to make an amphibious campaign against Britain. However, not being under any imminent threat themselves, these powers had plenty of time to conclude the negotiations with Britain on the Third Coalition. Conversely, to William Pitt the Younger (1759-1806), the British prime minister since 1804, the time pressure was obvious. He had to finalise a treaty of coalition with the other great power enemies of Napoleon quickly enough to be able to draw the French emperor's attention away from the English Channel before France would be able to equip herself for launching an invasion. Thus time was of essence and the coalition's military preparedness and build-up of strength were central to taking the struggle to the opponent.[19]

Of course, success and disaster matter as much to coalitions struggling to achieve a common goal as they do to individual fighting entities. If ultimate disaster strikes the lead-nation, this can spell the end to the whole coalition as it did to the Central Powers following the German setbacks on the Western Front in 1918. The coalition ceased to function and

---

[18] Alex Danchev and Daniel Todman (Eds), *War Diaries 1939-1945: Field Marshal Lord Alanbrooke,* (London: Phoenix Press, 2001) p. 306.

[19] Schneid, *Napoleon's Conquest of Europe,* p. 83.

disintegrated; and the three constituent empires fell apart. On the other hand, the defeat of the French Mediterranean Fleet at the Bay of Aboukir in 1798 and the French naval – and vain – attempt on invading Britain merely constituted temporary downturns. In the war in Afghanistan since 2001, the coalition-of-the-willing has had no outright major disasters, but the continuing lack of success in their attempts on ridding the country of the Taliban insurgency has had important repercussions for the allies. It has necessitated a change of strategy, an exit date in 2014-15 agreed NATO-wide, and a dawning realisation that a negotiated solution to the conflict will have to include some moderate elements of the Taliban movement.

No commonly accepted doctrine for coalition warfare appears to exist today, although the coalition fighting in Afghanistan seems to agree on aligning with the concepts laid down by changing US commanders-in-chief.

Any multinational operation will require planning by all the participants, interoperability, shared risks and burdens, emphasis on commonalties, and diffused credit for success.[20] In this context the absence of an officially approved common doctrine is an impediment to smooth collaboration. However, coalition partners sharing the same level of technological expertise and equipment as the lead-nation stand a reasonable chance of performing close to seamless co-operation at the operational and tactical levels. This in turn will ease the path to a commonly agreed doctrine as the basis for their combined operations.

While shared notions on doctrines are highly advantageous, a common planning process on the operational and tactical levels is essential, too. The degree to which national commanders and staffs understand each other, and are able to participate in common planning, impacts on the time required to plan and the sharing of knowledge of every component part of the operational forces.

Today, the command and control ability depends heavily on technological compatibility and integration and it rests largely on one notion – unity of command. This is one of the most fundamental principles of warfare in general and the single most difficult one to practice in combined warfare – and the more numerous the participants the more complicated this becomes. It is dependent on many influences and considerations. However necessary, because of the severity and consequences of war, relinquishing national command and control of forces is an act of trust and confidence that is unequalled in relations between

---

[20] The following considerations on doctrines for coalition warfare is largely based on Robert W. Riscassi's article "Principles for Coalition Warfare." From Internet, URL http://www.dtic.mil/doctrine/jel/jfq_pubs/jfq0901.pdf accessed on 11 August 2010.

nations and therefore done only cautiously. Thus, just as in 1813 the Danish King Frederick VI accepted the placement of his expeditionary corps in Holstein under the French Marshal Davout's, *Prince d'Eckmühl*'s, command, Danish forces now fight in Afghanistan under a British brigadier, who in turn is subordinated to the commander Regional Command South-West – an American Marine Corps general. As the unfettered collaboration within this chain of command depends heavily on compatible information technology – in particular the command, control, communications, computer and information systems, or C^4I in military parlance – and as large economies like that of the USA easily place themselves outside competition from lesser partners, collaborative dilemmas are bound to crop up.

In a way the United States has created a considerable problem for itself through spending vast resources on generating technological disparity between itself and potential coalition partners – with the possible exceptions of the UK and France. However, coalition partners are critical commodities in the ongoing fight against terrorism, and if this disparity is not put right one way or the other the political fallout from fratricide, collateral damage and sheer economic exhaustion might eventually make partners defect. In that case, in the end the coalition project will collapse.

A reasonable degree of commonality of technological levels amongst coalition partners is important to ascertaining that co-operating militaries can actually work harmoniously together and, hence, are capable of agreeing on a common doctrine. Moreover, if there is one requirement which is universally important in war, it is awareness of what is going on. Thus operating without the situational awareness, which is shared through the information network, will make units 'disappear' and they might easily become liabilities to the combined effort and dangerous to themselves as well as to neighbouring formations.[21]

Constraints

While mutual confidence amongst partners plays a vital rôle in coalitions, distrust of allies or individual commanders jeopardises combat efficiency and operational success. Neither the politician nor the military commander acts in a vacuum. Political circumstances and aspirations as well as strategic and operational necessities fence in their freedom of action and the practical, and often complex, conditions of actual war further restrict their leeway.

---

[21] Mitchell, *Network Centric Warfare and Coalition Operations*, p. 117.

Prior to agreeing on the establishment of the Third Coalition against Napoleonic France, Russia as well as Austria had serious doubts about Britain's ulterior motives for offering the huge subsidies she did to her prospective allies, and for that reason they tried to procrastinate treaty negotiations in order to see what might happen and to extract as much as possible from the British coffers.[22] On the personal level distrust between political and military decision-makers, as it happened in World War II between Churchill, Dill, and Brooke, contributed to straining the British effort during the early war years.[23] Churchill despised Dill and he frequently annoyed Brooke to the point of despair. And we may rest assured that animosity is a predominantly reciprocal phenomenon. Moreover, disagreements on realities and opportunities in war as well as on working habits are bound to erupt occasionally between politicians and military officers. Thus, in October 1944, Dill's successor as chief of the Imperial General Staff, General Sir Alan Brooke, later to become Field-Marshal Lord Alanbrooke, confided in his diary that he loathed these 'drunken, self assured politicians and diplomats [who] waste precious time.'[24]

Equally as destructive to coalition cohesion is the national self-importance shown over and over again by great power leaders, and this, too, is of no recent origin. Lamenting on the slow progress after D-day, on 27 July 1944 Alanbrooke told his diary that 'Ike knows nothing about strategy and is quite unsuited to the post of Supreme Commander... Bedell Smith on the other hand has brains, no military education in the true sense and unfortunately suffers from a swollen head... With that Supreme Command set up it is no wonder that Monty's real high ability is not always realised; especially so when 'national' spectacles pervert the perspectives of the strategic landscape.'[25] It is clear that Brooke was deeply offended by some of the American generals' condescending attitudes now they had become the de facto lead-nation of the western part of the coalition, and it may be difficult to find a more destructive constraint on coalition collaboration.

At the individual level self-importance frequently assumes the form of personal envy, and Brooke was not the only one to feel apprehension. Thus relations between British and Americans fighting in Northern Europe in 1944-45 were strained by mutual suspicion between Montgomery on the

---

[22] Schneid, *Napoleon's Conquest of Europe*, pp. 83-84.

[23] Danchev and Todman, *War Diaries 1939-1945: Field Marshal Lord Alanbrooke*, p. 199.

[24] Ibid., p. 603.

[25] Ibid., p. 575.

one side and the American commanders on the other. Rightly or wrongly, referring in his memoirs to a statement by former Commander-in-Chief of the Allied Expeditionary Forces in Europe, General Dwight D. Eisenhower, Montgomery claimed that "the impression was left that the British and Canadians had failed in the east...therefore the Americans had to take on the job to break out in the west... This reflection...is clear indication that Eisenhower failed to comprehend the basic plan to which he had himself agreed."[26] Much to his chagrin, Montgomery had been passed over by Eisenhower, who in addition to his job as the supreme tri-service chief also took upon himself that of land component commander, for which Montgomery – and Brooke for that matter – believed him utterly unsuited. Hence, Montgomery tried to mend his hurt feelings by claiming operational ignorance on the part of his American superior and, whether his appreciation was correct or not, it definitely contributed to a strained collaborative atmosphere.

In matters where honour, trust and reliability mean as much as they do in war it is hardly surprising that personal as well as national sensitivities are easily hurt. Allies do not want to be openly treated as a *tertium quid* (a third something), which apparently was exactly what Canadian Prime Minister William Lyon Mackenzie King (1874-1950) felt in the context of the Anglo-American co-operation during World War II in general and of Churchill's lack of attention in particular. Instead of engaging King in negotiations on an equal footing, Churchill developed a bad habit on merely informing him on what he himself had agreed bi-laterally with the Americans. Viewed from the perspective of the junior partner to a coalition, this kind of behaviour will invariably be seen as lead-nation arrogance.[27] Similarly, personal sensitivities amongst military officers like Canadian generals McNaughton and Crerar on the one hand and Montgomery on the other, prompted Brooke to persuade Monty, with whom empathy was no dominant feature, to tread carefully.[28]

Although the touchiness of alliance partners – whether nations or individuals – should be taken seriously or at least taken into account, there are examples of military top-brass prima donnas, whose petulance may be

---

[26] Bernard Law Montgomery, *The Memoirs of Field-Marshal Montgomery* (London: Fontana Monarchs, 1958) p. 262.

[27] Roy Jenkins, *Churchill* (Basingstoke and Oxford: Pan Macmillan, 2002) pp. 740-41 and 752.

[28] Nigel Hamilton, *Monty: The Making of a General 1887-1942* (Toronto: Fleet Books, 1982) p. 506.

difficult to handle. Coalition senior commanders like Montgomery and Patton were poles apart in their strategic outlooks and equally convinced of their own infallibility as well as the shortcomings of the other. Therefore, clearly offended, Monty wrote in his Memoirs that Patton, when in the summer of 1944 he was stopped during his advance, exclaimed 'let me go on to Falaise and we'll drive the British back into the sea for another Dunkirk.'[29]

However important personal and inter-ally relations certainly are, the key threat to coalition warfare is disintegration. Defection of allies remains an abiding problem in coalition warfare. Throughout history this has happened over and over again for reasons of receiving tempting offers from the opponent, from decline of zeal, from sheer exhaustion of resources and manpower, or simply because the war has taken a turn prohibiting continued fighting for one or more coalition partners. While in 1813 Denmark and Saxony defected from the Napoleonic cause, because the opposing coalition had pushed the French armies so far west that defence of own territory became incompatible with staying with the Emperor, in 2007 the UK and Denmark withdrew from the coalition-of-the-willing in Iraq although the desired end-state still had not been achieved. Similarly in 2010, the UK, Denmark and various other coalition partners chose to announce that as of January 2015 onwards they would no longer retain combat troops in Afghanistan. However, this intention was ameliorated by two facts, namely that at a conference in the summer of 2010, the Afghan President Hamid Karzai agreed that as of that moment the Afghan army and police forces would be capable of taking over the main responsibility for security in their own country, and at the NATO summit in Lisbon in November 2010 the alliance concurred that withdrawal of combat troops by the end of December 2014 would indeed be alliance policy. Nevertheless, this unison accord detracts little from the fact that this is precisely what the opponent might regard as most useful to his cause, not least seen in a propaganda perspective, and he will do his best to escalate the decomposition process.

Aim

The various parameters set out above, which determine the conditions for coalition warfare, will underlie the analysis of the historical as well as the contemporary facts, events and deliberations that will be described in the

---

[29] Montgomery, *The Memoirs of Field-Marshal Montgomery*, p. 270.

following chapters. Having acquainted ourselves with the history of 19th and 20th century foreign interventions in Afghanistan undertaken primarily by Britain and Russia, we shall look at the mission of and the solutions by the Danish ISAF Team 10 from August 2010 to February 2011 and its collaboration with the British lead-formation of Task Force Helmand. Doing this I shall aim at explaining how coalition warfare happens today. The ISAF Team 10 was deployed to the Helmand Province in Afghanistan to become one of more 'combined forces', i.e. a multi-national all arms formation, which together constituted Task Force Helmand, whose lead-formation was the British troop contributions of Herrick Teams 12 and 13 primarily constituted by 4th Mechanised and 16 Air Assault brigades respectively. Subsequent to dealing with the modern aspects of coalition warfare in Helmand, we shall cast a glance at the past viewing coalition warfare in a broader perspective. Employing examples from across military history, I intend to use the historical overview to present the mechanics, difficulties and benefits of coalition warfare in general as these have been perceived by politicians, commanders and forces over the years. Subsequently, a chapter on the theoretical aspects will attempt to merge the historical facts with military theory. From these three perspectives, the present, the past and the theoretical, I will endeavour to extract History's message to all those who wish to understand coalition warfare's advantages and complexities.

Chapter I – Afghanistan and Foreign Power Projection

LYING AT THE cross-roads of north-south and west-east, Afghanistan has seen – over the centuries – repeated interest by foreign and neighbouring powers looking for advantages in trade and security. It is not surprising, therefore, that Afghanistan's modern history is interwoven with British colonial aspirations and Russian southwards expansion, both being elements of the quest for trading routes, secure borders and wealth.

In 1611, signing a treaty with the Mughal Emperor the English trade corporation, the Honourable East India Company, began its operations in southern Asia. Over the centuries, which followed, the Company expanded its influence in India through a mixture of negotiations and persuasion buttressed by its own military muscle. Militarily the Company was vastly superior to the various Indian rulers and conflicts between the company and local forces were, to use a modern expression, always asymmetric. Thus, in 1757, on the battlefield of Plassey near Calcutta a force of 50,000 men under the ruler of Bengal, Suraj-ud-Daula, was defeated by an Anglo-Indian force of 3,000 under command of Robert Clive of the Company. In the early to mid-19th century the Afghans were squeezed from four sides – from the west by the Persians who had conquered Khorsan, from the east by the Sikhs who had taken the Punjab, including the Afghan winter capital of Peshawar, from the north by the Russians, and from the south by the

British.[1]

If one wants to comprehend the Anglo-Afghan wars, which were fought from the mid-nineteenth century until the early twentieth, one must consider the considerable commercial interests that, little by little, led to establishment of the British suzerainty over India – the British Raj. From the beginning of the nineteenth century onwards the British influence in southern Asia underwent steady expansion. This continued until at least 1876 when, on behest of Prime Minister Benjamin Disraeli (later 1st Earl of Beaconsfield, 1804-81), Queen Victoria was crowned Empress of India. From 1526 up until 1858 this land had been the realm of the Mughal Emperors, whose power steadily declined over the period while that of the East India Company increased. The British Raj covered the territories which today include the states of India, Pakistan, Bangladesh and Burma (Myanmar). Moreover, Nepal, Bhutan and Ceylon (Sri Lanka) were British dependencies, although they were not parts of the Indian Empire, which was eventually dissolved when, in 1947, all these territories emerged as independent states.

In the eighteenth century, France had been Britain's primary competitor in the business of acquiring colonies. During the Seven Years War (1756-63) she had lost most of her American colonies as well as some of her Indian trade stations, and all India except Sind was henceforth under the over-lordship of the Company.

However, the Franco-Russian treaty agreed in 1807 on the raft off Tilsit by the French and Russian emperors, Napoleon I and Alexander I, included a secret clause on a joint enterprise of invading British India from the west. Though France started negotiations with Persia to obtain her co-operation in this endeavour, it never materialised. One of the reasons that the Franco-Russian plan came to nothing might have been India's military preparedness. There were at the time reasonably detailed defence plans for India, which already had been in place for some years. As early as in 1804, while serving in India general Arthur Wellesley, later to become the 1st Duke of Wellington, 1759-1852, had managed to provide the British Raj with practically secure borders. Following the French defeat in 1815, France disappeared completely as a colonial competitor on the Indian sub-continent, and the Dutch and Danish trade stations also lost their relevance during the first half of the nineteenth century.

Until the Sepoy Rebellion (the Indian Mutiny) in 1857-58, the East India Company conducted its activities in most of what is today India,

---

[1] Tanner, Stephen, *Afghanistan: A Military History from Alexander the Great to the War Against the Taliban* (Revised Edition. Philadelphia: Da Capo Press, 2009) p. 129.

Pakistan and Bangladesh through three presidencies, each of which had its own army. The Company was a commercial corporation and, basically, it was the security provision for its business interests that constituted the military policy leading to the three Anglo-Afghan wars. Foreign policy, however, remained the business of the Foreign and Colonial Offices in London, whose interests and influence in the Company's dealings gradually increased over the years until London finally took over sole responsibility in 1858.

The Company's armies of native sepoys (Infantrymen, from Persian *sipâhi)* and sowars (troopers, from Hindi and Persian *suwar)* were led by British officers graduated from the Company's own training facility, The East India Company Military Seminary.[2] Known as Addiscombe Seminary, Addiscombe College, or Addiscombe Military Academy, it was situated at Addiscombe in Surrey, in what is now the London borough of Croydon. It was established in 1809 and closed in 1861. Its purpose was to train young officers for a career in the Company's armies in India, and it was a sister institution to the East India Company College in Hertfordshire, which trained the Company's civilian administrative personnel. In military terms, it was a counterpart to the Royal Military Academy at Woolwich – a forerunner to the Royal Military Academy Sandhurst. The Company was a commercial corporation and, basically, it was the security provision for its business interests that constituted the military policy leading to the three Anglo-Afghan wars. As the Company's activities had considerable implications for British foreign policy, a governmental board was set up to monitor its military activities.[3]

Neighbouring Afghanistan was never considered an objective for further expansion, but it was important to the Company because the land trade routes towards the West passed through that country. It was feared that Russia and Persia (Iran) as well as various Caucasian nations might exploit Afghanistan as a thoroughfare and a staging area for aggression against north-western and central India. Already shortly after the end to the Napoleonic Wars this menace seemed to materialise as Russia began moving south. From a British point of view this advance was a precursor for aggression. Through an alliance with Persia, Russia could threaten the British traders having to pass through Afghanistan. Therefore it became important to the Company and to Britain to achieve a good standing in Caboul – as the capital of Afghanistan was called by the British at the time.

---

[2] Ibid., p. 131.

[3] Ibid.

The East India Company foresaw two threat scenarios, namely one that indicated potential Russian invasion through Kabul and the Khyber Pass, and another materialising as a Persian incursion via Herat, Kandahar and the Bolan Pass. Thus, British interests in Afghanistan were of an entirely negative character, viz. denying others powers entry into India through the Afghan passes.[4]

At the time, Afghanistan was ruled by local tribes or clans in areas which were difficult to access and control, and by *sirdars* – the big landowners – in landscapes which were closer to big cities and population concentrations. The *sirdar* would be elected from amongst the scions of the local noble family and be responsible for raising troops from his fief. The administration was in the hands of the *jagirdar,* who was the government's representative, or lord lieutenant, who would be remunerated by means of his *jagir* – liens on land revenue – and he would be the *de facto* administrative authority in the area. He would then pay parts of the revenue to the *sirdar* and he would also enforce law according to a combination of Islamic scriptures and local tradition. The *jagirdar* might or might not be the same man as the *sirdar* and, in the cases where he was, his authority and esteem would be considerably enhanced. Although the *sirdars* were responsible for raising, maintaining and arming troops, after a specified time in the field the obligation to feed and pay them would fall on the central government – an arrangement which tended to limit the campaigning periods.

Central power was formally wielded by the princely ruler whose influence, however, was limited by the traditional atomised set-up of the country. In 1809, a rebellion took place against the sovereign of Afghanistan, Shah Shuja ul Mulk, while he was away in Peshawar, and Mahmud Shah took over the throne. In 1814, Shah Shuja ul Mulk settled down in British India, where he was granted a state pension allowing him to keep up his household and his life style.

In 1818, there was again rebellion and disorder in Afghanistan and in 1826 yet another ruler, Dost Muhammad, ascended the throne. As the kingdom had shrunk considerably during the previous years he styled himself Khan in lieu of the more impressive designation of Shah. In 1834, from his dwelling in British India Shah Shuja attempted a comeback, but he was defeated by Dost Muhammad in a pitched battle near Kandahar. He celebrated his victory by promoting himself to Amir. However, a few years later, in 1837, the neighbouring Sikh kingdom of Punjab under Ranjit Singh

---

[4] Ibid., p. 132.

took Peshawar, reducing Dost Muhammad's land anew.[5]

As foreign powers began to show interest in India – first Napoleon, then Russia, then the Germans – Britain found it necessary to control the overland route to India which ran through Afghanistan. As, in 1830, Russia approached both Turkey and Persia, the British Governor-General in India was instructed to initiate new diplomatic initiatives towards the states of north-western India and central Asia. This opened a new route for British trade: by sea to Karachi and up the Indus, then either over land via the Bolan Pass to Kandahar or via Peshawar through the Khyber Pass to Kabul. On the central Asian markets this would make British goods cheaper than those of their Russian competitors.

From 1834 onwards, the Russians urged the Shah of Persia (Iran) to claim ownership of the Herat Province, which had previously been in his possession. This would keep the British busy while Russia slowly but surely crept southwards across the Afghan border. However, the British envoy to Tehran did his best to persuade the Shah that this would be the ruin of his army as well as his treasury, as it would go against British interests. Nevertheless, in 1837 the Shah of Iran marched on Herat looking for an easy conquest, but the then local Afghan ruler, Kamran Shah Durrani, had no intent of giving up without fighting. As a Russian auxiliary force was defeated and the Shah faced a British ultimatum, the Persians withdrew, accepting the crude realities of the military force ratio. Subsequently to this Persian debacle, for some time the three states between Iran and British India – Sind, the Punjab and Afghanistan – were either friendly towards or dominated by the British. To exploit this window of opportunity Britain decided to try to negotiate an official agreement with the regime in Kabul, and in 1837 Alexander Burns was sent there as British envoy to solicit the ruler's, Amir Dost Muhammad's, possible interest in British support for keeping Persia at arm's length. While the amir appeared willing to accept such support, he demanded that Britain helped him restore Peshawar to Afghan rule. A coalition with two countries of conflicting interests was obviously unrealistic, and the British were presented with the choice either to go with the amir or to stick with the Sikh Kingdom of Punjab. As the Punjab had conquered Peshawar in 1834 and had been a good ally of British India for some time, the British chose the latter. Rejected by the British, Dost Muhammad turned to the Russians and received, in December 1837, a Russian envoy by name of Vitkevich at his court in Kabul.

Lord Auckland had been appointed Governor-General of India in 1835, and arriving in 1837 he realised that the benefit of British trade transit was

---

[5] T.A. Heathcote, *The Afghan Wars 1839-1919* (London: Osprey, 1980) pp. 17-18.

no lever for persuading the amir of Afghanistan, Dost Muhammad, whose primary, if not only, interest with respect to foreign co-operation was to obtain assistance to recover Peshawar from the Sikhs. Auckland was instructed to keep an eye on the Russians and counteract any progress in their influence, because, since the 1820s, Russia had explored all possible options for expanding southwards through the Caucasus. While Russian influence had increased in the early 1830s, she also managed to use Persia as her proxy prompting the attempted conquest of Herat. A Persian held Herat would have opened the way to Russian advances towards Kandahar, from which strikes might have been carried out against India via Kabul and the Khyber Pass or south-east through the Bolan Pass. The key to preventing such a development would be an amir in Kabul who was well disposed towards British Interests. In 1837, Lord Auckland's task was to find out whether or not Dost Muhammad might fulfil that description.[6]

This development challenged the Company's security interests and the policy of Lord Auckland's Indian government. It became apparent that there would be an obvious advantage to be gained by replacing the Russia friendly Amir Dost Muhammad with one more likely to please the Sikhs. Being the one who had once ceded Peshawar to Ranjit Singh, Shah Shuja ul Mulk seemed to be the right man for this purpose. He still lived on a British state pension in India and was available for the task. With this aim in mind Auckland hoped to be able to persuade the Sikhs to form a tri-partite coalition against Dost. Alexander Burns retraced his steps, leaving Kabul in April 1838, and Lord Auckland now offered Ranjit Singh, the Sikh king, support for his continued claim on Peshawar on the condition that he joined the Punjab with India in its efforts to assuage Afghanistan.

The agreement on the war coalition paved the way for a new and pro-active foreign policy, and in February 1839 the First Anglo-Afghan War started. With it 'The Great Game,' i.e. the Anglo-Russian race for acquiring strategically important positions in Afghanistan, began.

In the beginning of 1839, a task force named 'the Army of the Indus' crossed from Sind into Balochistan. This force comprised one division of the Bombay Army and one from the Bengal Army under Major-General Sir Willoughby Cotton, plus a large formation commanded by Shah Shuja. General Sir John Keane assumed supreme command of the expedition. In terms of overall numbers, the force consisted of 9,500 men from the Bengal Army, 5,600 from the Bombay Army, and about 6,000 men forming the contribution by Shah Shujah. There were well over 30,000 camels, around

---

[6] Michael Barthorp, *Afghan Wars and the North-West Frontier 1839-1947* (London: Cassel & Co. 1982) p. 29.

8,000 horses, and hundreds of oxen pulling carts. One part of this army advanced through the Bolan-Pass towards Kandahar where they arrived in May 1839 and proceeded towards Kabul. On 20 July the army reached the seemingly impregnable fort of Ghazni. There were now three options: to wait for the siege train coming up from Kandahar, to mask the fort and proceed with the main force, or the take the fort by a *coup-de-main*. The general-officer-commanding, General Sir John Kean, chose the latter. During the night, loads of explosives were clandestinely applied to the Kabul Gate, and by the efficient and daring action of Lieutenant Henry Durand in the early hours of 23 July the gate was blown open, and the advance and main storming parties rushed in. At the same time on the northern axis of approach Shah Shuja's forces and the Sikhs advanced through the Khyber-Pass towards Kabul where they arrived in early August.

On 7 August 1839, as Shah Shuja entered Kabul, Dost Muhammad made a fighting withdrawal north towards the Hindu Kush mountain range. He was captured in November 1840 and went into exile in Calcutta, where he subsequently lived on a British pension and under British protection in the same mansion that had once been occupied by Shah Shuja.[7]. By the end of 1839, the Army of the Indus retired from Afghanistan. It had been formed as a task force, and upon arrival back in India on 1 January 1840, as its mission was complete, it was disbanded. The First Anglo-Afghan War appeared to be over.

Thus, following thirty years of exile Shah Shuja was reinstated as amir of Afghanistan and a minor British garrison was stationed in the country to prop up his regime. Afghanistan remained quiet for the most of 1840-41, not least because the British wisely subsidised the *sirdars*. However, in October 1841 the Honourable East India Company found this practice to be too expensive and decided to terminate it. This resolution was not well received amongst the *sirdars* and others in Afghanistan, who had profited from the Company's erstwhile lavish support. Signs of uneasiness began to show – a revolt was in the offing. In November 1841 trouble began to stir and the British garrison in Kabul came under attack. This insurgency was not to be quelled until August 1842.

The British garrison in Afghanistan was commanded by the commandant of Kabul, the Waterloo veteran General William Elphinstone, who was now ageing and infirm and a far cry of the gallant officer of 1815. He did not have the necessary control of the situation, lacking full situational awareness as well as mental agility. Moreover, his political

---

[7] Tanner, Stephen, *Afghanistan: A Military History from Alexander the Great to the War Against the Taliban* , p. 150.

advisors, the envoy Alexander Burns and the political officer William Hay Macnaghten, were reluctant to gauge the situation and unwilling to accept that their political wheeling and dealing to promote friendly relations between Afghanistan and Great Britain had all been in vain. Then, on 2 November 1841 the expected revolt broke out when Alexander Burns and his family were murdered. A similar fate befell the paymaster, Captain Johnson, while his safe containing £17,000 was lifted by the assailants.[8] Macnaghten initiated negotiations with Akbar Khan, a leading rebel and son of Dost Muhammad, securing a promise of free passage if the British troops would leave Afghanistan. That was agreed upon, and soon after the troops filed out of the gates of Kabul heading towards India. Nevertheless, Akbar Khan's promise was a trap. On 6 January 1842, unaware of what lay in store for them 4,500 troops and 12,000 camp followers left Kabul. However, barely were they outside the city walls than local militias and bandits fell upon the column with the utmost ferocity and only very few of them managed to get safely back to British-Indian territory. Macnaghten, like Burns, had been murdered. However, Brigadier-General Sir Robert Sale successfully held out Jalalabad in the north-eastern part of the country close to the Khyber Pass and, similarly, Major-General Sir William Nott held Kandahar in the south. These garrisons had to wait a long time for reinforcement, since it was not until the beginning of May 1842 that a relief force – The Army of Retribution, commanded by Major-General Sir Georg Pollock – would assemble near Peshawar. On 5 June 1842 Pollock led his troops through the Khyber Pass. By cleverly dispatching some highly agile and fast moving covering parties to the mountain tops along the axis of advance, he deprived the insurgents the opportunity of ambushing the relief column. Thus Pollock's force managed to annihilate an enemy force of around 10,000 with a loss of a mere 135 dead and wounded.

But even before the Army of Retribution began arriving in Afghanistan, serious internal strife erupted in the country. Shah Shuja might have believed that he would benefit from the British withdrawal, as the presence of foreign troops was, and still is, a delicate issue in Afghanistan. However, rebels set up Nawab Khan Zaman Baraksai, a nephew of Dost Muhammad, as a rival ruler, whose son managed to assassinate Shuja. Moreover, fighting amongst the various chiefs had broken out in April threatening to tear the country apart. But, in June a compromise was found, and Fateh Jang, one of Shuja's sons, was recognised as shah with Akhbar Khan as Wasir, his captain-general.

On 19 April, the recently appointed Governor-General of India, Lord

---

[8] Ibid., p. 161.

Ellenborough, had ordered the Commander-in-Chief India, Lieutenant-General Sir Jasper Nicolls, to give up his post in Kandahar and withdraw to Quetta and Sukkur on the Indus. Moreover, he had ordered Brigadier-General Sale's and Major-General Pollock's formations to move from Jalalabad to Peshawar. The generals rejected this order and Ellenborough therefore submitted the dispute to the Duke of Wellington – an old Sepoy-general himself – who had, earlier in 1842, taken office as commander of the British army for the second time. Wellington was aware of the disastrous retreat from Kabul. He had informed the queen and the government of his assessment that now the enemies of Britain – France, the United States and Russia – were delighted to see their competitor humiliated; Britain's reputation had to be restored effectively and as quickly as possible.[9] He decided, thus, in favour of the two commanders and between 15 and 17 September 1842 Nott's formation from Kandahar and Pollock's from Jalalabad were joined in Kabul, where the Union Jack was now hoisted over Bala Hissar, the amir's fortress residence. Pollock had succeeded in rescuing all British prisoners from Afghan imprisonment including Lady Florentia Sale, the brigadier's wife, their daughter and baby granddaughter.

As mentioned above, Shah Shuja had been murdered during the rebellion and by the end of September his son Shapur ascended the throne. Again, the bulk of British forces were withdrawn to India and it was not until this moment that one could reasonably claim that the First Anglo-Afghan War was over.

As after a brief spell in office Shapur was murdered too, Dost Muhammad came back from his exile in Calcutta to resume his throne in Kabul. He reigned until his death in 1863 and he remained a loyal ally of Great Britain and representatives in India.

After the end of the First Anglo-Afghan War, British India, still ruled by the Company, had a neighbour who was an ally and a buffer to the west and north-west against Russia, Persia and Turkey. Afghanistan had to make some concessions as to her sovereignty, but she was not annexed into British India and therefore, to a large extent, retained her independence. In 1849, British India annexed Sind, Lahore and the Punjab and came to enjoy a common border with Afghanistan. Since Britain had suffered considerable losses during the First Anglo-Afghan War no unnecessary conflict was called for and, until 1876, she conducted her policy in India as so-called 'masterly inactivity' emphasising quiet and peaceful development of Indian affairs as long as nothing more was required.

---

[9] Ibid., p. 190.

Regardless, the common border was the object of frequent clashes, and British garrisons on the frontier had to keep a watchful eye on the development. If all else failed, punitive expeditions would be sent into Afghan territory. Between 1849 and 1857 fifteen such expeditions were mounted all along the Frontier, from the Bazais of the Swat Valley and the *cis*-Indus Yusufzais of the Black Mountains in the north, down through Mohmand and Afridi country, to the Wazirs and Shiranis round Dera Ghazi Khan in the south. Bound up with the question of tribal control was the defence of the frontier against external aggression. An invader from the north would have to negotiate the two large natural obstacles: the mountain range between Afghanistan and the Punjab, and the River Indus. The Russian threat had dwindled, but it re-arose briefly in 1853 when the Crimean War broke out.

The Crimean War began in 1853 as a consequence of years of conflicting interests between France and Britain and to some extent the Habsburg Empire on the one side and Russia on the other. While Russia liked to see herself as the protector of the orthodox Christians in Constantinople, she wanted unencumbered access to the Mediterranean Sea and wished to weaken the Ottoman Empire as much as possible. At the same time, the western powers, headed by Napoleon III, emperor of the French, believed it their duty to prevent exactly that. During the Crimean War, which was to last until 1856 as a consequence of this antagonism, Anglo-French interests in south-eastern Europe and central Asia collided with those of Russia. After three years of war Russia gave up. The eventual conclusion of this conflict was adverse to Russian aspirations, but it made her accept realities as these materialised at the time. This restraint, however, did not last forever.

Although British interests in the area were not seriously challenged by outsiders in the late 1850s, a menace materialised from within the army. On 10 May 1857, something happened which would eliminate the Company's government forever – the Sepoy Rebellion (or Indian Mutiny as it is also called). This rebellion had partly religious, and partly cultural origins. One of its consequences was that the British crown took over responsibility for governing India, and while the political leadership would henceforth lie with the viceroy appointed by Whitehall, a single military commander would be responsible for the armed forces of the whole of British India.

The treaty with Dost Muhammad ensured that Afghans did not join the rioters during the Sepoy Rebellion, which had ended the rule of the Honourable East India Company in British India. Since Dost Muhammad and his Afghan lieges had remained neutral, he was accepted by the British as ruler of Herat and Kandahar, too. He took Herat by storm in 1863 and died shortly afterwards. Although he had managed to make his country a peaceful place, his death was followed by years of power struggle amongst the contenders for the throne. The election of a new amir was put off for five years.

In 1868 and 1869 Bokhara and Tashkent were annexed by Russia, which had thus reached the Upper Oxus bordering with Afghanistan. A number of quiet years followed, but in 1876 the threat to the north-western frontier region and the trading routes re-emerged. In Britain the Tories were once more in office, and the government of Benjamin Disraeli found that the Afghan regime had become too pro-Russian, as opposed to its former friendship with Britain. As the trading routes through Afghanistan were essential to the prosperity of British India and because the Russians had repeatedly shown indications that they were willing to go out of their way to obstruct Britain's relations with the Afghans, the British government's evaluation of the security situation was gloomy. The development convinced the prime minister that not all was well.

Anxiety that the Russian advance should continue engendered two schools of thought: that of Masterly Inactivity (remaining close to the Indus), and those advocating a Forward Policy (to Hindu Kush or even the Oxus). In 1865, Sir Henry Rawlinson, member of the Council of India, had advocated re-establishment of an Afghan quasi-protectorate to protect British arms transports to Kandahar and Herat, but this did not occur. Since the death of Dost Muhammad, Afghanistan had been ruled by his son, Sher Ali. During the Russo-Turkish War, 15,000 Russians were on the move between the Oxus and the Hindu Kush. Then, the Berlin Congress of 1878 settled the matter and, once again, the Russians pulled back.

While Saint Petersburg tried to vindicate Imperial Russian expansion for the sake of spreading civilisation, in the late 1860s and early 1870s British India, personified by Viceroy John Lawrence, was less ambitious and carried on with the 'masterly inactivity' policy.[10] Lawrence claimed that experience showed that 'the Afghan will bear poverty and insecurity of life; but he will not tolerate foreign rule. The moment he has a chance, he will rebel.'[11] As if echoing these words, Islamist fanatics – the Ghazis – had begun an extravaganza of assassinations targeting British high profile officials such as Viceroy Lord Mayo in 1872 and the acting chief justice of Calcutta in 1871. The Ghazis were Wahabists believing in a seventh century interpretation of the Quran – in holy war against infidels and that the English language was destructive to pure religiosity. The Ghazis opened

---

[10] David Lyon, *Butcher and Bolt,* p. 82.

[11] Viceroy John Lawrence quoted in David Lyon, *Butcher and Bolt,* pp. 79 & 83.

their own schools across British India.

In 1873, Britain and Russia signed an agreement on Afghanistan. The Oxus and the mountains including a long stretch of land leading north to China was defined as belonging to Afghanistan, to make sure that a buffer zone existed between the two.

After two wars with the Sikhs following the death of Ranjit Singh, Britain had taken over the Punjab and the land west of the Indus up to the Suleiman Mountains, including the former Pashtu capital Peshawar.[12] Total British disengagement from Afghan affairs was no longer possible.

In 1876, Lord Lytton was appointed viceroy by the Tory government, which, realising the aggravated security situation in British India, gave him a more aggressive mandate than those of his predecessors. He was faced with a considerable challenge as Russia wished to move her presence forward to the Hindu Kush. Russo-Turkish tensions were solved at the Berlin Congress in 1878, but on 22 July 1878 Russia sent a diplomatic mission to Kabul persuading Amir Shir Ali to side with her, promising in return Russian support for taking back all Pashtunistan including Peshawar and the regions east of the Suleiman Range. On 14 August, Britain demanded to have a similar diplomatic mission set up in Kabul but was refused. Lord Lytton, who was a bit of a Hotspur, seized this diplomatic impasse as a welcome opportunity for initiating military action. Lytton was a high-handed ruler with great aspirations for India, but he had tense relationships with the military commanders. A Social-Darwinist, he had been accused by some critics that through his policy he aggravated famine that weighed down India during the early years of his reign, when between six and ten million Indians died. On 21 September 1878, he ordered an envoy and a military force through the Khyber Pass, but they were refused entry. Lord Lytton declared the triangle of Kabul-Jalalabad-Ghazni to be ideal in case of having to meet the Russians, and the passes across the Hindu Kush to have immense defensive as well as offensive value. The Second Anglo-Afghan War was declared.

On 20 November 1878 a British ultimatum expired. It had demanded Sher Ali to accept a British mission to Kabul and apologise for having refused the entry of British forces through the Khyber Pass.[13] The amir's reply, however, did not arrive until it was too late and, on the 21st, Lord

---

[12] On 21 February 1849 the Sikhs were defeated at the battle of Gujrat by the British, who took over the Punjab. The Punjab was annexed on 2 April 1849 and became part of British India.

[13] Tanner, Stephen, *Afghanistan: A Military History from Alexander the Great to the War Against the Taliban* (Revised Edition. Philadelphia: Da Capo Press, 2009) p. 204.

Lytton dispatched three military formations into Afghanistan, where they defeated Afghan forces in the Khyber Pass, at Ali Masjid, in the Kurram Valley and at Peiwar Kotal. Simultaneously, British troops managed to re-occupy Kandahar without a shot being fired.

The Amir Sher Ali fled north from Kabul with the Russians, who did not give him the support promised. He died shortly afterwards a bitter man, and his son Yakub Khan ascended the throne and sued for peace. Britain accepted dealing with him, and by The Treaty of Gandamak of 26 May 1879, she achieved full control of the passes of the Suleiman Range and tribal areas connected with them. Afghanistan henceforth would have to accept a permanent British diplomatic mission in Kabul.

Subsequent to the signing of this agreement a diplomatic mission was set up headed by the recently knighted Sir Pierre Louis Napoleon Cavagnari, who wisely, it seems, abstained from using his more francophone Christian names. He moved to Kabul with a sizeable entourage to set up the legation. Yakub Khan's authority, however, was limited outside the walls of his capital. Sir Louis Cavagnari believed that Yakub Khan might turn out a good ally, but he was as deceitful as many of the other Afghan leaders whom the British had had to deal with. Peace was short-lived and, on 3 September, fighting began around the Bala Hissar, while the Amir remained passive. From 9 o'clock onwards four sallies were made by British officers, though in vain, and on the same day Cavagnari was murdered along with his bodyguards and a number of other European residents.

The British Indian government had to move quickly, but all that was immediately available was the 'Kurram Field Force' under Major-General Sir Frederick Roberts (1832-1914), later Field-Marshal Lord Roberts, 1st Earl of Kandahar. This formation was set in motion at once and Roberts took it through the pass of 'Camel's Neck', as he had intelligence that tribal warriors made the Khyber Pass an unwholesome place to amass large military forces. During the advance towards Kabul the pressure on Roberts and his Kurram Field Force increased by the day and it turned out that the force ratio was 5:1 in the rebels' favour. Nonetheless, thanks to high mobility, units of the Charasiab Highlanders and the Ghurkhas managed to rout the Afghans, and in October 1879 General Roberts seized Kabul. Yakub Khan was arrested under suspicion of complicity of the murder of Cavagnari. Forty-nine Afghans were hanged as guilty of mass slaughter, and the fortification of Bala Hissar, the amir's residence, was blown up. The building material thus freed was used for reinforcing the fortified position of Sherpur north of Kabul.

This, as it turned out, was a wise precaution, because in spite of the successful operation and conquest of Kabul, two months later a public uprising forced General Roberts to withdraw to the Sherpur position, where he remained under siege for three weeks. On 23 December, however, he repelled an Afghan attack and routed the attackers, whereupon he re-occupied Kabul.

At this time, one more formation was operating in Afghanistan. In late 1878 Major-General (later field-marshal) Sir Donald Steward (1824-1900), 1st Baronet, had moved through the Bolan Pass with a force of 13,000 men and advanced towards Kandahar. The Afghan occupying force, however, fled the city in January 1879 before the arrival of the British troops. Then, in May 1880, Steward's formation left Kandahar marching on Kabul. Having beaten an Afghan unit under Ahmed Khel en route, Steward entered Kabul and took over command from Roberts.

By now, the attempts to create a manageable buffer state in Afghanistan had cost Britain two wars, huge disbursements and considerable losses of equipment and personnel; and every time the accords which had been struck up had been broken by the Afghans. From a British point of view, there seemed to be no other option than to break up the country into controllable minor territories. To facilitate that, Abdur Rahman Khan, a nephew of Sher Ali and for some time an Afghan expatriate in Russia, was accepted by the British as new amir in Kabul. With the exception of Kandahar, he was allowed to govern as much of the country as he might feel capable of. Kandahar would though remain under British protection.

In July 1880, Ayub Khan, a younger son of Sher Ali, tried to usurp the throne in Kabul, and on 27 August he managed to defeat a British-Indian brigade at Maiwand, which is situated between Kandahar and Musa Qala. The survivors held the fortress until 1 September, when General Roberts – having marched 334 miles in twenty-three days – arrived and dealt Ayub Khan a decisive blow outside Kandahar.

Nonetheless, in the autumn of 1880 the British cabinet finally decided to give up Kandahar and in May 1881 the British troops left Afghanistan completely. By then, the notion of carving up Afghanistan had already crumbled. General Roberts, upon his return to London, expressed the view that:

> We have nothing to fear from Afghanistan, and the best thing to do is to leave it as much as possible to itself. It may not be very flattering to our *amour propre*, but I feel sure I am right when I say that the less the Afghans see of us the less they will dislike us. Should Russia in future years attempt to conquer Afghanistan, or invade India through it, we should have a better chance of attaching the Afghans to our interests if we avoid all interference with them in

the meantime.[14]

Abdur Rahman, now in command of a sizeable army, wished to try, after all, to re-unite the country under his rule. In the spring of 1881 he succeeded in conquering Herat in the west as well as Kandahar in the south, assuming overall control of the whole country, which he kept together by ruling it with an iron fist for the rest of his life. By his death in 1901, he had not only united the country anew, but also re-established friendly relations with the British Empire. During his twenty years on the throne, for once Afghanistan experienced having a strong central power which wielded a consistent and hard-nosed policy, enforced the law and prevented the *sirdars* from once again tearing the country apart. Upon his accession to the throne, he had conceded to leading a foreign policy in harmony with the interests of Great Britain and India. This was a limitation to Afghan sovereignty, and it was not to be abrogated until the end of World War I.

In 1892, Amir Abdur Rahman had granted an audience to Sir Mortimer Durand at his court in Kabul. The reason was that Sir Mortimer, who was foreign secretary of the Indian government of Lord Lansdowne – Henry Charles Keith, 5th Marques of Lansdowne – wished to negotiate a permanent and stable border between Afghanistan and the territories under the British Raj. Contrary to expectations he succeeded, and in November 1893 the agreement was signed. The border, which has since been known as the 'Durand Line' was drawn in such a way that it included in British India the territories of Citral, Bajaur, Swat, Buner, Dir, Khyber, Kuram and Waziristan. This border proved stable, but its legality was challenged by the Pastun tribes right from the beginning. Moreover, conflicts in the area had assumed new and ominous dimensions. In 1897, for the first time, the British saw local religious leaders, the mullahs, encouraging uprisings, which turned out to be difficult to pacify by military might alone.

In his 21 years in power, Abdur Rahman strove to centralise and modernise Afghanistan, ruthlessly suppressing tribal uprisings, but he lacked the military might to affect detailed central governance of the regions. Headstrong *sirdars* continued to administer their fiefs according to time honoured codes and traditions, and centralisation was therefore far from complete when the Amir died.[15]

The acknowledgment of the Durand Line was to be the last major

---

[14] Ibid., p. 217.

[15] Peter Tomsen, *The Wars of Afghanistan: Messianic Terrorism, Tribal Conflicts, and the Failures of Great Powers* (New York: PublicAffairs, 2011) p: 64.

official act performed by Abdur Rahman. After having re-established peaceful and friendly relations with the British he died in 1901. On his death bed this hard-nosed ruler warned his successor to thread carefully when trying to modernise:

> My sons and successors should not try to introduce reforms of any kind in such a hurry as to set people against their ruler, and they must bear in mind that in establishing Constitutional Government, introducing more lenient laws, and modelling educational systems upon the system of Western universities, they must adopt these gradually as the people becomes accustomed to the idea of modern innovations.[16]

Taking this advice, his son Habibulla, who took over as amir, continued Abdur Rahman's policy. He avoided upsetting tribal and religious leaders, cultivated good relations with British India, and managed to keep receiving subsidies from the British right up to his death in 1919. In turn he acquiesced with the Durand Line and British suzerainty in foreign affairs. Moreover, he invited back some of those exiled by his father and placed several of these in high positions. One, Mahmud Tarzi enthusiastically advised Habibulla on reforms believed to be conducive to progress and formed a group of reformers – an Afghan version of the 'Young Turks' called the Young Afghans – which among others included the future king, Amanulla.[17]

In the early years of the 20th century, two men were the primary decision-makers on Anglo-Indian-Afghan affairs. In 1901, Lord Curzon became Viceroy of India, but it was not until 1908 that Sir George Roos-Keppel took office as governor of the *North-West Frontier Province* – the region which bordered with Afghanistan. He spoke the language of the Pathans (later to be called the Pashtuns) and he was well read as to their culture and traditions. He firmly believed that sooner or later there would have to be another war between the Empire and Afghanistan in order to – once and for all – create peace and stability in the country. Nonetheless, many years would pass before the third and final Anglo-Afghan war broke out. Moreover, during the entire World War I, when the military forces available for local skirmishes were few and far between, Roos-Keppel managed to keep the peace in the border region.

At the same time, Habibulla managed to stay aloof from the warring parties and preserve Afghanistan's neutrality throughout the First World

---

[16] Ibid., p: 63.

[17] Ibid., p: 66.

War. Even as the Ottoman Empire joined the war on the side of the Central Powers and the caliph-sultan, who was the nominal head of all Muslims, declared holy war against the Entente and did his utmost to persuade the Afghans to join as well, Habibulla remained neutral. Not even the German rapprochements with the Islamic peoples and the continued gossip that the Kaiser had secretly converted to Islam made him change his foreign policy. However, he did allow the Germans to open a consulate in Kabul, though he tolerated a so-called provisional Indian government, supported by the Germans, to be set up in his capital. However, his unwillingness to cash in on British difficulties and attempt full Afghan sovereignty and independence of the United Kingdom and India alienated the Young Afghans. In 1918 a student of one of the secular grammar schools he had himself established, attempted the assassination of the Amir.[18]

Eventually, Habibulla was assassinated during a shooting party in 1919, and his third son Amanulla endeavoured to take over the amir's responsibilities. There may have been some likelihood that the Third Anglo-Afghan War, which commenced shortly after, was initiated by Amanulla, in order to bolster his popularity. For this purpose he had probably planned to invade India, re-conquer the lost provinces west of the Indus and claim full sovereignty for his country. In any case, Amanulla publicly declared that Afghanistan was now an independent state and sovereign in internal as well as in foreign matters, and he made sure that his message was heard by the British-Indian viceroy. Ignoring the royal missive the viceroy responded that the existing Anglo-Afghan treaty was still in force and that Afghan foreign policy was therefore a matter for the British government to decide.[19] The invasion theory was supported by the fact that there were Pathans/Pashtuns living east of the Durand line who shared the same religious creed as those in Afghanistan. Moreover, the British Indian army was being demobilised after the First World War, hence the force ratio might have been favourable for the Afghans.

However, to others it seemed more probable that the unrest was a fall-out from Brigadier-General Reginald Dyer's bloody quelling of public unrest at Amritsar on the 11 April 1919. Regardless of the reason, Amanulla moved his troops forward to the Durand Line and declared holy war against the British. On 3 May, some of his formations crossed the border at the western entrance to the Khyber Pass, where they occupied the town of

---

[18] Ibid., p. 68.

[19] Ibid., pp. 69-70.

Bagh. From this position it was possible to cut the water supply for the forward British posts at Landi Kotal. Simultaneously, the Afghan army chief, Nadir Shah, moved his forces through the Camel's Neck Pass into the Kurram Valley, where he launched a propaganda campaign amongst the local Wazirian tribes to turn them against the British. Pathans/Pashtuns from both sides of the border now joined the Afghan forces, and the important British posts in northern Afghanistan had to be abandoned. The British therefore concentrated their efforts around Thal at the entrance to the Kurram Valley. Here a British brigade was encircled by Afghan troops until General Dyer managed to rescue them. The Afghan forces were slowly pushed back across the border, and little by little the hostile tribes were suppressed.

The war had been declared on 6 May and, in spite of difficult conditions for logistics and communications as well as the Imperial forces' war weariness and lack of enthusiasm, the British rather smoothly managed to drive the Afghans out of the Khyber Pass and make a resolute attack on Bagh.

Thus, the Third Anglo-Afghan War was now in full progress and, for the first time since the Great War, Royal Air Force aircraft were employed in a bombing campaign. This came as a complete surprise to the tribal warriors of the Dakka area of Afghanistan, and they were quickly dispersed. However, the planes' engines were not powerful enough to allow them to fly over the mountain tops. Consequently they had to use the Khyber Pass when flying between India and the area of operations, making them easy targets for the Afghans shooting down at them from the mountain ranges. After the seizure of Dakka the attack was meant to proceed to Jalalabad, but because of troubles further south and the Indian government's firm conviction of the hopelessness of occupying the whole of Afghanistan this never happened.

In the south in Balochistan the Afghan invasion was checked by a successful British attack against the fortress of Spin Baldak, from which the road to Kandahar could be dominated. The fortress was stormed and taken 27 May 1919.

On 3 June 1919 the fighting stopped and on 8 August a peace agreement was signed in Rawalpindi. The Afghans achieved full independence, including the ability to conduct their own foreign policy. However, an insurgency amongst the tribes in the frontier region followed in the wake of this war, and a prolonged armed resistance took place in Waziristan from 1919 to 1921. This was probably the most critical campaign the Indian army ever fought over the border region.

The Afghans now settled down to normal everyday pursuits making their country functioning anew. The amir sincerely wished to modernise the country and launched a glut of development programmes to make this happen rather more quickly than his grandfather Abdur Rahman would have liked. Some instability materialised along with his endeavours to westernise Afghanistan on the model that Mustafa Kemal (Atatürk) was successfully employing in the Turkish remnants of the Ottoman Empire. From about 1911 the Turkish secular Muslim movement had had considerable influence in Afghanistan, and nationalism, religion and modern technology were tied together in a way never seen before. A fortnightly magazine published in Kabul propagated the blessings of schools, road construction, equal opportunities for women etc. This magazine, however, was a forerunner of the worldwide movement of later years and it was seriously anti-imperialist as well as anti-British. Amanulla's ascension to the throne boosted the spread of this magazine, and still its models were the developments happening in Turkey and Persia. Everything, which the Taliban would later ban, now became greatly in vogue. The bowler replaced the turban, males were encouraged to wear western lounge suits and women persuaded to appear unveiled in public. Western norms of behaviour were *de rigeur* at court, the amir started to call himself 'king' and electricity was introduced. An educational programme for Mullahs was set up in order to prevent the Mosques from unsolicited influence, and a secular education of lawyers was initiated. Moreover, women were now free to choose their own spouses and the minimum age for marrying was fixed at eighteen years.

However, sadly, in the long run, Amanulla's plans and projects came to nothing. Over the next decade they clashed with tribal traditions, which were deeply rooted in Afghan society and not easily removed by a single far-sighted ruler. Amanulla was exiled to Italy where he died in 1960, and after a short interregnum Nadir Khan ascended the throne. Then, from 1930 onwards, the reform tempo stalled and to some extent Afghanistan slipped back into the ways of the past. However, reforms did not totally cease. Nadir Shah's Musahiban dynasty heralded more than four decades of quiet and gradual modernisation just the way Abdur Rahman had advised his sons in 1901.[20]

After the war, the British had been forced to focus their efforts on containing the uprising in several border districts and therefore accepted a fully independent Afghan foreign policy. The Afghan side repeated its acceptance of the Durand Line, but tribal rebellions continued from northern Balochistan all the way to Chitral in the far north.

During the winter and spring of 1920, regaining a mere minimum of control of the very difficult mountainous Waziristan had required the

---

[20] Ibid., pp. 75-76.

employment of 83,000 troops and supporting civilians, modern artillery, plus a significant air force of light bomber aircraft to pacify the well-armed Pashtu fighters. More limited operations in Waziristan, mainly using aircraft, continued until 1925. During the next 10 years, the pacification of the rebellious tribal districts succeeded by a balanced combination of military, economic and military means; and simultaneously the centre of tribal trouble migrated north to the Khyber Pass area. In the meantime, the local militias were thoroughly reorganised to facilitate re-deployment of regular troops stationed in Waziristan to new garrisons further east. These garrisons supported local economic development and provided reaction forces backed up by air power, doing so called 'air policing,' which could be deployed forward should the necessity arise. However, this system started to collapse in the mid 1930s, when political opposition emerged against the British Raj. At the same time, the 'air policing' was losing credibility. Moreover, as the traditional application of punitive action became unacceptable to the increasingly well informed and pacifist British public opinion at home, trouble began to stir anew in Waziristan.

Post-British Afghanistan

What was once the north-western part of British India bordering on Afghanistan is the present day Pakistan. The Durand Line, which still exists as the international border between the two, was drawn with little regard for tribal, traditional or religious layouts of the land. Today's insurgencies in Afghanistan – and to some extent in Pakistan – are reminiscent of that divide; they are Pashtu borderland unrests, and they should be seen and treated comprehensively as one and the same security issue.

In 1947 the British Raj ceased to exist. British India split up, initially into two states of Hindu India and Muslim Pakistan – the latter to be divided, from 1971, in the western and eastern independent states of Pakistan and Bangladesh – and for a few decades Afghanistan was left alone.

The last King of Afghanistan, Mohammed Zahir Shah (1915 – 2007), reigned for four decades, from 1933 until he was ousted by a coup in 1973 and went into exile in Italy. In 2002, following his return to Afghanistan he was given the title 'Father of the Nation,' which he held until his death.

Afghan society had always been based on extensive local autonomy with only marginal power vested in the central government, except from a few brief interludes. However, as during the 1960s and '70s many Afghanis had been educated in the West and in the Soviet Union, an intellectual élite began to form. In the early 1970s, these educated classes pressed for centralisation of power to buttress societal and economic reforms. Consequently, in 1973 the legitimate ruler of the country, King Mohammed Zahir Shah, whose reform tempo was seen as being inadequate, was deposed and replaced by his cousin Mohammed Daoud Khan. While Daoud tried to boost his popularity by supporting independence claims by Pakistani Pashtu and Baloch groups, his internal position remained weak. To consolidate his power, he sought the support of the competing Communist Party factions of 'Khalg' (masses) and 'Parcham' (banner). Backed by the Communists, he moved against the fundamentalist Muslim leaders, including the later Mujahidin chiefs Gulbuddin Hikmatyar and Ahmad Shah Massoud, who fled the country and were warmly welcomed by Pakistani Islamic politicians as well as by the Pakistani military. In early 1978, Daoud was killed in a coup d'état and replaced by the Khalg leader Nur Muhammed Taraki, who in his turn was assassinated the year after on the initiative of his rival, the Prime Minister Hafizullah Amin. Although in 1975 local rebellion against Daoud had been smashed quickly, the communist takeover in 1978 led to a general uprising against the central government. This rebellion was orchestrated from a political and military headquarters in Peshawar, Pakistan, where hundreds of thousands of exiled Afghan warriors were amassed in refugee camps.

In early 1979 the situation spun out of control, and Taraki applied for Soviet military intervention. Soviet military advisory and material assistance had increased in stages since 1972, and limited Soviet forces were already in place. After some months' contemplation the USSR decided to stabilise the situation with an operation modelled on the 1968 intervention in Czechoslovakia. In late December 1979, the operation went ahead with air landings and over land operations happening smoothly in a matter of a few days. Amin was killed by his Soviet Special Forces guards and replaced by Babrak Kamal, the leader of the less hard line Parcham faction of the Afghan Communist Party. Making the Afghan position even more complicated, in December 1979 Pakistani president General Zia ul Haq made a public declaration of the intention to move Pakistan towards Islamisation, thus setting the scene for a new 'Great Game' over the Durand Line – this time in the form of Communists versus Muslims.

The under-developed Pashtu districts on both sides of the Durand Line border soon turned out to be the main obstacle to the Soviet forces' supremacy in Afghanistan. The invading army had been trained exclusively for tank and mechanised warfare in Central and Western Europe and was completely unprepared for counter insurgency operations. Moreover, the Soviet forces had come to support the army of the Afghan government, but to a large extent this army's soldiers had deserted during the days of Taraki's and Amin's regimes, leaving the Soviet military having to learn how to do the job themselves. Employing Special Forces and other élite units they gradually developed a certain faculty for carrying out tactical 'cordon and search,' raids, ambushes and convoy operations combined with punitive actions against civilians. Soviet brutality and indiscriminate use of firepower forced around five million Afghanis to flee to Pakistan and Iran and turned the rest of the population against the Communist government in Kabul.

The Pakistani and Iranian refugee camps supplied recruits for the Mujahidin forces that operated from training bases beyond the Durand Line, entering Afghanistan to conduct hit and run attacks in support of local resistance groups. At the same time, Zia ul Haq's Islamisation programme aimed at propping up his political position fitted hand-in-glove with western wishes of exploiting Soviet over extension and the Saudi support of their fellow Muslims. Support for the co-religionists amongst the refugees was channelled through the Pakistani Inter Service Intelligence Agency (ISI), who decided who to assist. While seven Islamic resistance groups were selected as support worthy; secular, nationalistic and democratic ones were sidelined, with substantial support going to the brutal extremist Gulbuddin Hikmatyar's organisation. In 1986, after a surge of forces in order to achieve quick military victory the Soviet strategy changed to one of transferring operations to the Afghan military. A new Afghan government under Mohammed Najibullah decentralised the armed forces, giving a greater rôle to structures based on local militias and opened up the political system to the moderate part of the Islamic opposition. He also tried to find ways to preserve parts of the reforms already in place within a constitution of a slightly Islamic leaning. Having had a rough time in Afghanistan, after ten years the Soviet forces left in the winter of 1988 1989, the last tank passing the bridge into the USSR on 15 February.[21]

It is easy to believe that, had the US provided wholehearted support of the United Nations' effort to create a stable transition in the wake of Soviet withdrawal from Afghanistan, much would have been different.

Instead, the Pakistani Inter Service Intelligence (ISI) seized the opportunity to support the Pashtu paramilitary Frontier Corps in a double offensive to improve Pakistan's geo strategic position vis-à-vis India. Following a twin-track approach, Pakistan clandestinely infiltrated guerrillas into Kashmir to initiate and fuel an insurgency lasting from 1989 up until the limited war in 1999 between nuclear armed militaries of Pakistan and India, while at the same time endeavouring to replace Najibullah's regime with a Pakistan-friendly Islamist government under

---

[21] Tomsen, *The Wars of Afghanistan*, p. 322.

Hikmatyar. These attempts ended in defeat, and it took another three years before the Soviet Union's collapse and a worsening economic crisis led to Najibullah's fall.

The apparent lack of American interest and the competition amongst the former Afghan administration and resistance leaders, who were now mired in the traditional predicament of local warlordism, meant that no stable government was created. In January 1993, Hikmatyar's bombardment of Kabul started full scale civil war. Nearly all modernisation, national cohesion and infrastructure developed during the previous century were destroyed. In the civil war that followed, the puritan Pashtu Taliban creed spread in the refugee religious schools in the Quetta area of northern Balochistan. With its success in battle and backing by a war weary Afghan population, in a matter of five years the Taliban assumed control of nearly all of Afghanistan. By the end of the 1990s, the Pakistani military leadership under General Pervez Musharraf and ISI had shifted their direct support and co operation from Hikmatyar to the Taliban. There is some reporting to suggest that, indirectly, elements of the ISI may have supported the sophisticated leaders of Al Qaeda who had moved to Afghanistan and increasingly defined and dominated the foreign policy views and actions of the Taliban. With contacts and local knowledge gained from co operation with the Mujahidin in the 1980's, they quickly took over the fundamentalists' training facilities for Muslim militants, which used to be run directly by the ISI. Moreover, it merits some curiosity that when in 2011 Osama bin-Laden was found and neutralised, he was right under the ISI's nose.

Neither the British in the two invasions in the nineteenth century nor the Soviets in their intervention into the Afghan civil war in 1979 had expected to end up in demanding protracted counter insurgency operations, where, in all cases, the insurgents were supported from the other side of the Durand Line. However, Operation Enduring Freedom, the US invasion that started in November 2001, differed from both the British and Soviet invasions of central Afghanistan in its narrow military naïveté by not considering any significant post invasion deployment essential or even relevant for the regular American military forces. The very name of the operation highlighted the roots and depth of the pre invasion strategic analysis; a 'light footprint' would be sufficient. This was years before the crisis that led to the successful bureaucratic rebellion of senior officers of the American army and marine light infantry against the paradigm of winning by quick manoeuvre plus accurate strike and the introduction of a new counterinsurgency doctrine.²² According to the Pentagon's understanding at the time, its armed forces should smash the enemy military quickly with overwhelming, scientifically distributed, accurate firepower and then redeploy to bases ready for the next operation, leaving allies and civilian agencies to pick up the pieces and maybe providing some Special Force elements for a time to train the locals.²³ This economically attractive 'light footprint' was made dependent on the use of air power very similar in concept and motive to the British Royal Air Force's 'air policing' policy for the border zone in the interwar period. Due to the failure until 2011 to capture Osama bin Laden and continued low level problems on the border, a light division size army force remained, although most of the large country was cynically left to the re emerging warlords. Insufficient American example and merely moderate pressure was applied on the world community to live up to and realise the American president's spring 2002 promise of a 'Marshall Plan for Afghanistan' to support reconstruction of the demolished country. Pakistani generals used the US dependence on Pakistani bases to extricate and bring Taliban fighters, their Pakistani volunteers and ISI agents back to safety followed by reorganisation and re training in 'federally administered tribal areas' – or FATA – and northern Balochistan.

In 2005, following years of retraining and expansion from their bases in Pakistan, the different mainly Pashtu militant units under the 'Taliban umbrella' stepped up operations in southern and eastern Afghanistan. As early as the summer of 2004, the worsening security situation had led the *Médecins sans Frontières,* to leave the country.²⁴

Some of the insurgent organisations now employed to destabilise Afghanistan had actually been created in the early 1990s to fight the Indians in Kashmir. Many of the insurgent organisations are not religiously motivated Taliban and Al Qaeda warriors, but a new generation of tribal fighters striving to achieve control of territory, poppy production, mineral wealth, and smuggling routes. They include groups like the Haqqani

---

[22] It seems fair to suggest that this crisis was a direct consequence of the Bush-Rumsfeld administration's illogical defence policy of worldwide military engagement running parallel with cuts in budgets and personnel numbers.

[23] Moreover, nation-building, which would have been a logical follow-up on destroying an allegedly hostile regime and a country's infrastructure, was not on the US agenda during the first decade of the 2000s.

[24] *Médecins sans Frontières* is an international NGO of medical doctors and others trying to alleviate suffering in war-torn areas.

network, Hezb-e-Islami Gulbuddin and some smaller groups that have also joined the insurgency. The US has responded by sending armed unmanned aerial vehicles into Pakistan's Federally Administered Tribal Areas' airspace on cadre assassination missions, but the Taliban has managed to continue its attacks, including the bombing of the Indian Kabul embassy in July 2008, along with the rather limited facilities supporting US and other western operations. In response, the United States expanded into southern Afghanistan to contain and roll back the so called new Taliban offensive. The increasingly and understandably self confident militants in the border provinces that could get support from tribal members in the large Pakistani cities now started a terror offensive to destabilise the political system in Pakistan itself, as well as train disgruntled members of the vast Pakistani *diaspora* in the West. The murder of the opposition leader Benazir Bhutto in late 2007 did not change the situation and neither did the attack in Mombai (Bombay) one year later. However, the Pakistani military and ISI installations were now also targeted by the militants. Now that new governments had taken over in both Islamabad and Washington the destabilisation campaign directed against both countries from the border zone might finally be treated as the one indivisible problem it had been for more than a century after the Pashtuns were divided and left in poverty.

So far, the new Pakistani efforts against the Taliban have been concentrated first in the Swat Valley and then in Waziristan, and has been given the form of 'offensives' likely to have no lasting effect. The sanctuaries in the west have been left fully intact, making NATO and Afghani operations in the Helmand, Kandahar and Uruzgan areas both difficult and very risky. The only change has taken place in the Swat Valley and Waziristan, whence the US forces across the border has been attacked and from where the insurgents had started to move towards Islamabad. The fact that the US has now taken the leading role in the counterinsurgency in southern Afghanistan still has not diminished Taliban activities believed to be carried out from bases in northern Balochistan. One explanation for the lack of progress may be that the best regular infantry units in the Pakistani Army are the Balochi and Frontier Force Regiments, the latter being recruited amongst the Pashtuns. Other reasons might be lack of counterinsurgency training and the fundamental unwillingness on the part of the army to move forces away from the border with India.

However, there are few, if any, indications that can challenge the suspicion held by some that the driving force behind the operations against NATO and Afghan forces and institutions in Southern Afghanistan is the ISI's and some of the Pakistani armed forces' modern 'Great Game' to destabilise Afghanistan because their firm belief seems to be that only India stands to gain from a stable Afghanistan. So far, many believe that the US has not done enough to prop up the Pakistani civilian leadership in order to prohibit what increasingly looks like a Pakistani war by proxy.

There is nothing that indicates that the tribal Pashtuns in Afghanistan and Pakistan are fundamentally different from other human beings seeking learning opportunities for their children, better healthcare, improved infrastructure, full employment and fair and just treatment by the authorities – at the same time as they combine their aspirations with a cautious attitude to change and with a natural wish to be able to influence local development. The challenge is to be able to supply that little extra better government and the peace to achieve progress. It should not be impossible to compete with the Taliban and other groups that have nothing else to offer than a strict application of the Sharia law and regression into a perceived utopian past. A key obstacle to progress, however, is that the coordinated and combined effort cannot bring real progress if it remains limited to security (military, intelligence and police) efforts. To have a lasting effect progress must pervade political reforms, good governance and some economic development in the Pashtu tribal areas on both sides of the Durand Line.

Chapter II – ISAF Team 10

AFGHANISTAN'S PLIGHT WAS neither over with the departure of the Russians in 1989, nor was it easily ridded of the austere and dictatorial regime of the Taliban, who governed the country from 1996 to 2001 with an iron fist, religious dogma and Medieval outlooks on human rights and women's education. Moreover, towards the end the Taliban regime gave refuge to key elements of the Al Qaeda terrorist network, who, in 2001 perpetrated the attacks on the World Trade Centre and the Pentagon in the United States.

 In the aftermath of these attacks, the worldwide struggle against Islamist terror began. Shortly afterwards, the Taliban were overthrown by the United States' armed forces in Operation Enduring Freedom, and a period of relative freedom and social development commenced. Also in 2001, the multi-national military coalition, the International Security Assistance Force, or ISAF, moved into Kabul on a mandate by the United Nations' Security Council. In 2006 this force expanded its area of responsibility to covering most parts of the country and, from then onwards, the establishment and training of a reasonably efficient Afghan security apparatus gathered momentum. Since June 2007 – under the auspices of the British Task Force Helmand as the lead formation – a Danish-led battle group has been part of that effort.[1]

As we shall see in closer detail in Chapter IV, the concept of military coalitions is of no recent derivation and Denmark had participated in numerous wars as a member of such multi-national forces: in the Thirty Years War (1618-48), the War of Spanish Succession (1701-14) and the French Revolutionary and Napoleonic Wars (1792-1815) to mention but a few, and since the atrocities in New York and Washington perpetrated by the al Qaeda in 2001, once again, she saw herself a member of a large coalition of states combating the perceived evils of the time.

From 2003-2007 Denmark's main effort as a coalition partner was in Iraq, but since then Afghanistan – more specifically Helmand Province – has been the hub of her military exertion. Right from the beginning, Danish troops encountered heavy Taliban resistance and sustained many casualties, the death toll reaching forty-three in mid-2011. Nonetheless, over the years remarkable improvements of both security and community development have been seen, and in 2010 the Danish Battle Group 'ISAF Team 10' was able to launch the process of training and slowly but surely handing over responsibility to the indigenous security forces. In the following chapter we shall examine the challenges and gains of coalition warfare through the glasses of this battle group.[2]

On 14 June 2010, the Helmand Province became part of the area of responsibility of ISAF's newly established Regional Command South-West, under the overall direction of the commander of the 1st US Marine Expeditionary Force. The headquarters of this regional command was situated in Camp Leatherneck slightly west of Task Force Helmand's own operational headquarters in Lashkar Gah, 25 km west of the town of Gereshk. Task Force Helmand, comprising UK, Danish, US, Estonian, Lithuanian, Bosnia-Herzegovinian and Togan forces, remained, entirely under British leadership. Although the Afghan National Army's 3rd Brigade was also operating in the Helmand Province, this formation was not formally under British command. However, Task Force Helmand had a close co-operation with the Afghan 3rd Brigade, whose *kandak*s (battalions) and *tolay*s (companies) were permanently mentored by the task force's battle groups and sub-units. It was clear from the outset that the ISAF-Afghan National Army relationship was one of co-ordination, and that there was no formal subordination of Afghan troops under Task Force Helmand.

---

[1] Parliamentary decision B 161 taken on 1 June 2007.

[2] This chapter largely based on interviews with the commander of Task Force Helmand Herrick 13, the commanding officer and officers of the Danish Battle Group ISAT Team 10 and diaries kindly lent to the author.

For this reason Afghan units' and individuals' adherence to international law remained a national responsibility.

While from the arrival of the Danish Battle Group, ISAF Team 10, in August, 2010, and throughout the following two months the 4th (UK) Mechanised Brigade was the lead formation, from 10 October onwards the task force came under the aegis of the 16th (UK) Air Assault Brigade, commanded by Brigadier James Chiswell. Within the Helmand Province each battle group was responsible for a district or parts of a district. The Danish Battle Group's area of responsibility was the northern part of the Nar-e Saraj District, and, in order to underscore the multi-national – or combined – character of the activity, the Danish Battle Group with its collaborators were designated Combined Force Nar-e Saraj North or, in daily parlance, CF-NES (N). Essentially, this force was a Danish battalion of the Royal Life Guards with a number of Danish and foreign sub-units and staff officers attached.

As it was in the past so it remains true today that a country going to war – whether alone or as part of a coalition – must have a goal oriented strategy based upon the aims with the war. The ISAF Team 10 Battle Group was deployed to Afghanistan as a tool meant to advance the Danish official strategy in which the government had stated that, as a coalition partner it was:

> Denmark's objectives, from 2008-2012, to increase security and stability, state building and human rights, education, the improvement of living conditions and cross cutting issues involving women's rights, good governance and counter-narcotics. These objectives form the foundation of Denmark's long-term commitment to improving the safety, opportunities and every day lives of the Afghan people.[3]

Although the Danish Battle Group was mostly designated 'ISAF Team 10,' it also carried the nickname 'Valdemar 10' – a parallel to the British 'Herrick numbers' that were used occasionally and for purely national purposes.[4] The Battle Group and its parent British 'lead-formation' did not rotate simultaneously, which entailed both advantages and snags. On the one hand this was good for continuity, but on the other the Battle Group's

---

[3] Danish Ministry of Foreign Affairs, http://www.afghanistan.um.dk/en/menu/DenmarkinAfghanistan/DenmarksObjectives accessed 4 August 2011.

[4] Each 6-month-team of either British or Danish troops were nicknamed nationally as Herrick or Valdemar respectively followed by their number in the succession of teams.

commanding officer and his headquarters had to adjust twice to the changing task force commanders' operational priorities and preferences.

As the Danish contingent arrived in Afghanistan, Task Force Helmand was already carrying out its mission in close co-operation with the Afghan National Security Forces – the Afghan National Army, ANA, and the Afghan National Police, ANP. This tri-partite co-operation was to become known as 'the rule of three' and was the basis both for mentoring the Afghan forces and for the operational partnership with them. It implied that whenever possible, an ISAF company, an Afghan Army sub-unit and an Afghan Police sub-unit would work together planning and executing operations as a combined force. They would share the mutual obligations of protecting communities, creating conditions for the civilians' freedom of movement on main thoroughfares in the area, expanding the influence of the Afghan central and local governments and, identifying possibilities for re-integration of former Taliban and other insurgents in society.

The Danish Battle Group was expected to leave their home base for Helmand in mid-August but, as always when military forces are deploying, there was an advance party arriving early. By Sunday 1 August, the Chief of the Operations Branch (S3) Major Christian Bach Byrholt was already there and engaged in familiarising himself with the tasks he was soon to solve and the conditions on the ground. Although he was initially overwhelmed not only by the work load, but in particular by the intense heat – 35 to 40 Celsius in daytime – he engaged himself in various sports activities whenever time allowed, and though he managed to work long hours, he admitted that during briefings he frequently felt the sandman's presence. Thus, for all practical intents and purposes, the ISAF Team 10 Battle Group – i.e. the Danish part of the Combined Force Nahr-e Saraj North – had started arriving in early August, and on the 11th of August the official transfer of authority between out-going and in-coming teams commenced with a view to being completed on the 18th. The Combined Force was tasked with occupying, developing and controlling during autumn and winter 2010-11 its area of responsibility, the town of Gereshk as well as the thinly populated northern part of the Nahr-e Saraj District. It comprised staff, combat units, a headquarters and logistics company and support elements. The combat units were one troop of Leopard tanks from the Regiment of Jutland Dragoons, 'A Troop,' two Danish mechanised infantry companies, B and C Coys of 1st and 2nd Battalions Royal Life Guards respectively, two British infantry companies, D and F Coys, which were – since the arrival of the UK 16th Air Assault Brigade sub-units of 1st Battalion Irish Guards, a Light Reconnaissance Company, E Coy, (US later to be replaced by D Squadron, the Household Cavalry Regiment (UK)) and a Viking Group.⁵ The Combined Force's Headquarters and Logistics Company comprised the necessary command and control elements, engineers, artillery observers, an electronic warfare component, human intelligence resources and medical support. Moreover, with the Danish contingent there were various advisory bodies working with the Afghan National Security Forces (ANSF).

The Danish battle group was in Helmand to solve, primarily, military tasks, but they were doing so to fulfil political aims. The policy guiding their efforts was set out by the Danish government in its Helmand Plan published in the spring of 2010; a plan that would complement the Danish 2008-12 Afghanistan Strategy cited above:

> As early as 2008, Denmark had developed a long-term strategy for her engagement in Afghanistan. The strategy stated that Denmark would gradually shift the balance between civilian and military capacities moving towards an increased civilian effort and a more unobtrusive military rôle. Along with reaching the goals set for the military efforts, the Danish military engagement would be gradually reduced.⁶ The Helmand plan covers specific initiatives within the comprehensive Danish engagement in Afghanistan that entails political as well as military and civil economic assistance. The plan complements the 5-year strategy 2008-12. The Helmand Plan for 2011-12 involves adjustments of the Danish military engagement towards an increased focus on training of the Afghan security forces. The plan also entails a markedly strengthened police training effort as well as an increase in civilian and developmental contributions to Afghanistan. The gradual adjustments to the Danish engagement in Afghanistan - initiated in 2010 - continue. The Danish soldiers will over the coming years gradually assume a more retracted posture, allowing the Afghan security forces to take the lead. The changed focus from combat to training of the Afghan security forces means that Denmark can begin to reduce the overall number of troops. The reductions will continue and by the end of 2014, there will be no Danish combat units in Afghanistan, but Denmark will continue its active engagement in the country with an emphasis on development assistance.⁷

---

⁵ This sub-unit of the 2ⁿᵈ Royal Tank Regiment was nicknamed the Vikings after their lightly armoured 'Viking' personnel carriers used until 2010 and then being replaced by Warthogs.

⁶ Danish Strategy for Afghanistan 2008-12, from internet accessed 23 July 2011: http://www.fmn.dk/eng/allabout/Pages/ThedanishengagementinAfghanistan.aspx

⁷ http://www.afghanistan.um.dk/en accessed 4 August 2011.

Fiction and Reality

In early 2010, a film called "Armadillo" was shown in Danish cinemas and on television. It was fiction, but made heavy use of footage shot in the Danish camps and bases in the Helmand Province and during patrols with the infantry in the Gereshk Valley. Jeanette Serritzlev, M.A. had worked as a press and information officer with the Danish contingent, ISAF Team 8, shortly after the scenes had been shot. On 15 June 2010, she told the newspaper *Berlingske Tidende* that in her opinion this was "a film pretending to depict reality, but in fact it was merely fiction looking real." The film did use relevant footage of a number of actual and dangerous situations in the mission area and might well have been good entertainment for soldiers who knew about the underlying reality, but it would be a harsh misguidance to ordinary film audiences looking for information on Danish defence and security policies. The film, she claimed, showed a depressingly negative picture of army life in general and the development in Afghanistan in particular, which she could not vouch for. Admitting that she might be a little naïve, she related that walking around in the Streets of Gereshk she met grateful and obliging people, girls going to school for the first time in their lives, thriving trade in the bazaar and a generally positive development of the Afghan society. For these reasons she dismissed the film's claim to be a narrative of reality. Nevertheless, Jeanette Serritzlev observed some well-made portraits of the soldiers, their military background, conscientiousness, efficiency and zeal. Moreover, she was astonished that, while she was away with the ISAF Team 8 contingent, press clippings from home told of soldiers losing their enthusiasm and belief in the usefulness of their business, and she was as flabbergasted as the rest of them. Those around her did indeed believe in the value of their mission, she maintained.[8]

Regardless of the message of this film and similar accounts by authors, journalists and some politicians, as, in the spring of 2010, Colonel Lennie Fredskov – the commanding officer of ISAF Team 10 – prepared for deployment of his battalion to Afghanistan, the general attitude to Danish participation in the coalition-of-the-willing seemed to be confidence that this was indeed an undertaking contributing to a safer and better world.

Politically, military engagements in coalitions in far-flung countries serve Denmark's security indirectly, but they also send messages of solidarity with partners wishing to keep international law and order and combat terrorism. For this reason there is much political goodwill to be

---

[8] Jeanette Serritzlev, "Andre scener fra Afghanistan [Other Scenes from Afghanistan]" in *Berlingske Tidende*, 15 June 2010.

gained by coalition participation. Returning from Helmand in February 2011 and judging on the background of his previous experiences as the security advisor at the Prime Minister's Office, Colonel Fredskov expressed his serious belief that the effort in Helmand had yielded considerable benefits for Denmark in the form of eased access to and clout with decision-makers in Whitehall and Washington, as well as a high profile amongst NATO partners. Moreover, in recent years the military engagement in Afghanistan had engendered immense improvements of the technological level of the armed forces' equipment, training standards and the officer corps' insight and understanding of the requirements of international operations. Colonel Fredskov believed that never before, at least not during post-Cold War years, had Danish units been so strong, well-trained and well-equipped as during the recent operations in Afghanistan, although, he admitted, to some extent the progress overseas had over-stretched the forces to an extent that had reduced the army's ability to train and develop at home.

Training and preparation for deployment to the hot spots of modern anti-terror warfare should, however, never be taken lightly. The foundation for these endeavours will always be military history, as it is through the experiences of the past that the need for development and adaptation become clear. However, since the brief Danish co-operation with Napoleon's 13th *Corps d'Armée* under Marshal Davout in Northern Europe in 1813-14, Denmark had had little experience with coalition warfare.[9] As a result, in 2010 the Danes' comparison with the past and their ability to extract lessons for the present primarily concerned the most recent years' changes in outreach, equipment standards and *modus operandi*. Conversely, UK forces have a rich historical background for developing military doctrines and adapting to new conditions – colonial war, the world wars, Korea, the Suez crisis, the First Gulf War, etc. It is, therefore, pertinent to ask to what extent military history formed the basis for the British task force commander's comprehension of his rôle and his operational choices within the framework of this coalition. Although, in the case of Brigadier James Chiswell, there was no doubt in his mind that the primary historical foundation was cast while undergoing obligatory educational programmes on military history at institutions such as the Royal Military Academy at Sandhurst, Staff College and Higher Command Course, it was primarily his personal experiences from coalition operations in the Balkans in the 1990s and in Iraq in the early 2000s that provided the background for his

---

[9] Louis-Nicolas d'Avout (10 May 1770 – 1 June 1823), better known as Davout, 1st Duke of Auerstaedt, 1st Prince of Eckmühl, was a Marshal of France during the Napoleonic Era.

deliberations and actions as a lead formation commander in Afghanistan.

Historically, leadership of coalition forces has involved issues such as who makes the decisions and at what level direction should be given when putting officers and soldiers into harm's way. Traditionally, or at least since the mid-19th century, decision-making in war has happened at three distinct planes, viz. the strategic, the operational and the tactical levels. These levels were usually linked to the capitals or supreme headquarters as far as strategy was concerned, army groups or armies for operational decisions, and to lower headquarters in case of tactical dispositions. However, as far as management of the coalition warfare in Afghanistan is concerned the Danish doctrine has not been absolutely clear – the British perhaps more so.

Brigadier Chiswell saw decisions being taken as follows: the politico-strategic level of decision was to be found in national capitals of coalition partners, in Brussels as far as NATO was concerned and in Kabul with the Afghan president and the NATO Commander-in-Chief of ISAF. While the operational level would be the International Joint Staff in Kabul, tactical decisions concerning the Helmand Province were taken by the regional commander south-west and commander Task Force Helmand as well as the commanding officer Combined Force Nahr-e Saraj North. A slightly different, though not contrasting, view was offered by Colonel Fredskov: 'While the national capitals and the ambassadorial level in Kabul constituted the politico-strategic level, ISAF HQ would perform as the military-strategic decision-maker. The International Joint Staff in Kabul, then, was the operational, and Task Force Helmand and Commander Combined Force Nahr-e Saraj North the tactical levels.' However, Fredskov believed that, as the Danish commanding officer of Combined Force Nahr-e Saraj North, he was duty-bound to putting the Danish strategic intent into practise, and his battle group was in some respects to be found at the operational level. These two views are slightly disparate, though not mutually exclusive, but they do show some unavoidable intricacies in perception of coalition warfare.

The levels of decision-making were crucial to stating the aim *with* and the aims *in* the war, i.e. the political reason for committing forces to the war effort and, respectively, the military steps taken to ensure its fulfilment. The point of departure for any martial endeavour is the political aim <u>with</u> the war. Then, how was this aim perceived by commanders, what was their rôle in it and their contribution to providing long term as well as short term *indirect security* to their respective communities? The task force commander saw this aim as being the same as the central NATO intention, which must of necessity coincide with that of the Afghan government: denying Afghanistan to international terrorism and promoting a stable community and a reasonably well functioning state, giving the Afghans confidence in their government, and bringing about a sense of security and opportunity. And once started, one would have to follow this aim and see it through to the end. Taking his point of departure in the Danish Ministry of Foreign Affairs' official policy, Colonel Fredskov stated the war aim, as he saw it, slightly differently: supporting civil reconstruction, strengthening the Afghan National Security Forces, weakening the insurgents and thereby creating improved security and conditions for the Afghans to take over responsibility for their country. Although differently phrased, in essence there was no discrepancy between UK and Danish perceptions of the aims *with* the war, and, since these aims formed the basis for strategy, the agreement was indeed fortunate and remained unchallenged throughout the Danish ISAF Team 10 Battle Group's stint in Afghanistan.

Unlike the aim with the war, which was politically fixed, the aims in the war lay within the realm of military operations. Brigadier Chiswell saw two such aims, namely contributing to the struggle in support of civilian development, and supporting the development of the Afghan National Security Forces to allow them to take over gradually in order to be fully responsible for the safety of their country by the end of 2014. On the basis of Chiswell's operations orders and complying with the Danish Helmand Plan 2010, Colonel Fredskov decided that his primary aim *in* the war would be to train and improve the Afghan National Security Forces in his area of responsibility, with special emphasis on the police in the town of Gereshk, which ought to perform major progressive steps within a period of six months. Moreover, there were important aims such as reducing the violent activities in the Deh-e Adam Khan area and on and along the Bandi Barq Road in order to allow repair of Gereshk's hydro-electrical plant, reduce violence in the area close to the Patrol Base Line – the row of patrol bases running north to south about six kilometres east of Gereshk – to facilitate peaceful activity, and to improve safety on the main Afghan thoroughfare, Highway 1, in order to ascertain freedom of movement and smooth the progress of civil life and commerce.

In war, sometimes pressure is applied by political or military superiors on a commander in order to achieve aims which are politically opportune though irrelevant or outright detrimental to his objectives on the ground. The Swiss-French general and military philosopher Antoine Henri Baron de Jomini called such aims 'political decisive points' and he found them harmful to the conduct of military campaigns. One example of this phenomenon might be the American insistence on executing Operation Dragoon, the pre-planned landings in southern France in mid-1944, in spite of the dire need that summer to have as many forces as possible available for the operations in Normandy. Such maladroit political interference might easily have happened to Task Force Helmand in 2010-11 as well, but, fortunately, it did not.

Before the advent of international terror, the UK had had a counter insurgency doctrine based on a long history of combating colonial riots, but this had been made obsolete by the atrocities of 11 September 2001. However, for a year or so prior to the deployment of the Danish ISAF Team 10 a new doctrine addressing the challenges of the wars against Islamist terror had been in place. This doctrine seemed to tally nicely with the notions of the Danish Battle Group in 2010-11. Moreover, no difficulties arose vis-à-vis the Afghan forces, although the points of departure were certainly two very different value systems. Colonel Fredskov agreed with this observation telling that, prior to deployment, Danish officers had closely studied the UK Army's Field Manual *Countering Insurgency* as well as various independent authors' contributions such as John A Nagl's book *Learning to Eat Soup with a Knife: Counterinsurgency Lessons from Malaya and Vietnam*. However, while British and Danish approaches tallied almost completely, the Afghans showed some sign of having been taught by the Americans which made their views differ slightly from that of their Helmand collaboration partners. In particular, the way civil-military-relations were conducted, the rules of engagements, including the use of force, and battle procedures differed. Nevertheless, no major disagreement surfaced and, because of the need to solicit the support from the population, their understanding of counter insurgency action grew every day.

Throughout his period in Afghanistan, Colonel Fredskov conducted an offensive line of operations in order to seize and keep the initiative vis-à-vis the Taliban. His sub-units were composed of regulars, whose equipment and level of training allowed this approach, and, in particular in Patrol Base Line, this proved successful. Nonetheless, there were instances when slightly more defensive measures were called for. When taking the initiative through offensive action at one point, emphasis of effort (i.e. the majority of one's forces) was required there, necessitating a less forceful and more defensive action in other places and diluting one's troop presence, as for instance happened at Forward Operating Base Budwan.

In late September 2010 the newspaper *Berlingske Tidende* claimed that from then onwards Danish forces had changed their operational priorities – although the paper called this a change of strategy – no longer chasing the Taliban and exposing themselves to improvised explosive devices (IEDs) in the Upper Gereshk Valley, where Forward Operating Base Budwan was situated. The paper claimed that the bulk of the troops were to be withdrawn into the town of Gereshk to protect the population there.[10] As had been seen so often before, this was merely the simplistic understanding of a journalist knowing little of strategy, let alone operations and tactics. The actual intention was not to concentrate all troops in one town, but by 'unfixing' the subunits making sure that there would be fairly large mobile forces available to deploy quickly to where and when overwhelming firepower might be required simultaneously with keeping up sufficient armed presence anywhere the local citizens in the Gereshk Valley might need reassurance.

During ISAF Team 10's presence in Helmand, the intensity of the public debate at home concerning the utility or futility of keeping combat troops in Afghanistan grew. Researchers at universities and think tanks raised doubts about whether combating the Taliban had any positive effect at all. Nevertheless, with general elections coming up in the autumn of 2011 and well aware that most voters were sensible to any criticism of their dear ones risking their lives in Afghanistan, political parties were keen to emphasise that their Helmand policy was almost identical with that of their political opponents. In *Berlingske Tidende* 22 February, the leader of the opposition, Social Democrat Ms. Helle Thorning Schmidt, declared her unequivocal support of the efforts in Afghanistan until the end of 2012. However, as far as the period after 2012 was concerned she and her collaborators of the Socialist Popular Party were less clear about their intentions.[11]

As the Combined Force moved into Nahr-e Saraj North, there was a lot of media talk about the need to change strategies, and indeed the next six months turned out to become a period of alterations. However, the strategy remained the same. The primary goal for Colonel Fredskov and his Combined Force was to convince the local Afghans that the rule of the official government and its local affiliates was preferable to that of the Taliban and the insurgents, with all what this would imply of Sharia and subjugation of women. Achieving this aim would require novel approaches not least of which with respect to planning the way ahead in accordance with the needs of the Afghan security forces, who would eventually have to take over sole responsibility. Peace and progress were what the citizens craved, and the keys to such development were security, functioning power plants, and an intact infrastructure. Reactivation of the old hydro-electrical

---

[10] "Danmark ændrer strategi i Helmand [Denmark Changes her Strategy in Helmand]" in *Berlingske Tidende* 22 September 2010.

[11] "S: Helmand-bidrag star fast efter valg [Social Democrats: Helmand Contribution Remains Unchanged after elections]" in *Berlingske Tidende* 22 February 2011.

works in the area between Bandi Barq and Deh-e Adam Kahn north-east of Gereshk would be a boost to life and commerce in the town, but repair and running of the plant at its sluices required macadamisation of the road leading to it and protection of the employees during the construction work. However, this project was deemed immensely important as it would convince the citizens of Gereshk that the international effort was successful and to their benefit.

Photo taken by OC B Coy, Danish Battle Group: Sniper in position during closure of Patrol Base Zumbalay.

The newspapers' stories about changes of the Danish strategy for Afghanistan were grossly exaggerated. The fact of the matter, of which the newspapers had only superficial knowledge, was that Colonel Fredskov was adjusting his *modus operandi* so that henceforth his troops could act more swiftly when challenged by the insurgents. As mentioned, the repair and maintenance of the hydro-electric plant was essential and the Afghan district governor consequently placed great emphasis on securing the plant, the road and the ongoing repair activities. In a meeting between the District Governor, the Combined Force's deputy commander, Lieutenant-Colonel Thomas Funch Pedersen and the stabilisation advisors it was decided which development and security measures should be central to the battle group's campaign plan. Subsequently, a sketch for the development in Nahr-e Saraj was drafted stating that while the bases Budwan and Zumbalay, 15 km and 12 km east of Gereshk respectively, were to close, more forces could be employed in mobile rôles in the vicinity of Gereshk.

One clear advantage of this plan was that the companies operating east of Gereshk protecting the reconstruction works at the power plant, the companies in Patrol Base Line and the company in Patrol Base Rahim, situated 3 km north-east of the Line, were able mutually to support each other. Parts of this plan were immediately approved by the commander of Task Force Helmand, but others, such as the closure of Forward Operating Base Budwan, took a little longer. The reasons for this lay in the fact that while Colonel Fredskov wished to concentrate forces near Gereshk, Brigadier Chiswell was equally convinced that, in order to assure the local citizens of ISAF's will and ability to protect them, security had to be deepened in wider parts of the Upper Gereshk Valley.

Nonetheless, Brigadier Chiswell supported the closure of Budwan, but it took a fair time to get final agreement on this. Task Force Helmand had to confirm the Afghan senior decision-making level's agreement, which was eventually brought about by Chiswell personally speaking to Provincial Governor Mangal. Moreover, he needed to ensure that Headquarters Regional Command South West were completely happy because – at the time – they were developing plans for greater effort further north in the top end of the Upper Gereshk Valley. Chiswell didn't want to find himself closing a base they might subsequently find useful. Once it was confirmed that the Regional Command would shift its effort onto the east side of the River Helmand (relating to upgrading and securing Route 611 to the North), Task Force Helmand agreed to Colonel Fredskov's proposal on the closure of Budwan, which would henceforth have no further utility.

In November 2010, it became obvious that, with the much proclaimed surge of American troops, NATO had taken the initiative and the Taliban became increasingly stressed. Upon returning from a stint as ISAF's Regional Commander South, British Major-General Nick Carter visited Copenhagen and told the Danish press that considerable military progress was now seen ubiquitously and that the Taliban's morale was declining. Many insurgents had put down their weapons to enter formally the Afghan reintegration programme. Carter pointed out that in 2010 many places in the Kandahar and Helmand Provinces were now sufficiently safe for governors, civil servants and ordinary citizens to perform their respective tasks and to travel freely within these regions. Targeting Taliban leaders by Special Forces had worked well; across Afghanistan 235 leaders and 1,066 ordinary insurgents had been killed and another 1,673 had been captured during these operations. ISAF and the Afghan Security Forces were now so well-equipped and well-staffed that no major hiding place was a safe haven for the insurgents, and their freedom of movement had been seriously curtailed. The most difficult issue now, therefore, was the need to establish a well-oiled civil administrative machinery in all districts. The lack of education and of honest and decent officials were the main challenges.[12] Brigadier James Chiswell spoke along the same line as, on 14 February 2011, he told the BBC that there was reason for cautious optimism. Momentum was shifting and the opponent was under severe stress. But even more importantly, there was growing confidence amongst local citizens that Afghanistan might soon rid herself of the insurgents and that government authorities were to be trusted.[13]

While in the summer of 2010 the British and Danish prime ministers agreed on 2014 as the year in which to terminate the presence of combat troops in Afghanistan, the UK Ministry of Defence believed that mentoring of Afghan forces and partnering between these and the western military units might lead to immediate redundancy of troops in Afghanistan. However, since these activities would require more rather than fewer western mentors and partner units the British Army was adamant that all 9,500 troops would be needed – at the very least until 2012.

Along the same lines, in mid-2010 the NATO summit in Lisbon revealed that the alliance planned to start handing over responsibility to the Afghan National Security Forces in various regions within the following six month. However, fitting reasonably with the Anglo-Danish decision, the Helmand and Kandahar provinces were expected to remain ISAF responsibilities for at least two more years.

Co-operation within the Coalition

While coalition partners may share some interests, they do not necessarily do so to the same depth or for the same length of time as do true allies. Indeed, they may agree in some vital fields but be competitors in other areas. In Afghanistan in 2010-11, the UK and Denmark were allies as well as coalition partners, but very disparate associates as far as the sizes of contributions were concerned. To the lesser partner it appeared important to be clearly visible – internationally as well as domestically – and for this reason it was essential not only to participate with an independent battle

---

[12] "Det går fremad i Afghanistan [Progress in Afghanistan]" in *Information* 25 November 2010.

[13] "Britisk general i Helmand: Fjenden under massivt pres [British General in Helmand: The Enemy is under severe Pressure]" in *Berlingske Tidende* 15 February 2011.

group, but also to occupy key posts within the coalition's decision-making machinery.

Speaking in the early months of 2011 with centrally placed Danish staff officers, I realised that there was an apprehension amongst many of them that although co-operation on the battlefield was perceived as being excellent, Denmark ought to occupy more of the important posts in the lead-formation's and higher level bureaucracies' command structures on both the Task Force Helmand level and above. But how did this appear seen through British glasses? Would it be desirable and was it realistic? Which difficulties could be foreseen and which benefits might be reaped? In June 2011, Brigadier James Chiswell, kindly invited me to interview him in his charming house at the Royal Military Academy, Sandhurst and, as it turned out, his view was a decidedly positive one. In his opinion, the wish for more high level staff positions was a natural and positive ambition and, indeed, he thought that it should be implemented at all levels of command. Thus, coalition partners should be able to provide input to the operations planning processes at formation level, which would help in 'oiling the wheels' and 'ironing out problems further up.' For national reasons there might be reservations within the field of intelligence (J2), but wishes to occupy key appointments within the branches of operations (J3) and plans (J5) would be reasonable, and the UK decision-making hierarchy – so Chiswell assumed – would hardly resist a request to that end.

But co-operation was not merely an important bi-national, high-level matter. Officers had to act in a milieu of peers contributing as best they could to the mutual endeavour. But British and Danish officers attend different staff courses, the British nowadays mostly at the Joint Services Command and Staff College at Shrivenham, the Danes in Copenhagen at the Royal Danish Defence College, and for this reasons only NATO agreed procedures are taught in exactly the same way. Lots of other practical methods, aid memoirs and working procedures are different, and prior to collaboration in an overseas mission there is a need to take a look at these to see what might better be streamlined. To ensure that this happened, the Danish Battle Group introduced some procedures learnt from their British colleagues during combined training prior to deployment. They adopted the so-called 'back briefs,' which were basically examinations of subordinate commanders to ascertain that the orders received had been properly understood. Moreover, the concept of 'roc drill' was introduced – co-ordination meetings of commanders in order to discuss and pinpoint needs for harmonisation during an upcoming operation. As English was the common working language, but not the native tongue of the Danes or many other coalition partners, these procedures seemed particularly pertinent. While such novelties were absorbed gradually by Danish teams serving under British lead, the primary advantage of participating in courses in the UK run prior to deployment was that these familiarised Danish officers with UK staff work, improving their ability to contribute adequately to the common planning procedures.

However, harmony is not just a matter of shrewd practices and procedures; personalities of co-operating officers are of importance too. Amongst commanders there will always be some whose ambitions and self-importance stand in the way of smooth collaboration and amicable communication with their peers. As we shall see in Chapter IV, this had been a prominent quandary in the cases of, for instance, Württemberg and Marlborough during the Williamite campaign in Ireland (1690) and Montgomery and Patton during the operations in Northern Europe towards the end of World War II (1944-45). Why should there not be similar clashes of 'prima-donna egos' in the relationship between British and Danes in Afghanistan 2010-11? However logical this might seem, such bickering was not observed on the 2010-11 tour – at least not by Brigadier Chiswell. Nonetheless, there were differences of attitude to several single issues. In particular, there were discussions on the relative importance of being able to strike against insurgents in an unpredictable and surprising manner and of being present in as many places as possible. The Danish position was that, due to shortage of personnel, it was essential to be able to choose time and place oneself. The approach preferred by the Afghans was to man fixed positions and ask for ISAF's support whenever needed. This was not necessarily a bad idea, but it ran counter to Danish intentions of keeping the initiative vis-à-vis the insurgents by flexibility, speed, surprise and manœuvrability. On the other hand, the citizens of Helmand needed assurance that they might live in peace and be able to move freely and this, they believed, was obtainable only by troop presence. The task force was there to further exactly that, and there were, therefore, occasions when Commander Task Force Helmand was more in agreement with the Afghan brigade commander than with the Danish commanding officer. There were, in other words, a certain disparity of opinions as to task force outreach through fixed positions as opposed to flexibility and speed. One example of this occasional disagreement might be the operations in Deh-e Adam Khan, where the commanding officer was ordered to fix a platoon in Patrol Bases Shir Aga and Viking in spite of his advice not to do so. The Danish view was based not only on adherence to the old principles of flexibility and surprise in actions taken, but also on a perception of what would – in the long run – best advance the development of the independent capabilities of the Afghan forces.

Seen from the vantage point of Task Force Helmand, Colonel Fredskov's wish to close bases in the Upper Gereshk Valley might have been influenced by Danish political desire to consolidate further south pending a later shift from combat operations to training. Brigadier Chiswell agreed on the need for maximising manœuvre, but he was adamant that this had to be balanced with continuing efforts to extend the security bubble from a number of key fixed sites. Like the commanding officer of the Combined Force, he was concerned about the need to break the security deadlock in the Bandi Barq area. This was critical principally because instability there would undermine the confidence of the people in Gerershk, but also because the efforts in the Bandi Barq area were absorbing significant levels of security resources. In Commander Task Force Helmand's perspective, the proper solution was to deepen the security in the Deh-e Adam Khan neighbourhood as well as, critically, in the area to the east of the river. It was with this in mind that the task force directed the Combined Force Nahr-e Saraj North to put more resources into Hazrat. In terms of mission command, Chiswell was happy to leave the various subordinate units to deliver success in their own ways, though, on occasions, he stepped in where he judged a touch on the tiller was required, and the Hazrat situation was one of these.

The Danish deputy commander Lieutenant-Colonel Thomas Funch Pedersen observed that on occasions the British views on how to develop the Afghan troops' independence seemed somewhat cavalier with respect to long term perspectives. However, while such occasional disagreements entailed some tensions between leaders and subordinates, Colonel Pedersen was adamant that it never seriously damaged or even challenged the Anglo-Danish collaboration. Likewise, there were opposing views between the Afghans and the Danes, especially concerning fixing or unfixing troops. While Colonel Fredskov wanted to 'unfix' as many troops as possible to ascertain that a large manœuvre force be available at any time, the Afghans seemed more inclined to secure ground in their possession through placing troops in fixed positions.

Pre-Deployment Issues

Long before the deployment of ISAF Team 10, officers of previous Danish contingents had sharply criticised the preparatory training in Denmark.[14] In 2008, a captain, previously a tank squadron commander, claimed that private soldiers were inadequately trained at home and had to catch up when on the ground in Helmand. This, the captain claimed, was unfortunate

---

[14] Information, 25 June 2008.

and irresponsible as they needed fighting skills and dexterity with the equipment from the very minute they arrived. But they had not been trained properly to using the relevant equipment and weaponry because too little – or nothing at all – of this materiel had been purchased for training purposes. This allegation was correct, but when ISAF Team 10 trained a couple of years later, most of the shortcomings had been remedied. However, one field in which equipment for training was still inadequate was intelligence. Intelligence, Surveillance, Target Acquisition and Reconnaissance – in daily newspeak ISTAR – equipment was essential for keeping an eye on the insurgents, day and night. Such gear existed and, carried on various remotely directed platforms; it was used on location in Helmand. Unfortunately, during training in Denmark some of these paraphernalia were not always available.[15] Moreover, as Denmark does not have an intelligence corps, and as no proper dedicated intelligence training facilities and programmes are in place, the intelligence officers and personnel (S2 Branch) were inadequately prepared and had to undergo specific training immediately prior to deployment, thus missing essential parts of the common battle group pre-deployment preparation.[16]

Regardless of the few, although important, shortcomings, training was generally perceived as being adequate and relevant. And this was not the only kind of preparation undertaken by the battle group. Prior to deployment, the commanding officer worked out, in co-operation with the Army Operational Command, a survey of foreseeable tasks such as the provision of security during Ramadan and elections, changing of patrol patterns, and the manning of bases and support for the so-called 'Food Zone Programme', etc. As it turned out, more or less all these tasks materialised and were successfully solved by the Battle Group during its tour. Also before the actual deployment, sixteen members of the Danish Battle Group's staff exercised with 4 Mechanised Brigade in Warminster, UK, during a Computer Assisted Exercise (CAX), and a smaller team of three persons liaised with 16 Air Assault Brigade. In both cases Colonel Fredskov, found it a most useful experience and a booster of personal relations and networking which would become of importance when meeting again in Afghanistan.

ISAF 10 did not leave Denmark without having acquainted themselves with the milieu in which they were foreseen to operate. While

---

[15] ISTAR meaning the systematic combination of Intelligence, Surveillance, Target Acquisition and Reconnaissance.

[16] There are, however, a number of 2-3 week courses available at the Royal Military Academy in Frederiksberg.

reconnaissance prior to departure for Helmand was conducted with specific operational aims, the preceding fact-finding tour aimed primarily at creating a basis for the pre-deployment training programme for the Battle Group. The commander, deputy commander, chief of staff and the officers commanding sub-units took part in the fact-finding and had the opportunity to discuss ideas and suggestions with ISAF Team 8. As opposed to this rather limited group, the commanders' reconnaissance tour comprised twenty-three persons. Its main purpose was to prepare for the so-called hand-over/take-over procedure by which ISAF Team 10 was to assume sole responsibility in mid-August, and at the same time provide some basic input for the final pre-deployment exercise. A number of administrative documents were prepared during the reconnaissance such as rules for Annual Confidential Reports, directives for pay and salaries, etc. and, finally, the basis was laid for writing the *Commanding Officer's Tactical Guidance*. Moreover, because of the character of the upcoming tasks, the deputy commander and heads of the operations and information branches (S3 and S9) prepared a revised analytical model with increased emphasis on information operations – the 'modern version of intelligence' placing enhanced emphasis on civilian aspects and influence.[17] Unfortunately, the development of this model, which would in many respects influence the requirements of intelligence products, had to take place without the participation of the head of the intelligence (S2) Branch, because he and his team was away on supplementary intelligence training; and although the entire staff were subsequently briefed on the essentials of these analytical procedures, the impossibility of engaging all relevant officers in the preparatory work entailed considerable repercussions in the co-operation during the deployment.

Intelligence

While the primary purpose of this chapter is pinpointing elements of contention and collaborative challenges amongst coalition partners of different nations, it is of no little importance that within the Danish Battle Group there were inter-staff procedures which did not function optimally: those of operational exploitation of the work done by the intelligence people or, to see it the other way round, those concerning the intelligence

---

[17] Information Operations, although making use of technology, focuses primarily on the more human-related aspects of information use, including, amongst others, social network analysis, decision analysis and the human aspects of Command and Control.

branch's responsibility of providing timely, relevant and precise intelligence upon which to base operations. Within the Combined Force Nahr-e Saraj North, intelligence simply did not function adequately. On the one hand, the head of the S2 Branch, Captain Thomas Larsen, sensed that the Operations Branch (S3) as well as the commanding officer did not listen and did not use his thoroughly prepared Intelligence Preparation of the Battlefield traces, charts and situational appreciations.[18] On the other, while the head of the S3 Branch, Major Byrholt, found that though he did indeed listen, he did not get what was needed for the operational planning at hand. The commanding officer saw a reason for this serious impasse due to the immensely well-prepared intelligence material being of a complex nature ill-suited to meet low-level tactical challenges calling for immediate responses. What he had needed was intelligence digests explaining details pertinent to short as well as long term tactical planning. On the same note, S3 found the intelligence contributions too general and the intelligence collection plan unimaginative, and he put his finger on a crucial and very likely cause: the very brief and purely theoretical training of intelligence officers offered by the Danish army. This training emphasises production of network analyses, databases and link examinations rather than clear-cut pin-pointing of targets against which one might train one's guns or direct one's troops. From the commanding officer's and his chief operations officer's point of view, the value of the S2 Branch's very well processed Intelligence Preparation of the Battlefield products would have been greater had these been more detailed and more focused on the immediate threats – in other words, more actionable.

Across history we find examples that politico-military opportunism has forced commanders to undertake tasks which did not fit in with the reality on the ground as presented by their intelligence branches. This happened during Operation Market Garden in 1944, when Montgomery and Browning refused to accept last minute intelligence updates, and it might as well have happened in Afghanistan in 2010. However, according to Colonel Fredskov, this was not seen during his tenure as commanding officer of the Combined Force Nahr-e Saraj North. Authorities in Copenhagen laid out a policy for 2011 clearly aiming at preparing the Afghan forces for taking over responsibility of their country, but this had no bearing on the activities of ISAF Team 10, who were, after all, redeploying to Denmark as early as February 2011. Although the Danish Battle Group did take steps to shift

---

[18] 'Intelligence Preparation of the Battlefield' is a concept that allows analyses of terrain, enemy, weather, own forces and various other elements influencing one's combat options to appear in a logical and operationally easily manageable way.

efforts towards this goal by preparing the handover of the Patrol Base Line to the Afghan National Security Forces, this did not happen at the expense of the immediate security tasks at hand.

While the efforts of operations and intelligence did not interact optimally, there should be no doubt that during their six months in Afghanistan Captain Thomas Larsen and his S2 collaborators served all levels as best they could; and so did Major Byrholt and his S3 team, too. S2's customers ranged from the sub-units of Combined Force Nahr-e Saraj North to the Regional Commands and intelligence services at home in Denmark and in the UK. Their primary client, however, was their commanding officer, Colonel Fredskov, whose obvious need was accurate and timely information on what happened in the Combined Force's area of responsibility. Moreover, the intelligence branch supported the sub-units either generally, on a case-by-case basis or by detaching an intelligence officer to the sub-unit in need of support for a specific task. Finally, information was passed to Task Force Helmand as input for the 'Commanders Critical Information Requirement' process. Captain Larsen sensed that, while his own superiors rarely targeted anything on the basis of his recommendations, there was an obvious need to relay information to others in order to achieve focus on the situation at hand and provide a basis for addressing it. While, during our conversations, the commanding officer did not actually deny that Captain Larsen's perception was in many respects correct, he regretted that as far as he was concerned the S2 products, which were often quite impressive, were mostly too general for him to transform them into concrete operational tasks. Unfortunately, Larsen got the feeling – unwarranted, as it turned out – of being ignored by the commanding officer and the S3 Branch. Nevertheless, he was able to pass useful information to neighbouring formations who accepted it, and he sensed that, frequently, this led to fruitful results.

The differences between intelligence and the Combined Force's operational leadership, in many respects, seemed to be a matter of diverse viewpoints on what was the more important: short term insurgent activity or long term economic and security trends. A headquarters will normally lay down a number of 'Priority Intelligence Requirements' as a basis for the intelligence branch to prepare an 'Intelligence Collection Plan' which can be annexed to the operations order. In a number of conversations with the commanding officer, Colonel Lennie Fredskov, with the chief of operations, Major Christian Bach Byrholt, as well as with the intelligence chief, Captain Thomas Larsen, I realised that while the latter turned all his considerable talent to mapping the security, commercial and political landscapes with a view to visualising trends in the development of the Afghan society towards a more peaceful and mature state, the commanding officer and his operations branch were focussed on immediate needs such as insurgents' movements, manufacture of improvised explosive devices, government outreach and on solving short term tasks assigned by Commander Task Force Helmand, Brigadier James Chiswell.

It is likely that the intra-staff trouble was, at least partially, mired in a mental dichotomy not allowing the professional intelligence officer to see simple tasks like 'find out what is going on in Gereshk' as a Priority Intelligence Requirement, while at the same time, and with very good reason, the operations branch found that exactly that was a perfectly reasonable starting point for the 'Intelligence Cycle.'[19] Colonel Fredskov noticed that 'intelligence must support long term as well as short term planning, and S2 Branch, however excellent in the former, was somewhat remiss in the latter.' Captain Larsen saw accuracy and detail on, for instance, the ownerships of shops in the Bazaar, the trades and networks operating there and the volume of the turnover of the various businesses as the epitome of good intelligence allowing for long term predictions on security and economic development. Major Byrholt agreed that this was indeed needed and very useful, but he missed further analysis leading to brief conclusions and recommendations on actions to be taken – armed or otherwise. In order to smooth the progress of current operations there was a need for equally detailed but 'close to real time' insight into purely military matters, insurgents' dispositions, corruption within the security forces and immediate threats to own forces and local citizens.

Successful collection of information data depends on the sources as well as the means of collection. Apart from the obvious advantages of modern equipment and systems like those of the ISTAR concept, patrols, posts, convoys and others with the ability to observe and report were essential. Their usefulness depended on their being on the right spot at the right moment, and in this context the Combined Force was faced with at least two challenges: on the one hand, Colonel Fredskov's priority intelligence requirements focussed on the town of Gereshk, but the police and the police operational mentoring and liaison team, the P-OMLT, in that town were not very good at reporting. On the other, the military sub-units and platoons were mostly engaged in the Upper Gereshk Valley area far from the town, and therefore of little use as collectors of the intelligence most wanted by the commanding officer. It was a frustration that initially the Gereshk based police operational mentoring and liaison team did not

---

[19] The 'Intelligence Cycle' is the iterative process of direction based on needs for information, collection of intelligence data, analysis and comparison of material collected, and dissemination to users of the results of the process.

provide much useful information, but in the long run Captain Larsen managed to solve this problem. As to the troops in Upper Gereshk Valley, S3 recalls that they were excellent intelligence collectors, and that, in particular, the two Irish Guards companies provided very useful input – though not relevant to the priority requirement of knowing what was happening in town.

Since, in Larsen's view, the Priority Intelligence Requirements that were forthcoming from battle group headquarters were insufficient as a basis for his intelligence collection plan, he tried to deduct some from those of the Task Force Helmand level adjusting them to fit into the conditions of the Combined Force Nahr-e Saraj North's area of responsibility. In his opinion, this adjustment worked well.

The adjusted list, which he would now see as 'S2's own Priority Intelligence Requirements,' was then broken down into 'Requests for Information', which Captain Larsen believed that sub-units could act upon. Nonetheless, the S2 Branch found it difficult to deliver the right information precisely and pertinent to operational planning in a timely manner because, in their view, the battle group's Priority Intelligence Requirements – as stated by operations – were too vague and too general. Be this as it may, the accusation of being too broad and too little target oriented went both ways, and the operations branch did indeed pose some very precise questions, as did the commanding officer, to which, in some cases, they got no or inadequate replies.

As briefly described above, a systematic approach to intelligence called 'Intelligence Preparation of the Battlefield' has been developed over the last two or three decades in order to facilitate collection, analysis and presentation of battlefield intelligence. This is a lengthy and iterative process aiming at fusing all known features of terrain, weather, enemy forces' and own forces' situation together in a holistic situational picture. It requires co-operation amongst various functions within the formation or unit headquarters and must be updated concurrently as operations are being prepared and executed. The S2 Branch of the Combined Force Nahr-e Saraj North worked hard to achieve this, but frequently they felt that from staff members in the S3 Branch there was scant interest in the product as well as limited inclination towards contributing to its genesis.

Throughout their six months in Afghanistan, Captain Larsen and his collaborators strove to make sound intelligence preparations of the battlefields at hand. In particular, prior to the Combined Force Nahr-e Saraj North's operation in the town of Deh-e Adam Khan, on which no information was readily available, this work was time-consuming as it had to be done from scratch. All the details had to be meticulously worked out by intelligence in close co-operation with operations, S3 Branch, and information operations, S9 Branch, civil-military co-operation, CIMIC, the Military Stabilisation Support Team, MSST, and the Stabilisation Advisor, STABAD. However, Larsen was not satisfied by this process because he sensed that the co-operation fell far below the optimum and the battle group leadership was remiss in giving direction. Intelligence Preparation of the Battlefield is a mutual effort to be done by almost the whole headquarters and should be led, or at least initiated, by the Chief of Staff. Not surprisingly, this was seen differently by different eyes. Major Byrholt did indeed co-operate with S2, but he sensed that a kind of mental inattentiveness or insistence on preconceived notions prevented the latter from delivering products which could be readily exploited by companies and platoons. On the contrary, the intelligence products received by the S3 Branch were excellent for briefing higher headquarters on general trends, but hardly sufficiently detailed to be of any immediate use at the sharp end.

As to collection of intelligence data there was a plethora of means available – or mostly available. Generally, S2 might ask for support from any of the intelligence collectors in and near the Combined Force Nahr-e Saraj North's area of responsibility, whose assistance would be granted if the asset or capability asked for was currently available. These were:

- All kinds of unmanned aerial vehicles (UAV) with Full Motion Video, signals intelligence, stills capabilities
- Satellite photos
- Find, feel & understand teams (Human Terrain Mapping)
- Human intelligence
- Counter improvised explosive device teams
- Weapons Intelligence Teams
- PGSS balloons, i.e. cameras and sensors mounted on blimps
- Various camp surveillance systems
- Sub-units, platoons etc.
- Electronic Warfare (EW) Section
- CIMIC Support Teams
- Signals Liaison Officer

Normally, intelligence collection agencies could be asked to support as long as the requests were submitted early enough to allow Task Force Helmand or the Regional Command to prioritise the various missions. There was, however, one snag: Danish S2 and ISTAR personnel were not fully familiar with this abundance of assets. They had limited knowledge and skill in assigning tasks, which – in a perfect world – should have been anticipated and trained prior to deployment. Fortunately, UK personnel were there to help.

Information data – the raw material for producing intelligence – was mostly collected and processed within reasonable time frames, i.e. early enough to be acted upon, but there were certain limitations which had to be kept in mind. Flying ISTAR assets must be tasked in advance and the material gathered would have to be processed and interpreted, which at times was a time-consuming process. Moreover, lack of reports from sub-units – and in particular from the P-OMLT in Gereshk – hampered S2's endeavours to deliver timely and accurate estimates to those needing them. In Patrol Base Line, the Danish company there, C Coy, was largely negligent of reporting. Conversely, another Danish company, responsible for, i.a., Deh-e Adam Khan, co-operated smoothly with various other intelligence collection assets and provided much valuable information. Through diligent and goal oriented intelligence collection, the British F Coy provided useful information on the opposition's operational patterns and their organisation in the Rahim area. Nevertheless, it happened that there was insufficient time available for a 'find phase' before a given operation could be planned or launched. This meant that, occasionally, operational planning proceeded de-coupled from intelligence collection, which, according to Captain Larsen, constituted a number of potential hazards. The commanding officer, however, saw this from a different perspective: although planning normally happened with reasonable time for information to be collected and processed, there was always a need to weigh the necessity for perfect intelligence against time consumption in the overall operational planning and issuance of operations orders.

Humint, that is the use of human sources and contacts for intelligence purposes, was important and in Gereshk, for a long time, it was the only trustworthy way of collecting information on the insurgency leadership's capabilities, organisation, activities, intentions and determination. Although Combined Force Nahr-e Saraj North's personnel did not interrogate suspects, S2 had an intelligence officer at Task Force Helmand headquarters when interrogations were conducted and thereby gained vital insight into the challenges at hand.

The so-called Intelligence Processing Application (IPA) – a system automatically fusing, comparing and analysing information – was not used to any considerable extent by the S2 Branch of ISAF Team 10, because their access to the British secure Mission Secret data network was limited. There was, however, much to be gained through full access to this system and by the day of departure from Afghanistan all intelligence officer desks boasted a Mission Secret terminal. A purely Danish system – the I-base – operated in parallel with Mission Secret. It was accessible to S2 and the EW Section only and thus of little use as far as sharing information with neighbours and higher formations were concerned.

The battle group S2 Branch had a link to their national military intelligence service, which was useful in many respects but rather defunct in others. Although it will come as a surprise to no one that such co-operation took place, for reasons of personnel as well as operations security, details cannot be discussed in a book available on the open market.

The head of S2 had all the military interpreters of the Combined Force under his wings and was, therefore, in a position to 'decipher' most written material and translate all communication more or less simultaneously with its occurrence.

As for dissemination of intelligence, S2 issued daily intelligence summaries, INTSUMs, which were distributed widely to superior, co-ordinate and subordinate headquarters so that all potentially interested parties might know what was going on in the Combined Force Nahr-e Saraj North's area of responsibility. A weekly INTSUM was published every Sunday, summing up events amongst the opposition and trends materialising from daily summaries as well as estimates on the threat level. Moreover, whenever something of importance happened, S2 issued an intelligence report, INTREP, or a threat warning. Supplementary intelligence reports, or SUPINTREPs, were issued when a thematic analysis was called for, and during the six months of ISAF Team 10's stay two such supplements were produced. While one juxtaposed the improvised explosive device trends from 2007 to 2010 with the changes in Danish force presence during the same period; the other concerned the poppy cultivation in the Upper Gereshk Valley and its importance to the insurgency's ability to go on fighting. All reporting went to all those concerned locally, i.e. within Task Force Helmand, as well as to Danish authorities in the form of a monthly summary signed by the commanding officer. The latter became one more matter of contention between the Chief S2 and the commanding officer.

After a month or two, Colonel Fredskov realised that these reports had become too long and unwieldy and consequently trimmed them down to what would be of primary concern to readers at home. Captain Larsen, however, speculated that this might display a much too optimistic view of developments in the Combined Force' area of responsibility, which fitted better with what was politically desirable at the moment, relegating S2's diligent elaborations to obscurity. With the benefit of hindsight and that of having had access to both viewpoints it appears that, much like the top-brass prima donnas of the past, the Chief S2 failed to see the obvious – you cannot put everything into a report. S3 chronicled what was actually happening and the Colonel simply wrote in order to make sure that all points of importance to decision-makers at home were actually being read.

S2 got the impression that no co-ordination took place of the actions in neighbouring battle groups' areas of responsibility and that much effort was therefore wasted. This, one may wonder, might have been a rather defective observation, as there were numerous examples of the opposite as, for instance, the generally good collaboration with the Combined Force of Nahr-e Saraj South. Moreover, S3 – having made a similar observation – had taken the opportunity to suggest to J35, the Task Force's chief operations and plans, that the S3s of all the battle groups meet regularly to coordinate. This, unfortunately, did not happen during ISAF Team 10's tour.

Photo taken by OC B Coy, Danish Battle Group: No 1 Platoon during Operation Omid Panj 10 Dec 2010.

Nonetheless, S2 Branch contributed decisive input to several major enterprises undertaken by the combined force and its co-ordinate formations and units, and amongst the primary efforts made by S2 was the provision of intelligence estimates concerning:

- Taliban assault on the road construction company VICC on Route 611 19 August 2010
- Neutralisation of Qari Hazrat 29 August 2010
- Support of the general elections on 18 September 2010
- Human Terrain Mapping (HTM) of Deh-e Adam Kahn

- Operation Riverdance North of Combined Force Nahr-e Saraj North
- Operation Omid Panj in Deh-e Adam Kahn

Captain Larsen found considerable challenges in his day-to-day routines and in what he perceived as unclear directives, limited issue of Principal Intelligence Requirements, imperfect understanding by staff members of the intelligence cycle and an approach to operations based on a symmetric scenario which did not exist. However, his view ignores the fact that meticulous discussion and adjustment of analytical procedures had taken place during pre-deployment training although without participation by himself and his team. Moreover, reality did not support his speculation on these issues. Colonel Fredskov's direction of planning and execution of operations happened in a way nicely streamlined with the agreed Counter Insurgence Doctrine, and the Danish tactical approach employed in Afghanistan was in fact an *ad hoc* adaptation of the doctrine to the thoroughly asymmetric milieu prevailing on the ground.

Operations

As soon as intelligence has provided the necessary insight into the opposition's whereabouts and intentions, one can start planning and launching military operations aimed at constraining his freedom of action. A well-led coalition force, however, does not initiate operations just for the sake of being seen as pro-active – they do so as part of a well thought-out and nicely concerted campaign, as indeed did the Danish element of Task Force Helmand, the ISAF 10.

It is arguable that 16th Air Assault Brigade's stint in Afghanistan – or that of the Danish Battle Group for that matter – could be considered as being more or less one continuous campaign. The Danish Battle Group commander, Colonel Lennie Fredskov, believed that it could, and he planned ahead for a campaign running well into the next two contingents' periods of responsibility. The task force commander, Brigadier Chiswell, agreed that there was an obvious need for a manifest plan, because a clear plan was and remained the precondition for making sure that one's own rôle and aims can be fulfilled. His view was that continuity was critical between the task force teams – the 'Herrick Headquarters' – as well as between those of the combined forces. With that in mind, he was clear on the need for a single campaign plan which covered the period of the actual deployment and the years ahead. The basis of this was the Helmand Plan. The military part of the Helmand Plan called for gradual improvement of local structures and forces and transfer of all security responsibilities to

Afghan authorities leading to a complete handover by the end of 2014.[20] Another side of the same 'Helmand Plan' was of a more civilian nature and co-signed between the Provincial Governor, Head of the Provincial Reconstruction Team and Commander Task Force Helmand. Within its framework the so-called Provincial Reconstruction Team composed of British and Danish collaborators works to deliver stabilisation and development. The plan is structured around seven themes: Politics and Reconciliation, Governance, Rule of Law (Justice, Police and Prisons), Security, Economic and Social Development and Reconstruction, Counter Narcotics and Strategic Communications.[21]

Subordinate to this, there were the District Development and Security Plans, which sought commonality of aims and priorities between the District Governor and Afghan National Security Forces, Stability Adviser and combined force commanders. Against this backdrop, headquarters Task Force Helmand and the combined forces under its command needed goals for their six-month tenures – stepping stones on the campaign path. Each of the combined force commanders produced their unit-level Commander's Directive or Operational Orders on how these goals would be sequenced and achieved. Chiswell expressed some surprise that Colonel Fredskov had referred to his plan as a 'Campaign Plan', but he was happy with this so long as the objectives it set out conformed to the Helmand Plan, the Nahr-e Saraj District Development Plan and the task force's scheme of manœuvre.

Colonel Lennie Fredskov was adamant that a campaign plan for the Danish contribution in Helmand was essential, and in this respect he was wholeheartedly backed by his national authorities. Brigadier Chiswell recognised that, being a major national contributor, it was right and proper if Copenhagen would like to influence the aims and approaches of its forces. Fredskov, in his attempt at setting up a long-term arrangement, took his point of departure in what ISAF had laid down as a concept for the future development in Afghanistan. In November 2010, Regional Command South-West had transformed this concept into what he called 'RC (SW) Winter Campaign.' Initially, Combined Force Nahr-e Saraj North carried on with the plan made by their predecessor battle group, the ISAF Team 9, but during their first two months they ventured into drafting a 'campaign plan' of their own covering the period of their stint as well as roughly

---

[20] Danish Ministry of Defence, 2011, http://www.fmn.dk/temaer/afghanistan/baggrundforindsatsen/Documents/Helmandplan2 011_FINAL_web.pdf accessed 1 May 2012.

[21] See also the Danish political level Helmand Plan mentioned above, cf.: http://www.afghanistan.um.dk/en accessed 4 August 2011.

another year. As the basic idea of having Danish troops in Afghanistan was to help the Afghans to take care of their own security in the long run, this plan needed to be developed in a way commensurate with the perceived needs of the Afghan security apparatus. During his campaign planning, Colonel Fredskov negotiated measures and priorities with the District Governor Gulab Mangal, the commanding officer of 3rd *Kandak*, Lieutenant-Colonel Hamiyun and the Head of Police, Mr Ezmerai. Moreover, the draft was discussed with Brigadier Felton, the commander of Task Force Helmand during Herrick 12 (i.e. until mid-October, when this job was taken over by Brigadier Chiswell), with the Danish Army Operational Command and with representatives of the Danish Chief of Defence.

The notion of a one year plus campaign plan originated with the deputy commander Lieutenant Colonel Thomas Funch Pedersen. In his staff work in Colchester during the pre-deployment training with 16 Air Assault brigade the previous spring, the idea of a long term plan emerged as a logical solution to issues raised by preceding Danish teams. Pedersen sensed that both Danish and British plans showed a dearth of continuity, which, he believed, was likely to reduce the effect of the military efforts and hamper the long term development of the Nahr-e Saraj district. This led him to realise that the only viable method for creating a sustainable campaign plan was engaging the Afghan authorities in discussions on their long-term security needs and their wishes as to the transition of authority.

The task force commander, however, could not agree with Pedersen's view that there was a lack of continuity. He believed the continuity was there, though perhaps not entirely the way Pedersen would have liked. Nevertheless, much of the continuity for the wider effort in Task Force Helmand was derived from the fact that the task force conducted its pre-deployment training and conceptual thinking together. For example, Chiswell completed a 10-day commanders' reconnaissance to Helmand some five or six months prior to deploying taking all the combined force Commanding Officers and their operations officers with him. This provided an important forum for examining collectively the Helmand Plan in detail and to set the objectives for the coming period. By subsequently maintaining close links to those deployed in the run-up to the start of their tour they strived to hit the ground running with the long term campaign aims in mind. A similar approach was applied by the Danish teams, and, wherever possible, the work-up was performed on an Anglo-Danish basis. However, having the British and the Danish elements deploying on different rotations inevitably built in some managerial gaps, although these were compensated for by the advantage of not having the complete chaos of all units being relieved at the same time.

The Danish campaign plan was conceived during the first six weeks the Battle Group was in Helmand. Occasionally, however, to the Danish headquarters there seemed to be an absence of similar British long term planning, which created a challenge. This challenge was particularly felt when the Battle Group was ordered to solve missions in a manner that did not agree with the long term goals laid down in the Danish campaign plan, and it was aggravated by what was seen by the Danes as Task Force Helmand's sporadic imposition of specific tasks on sub-units under the Combined Force. Unfamiliar with this procedure, the Danish headquarters' staff perceived it as a *deus ex machina*-like interference and a violation of the principle of mission command, which states that a superior commander will merely order a given mission to be executed while leaving to his subordinate commander how to do so. This notion – the mission command way of leading military forces – lies at the centre of modern British as well as Danish military thinking, but at the time there were occasional disagreements as to its application. There were situations when adequate fulfilment of Task Force Helmand's overall mission necessitated certain dispositions which might be – and was – seen as interfering with subordinate commanders' freedom of method. This we shall deal with in more detail in the discussion of cultural differences below.

The Danish operations planning took its point of departure in, and was largely inspired by, the UK doctrine on counter insurgency. During pre-deployment training in England, the deputy commander, and the chiefs of branches for Operations (S3) and Influence Operations (S9) had developed a concept for the battle group's staff procedures which would secure an influence dimension in all operations.[22] These procedures also made the deputy commander responsible for the development of the campaign plan, oversight of the staff work and execution of the major operations, guaranteeing continuity. Moreover, as the deputy commander was the leading liaison person with the Afghan Security Forces and supervisor of the civilian stabilisation effort, the commanding officer was free to perform the genuine commander's tasks at the sharp end.

The long-term campaign plan materialised in the autumn of 2010, setting out some general goals, with the intention that the battle group would only take up and retain positions which the Afghan National Army could, and would, take over when Danish combat troops eventually left. Therefore, the locations of bases had to tally with the District Governor's

---

[22] With some approximation 'influence operations' may be seen as a modern, more civil version of psychological operations (PsyOps) in combination with what is now called civil-military co-operation (Cimic).

priorities, the force lay down should free troops for manœuvre purposes from positions where they were no longer needed and, in the long run, open up for reductions in the Danish troop presence so as to increase Afghan independent responsibility.

The Combined Force's area of operations was subdivided so that each sub-unit – apart from A Troop, which was held in Camp Price as a general mobile reserve – was responsible for a well-defined area of its own.

Photo taken by OC B Coy, Danish Battle Group: No 2 Platoon trains Afghan soldiers using the .50 Browning heavy machine gun; August 2010.

In August, the Danish B Coy still had not got an area of operations of their own, as they were split between three tasks. One platoon guarded Main Operating Base Price, one platoon was detached to Forward Operating Base Budwan, where it replaced the platoon from C Coy, and one was patrolling the area around Price partnered with the Heavy Weapons *Tolay* from the 3rd *Kandak*, which they were also training. The Danish C Coy operated from Patrol Base Line in the southern part of the Upper Gereshk Valley West. This company could move only as long as it remained close to the base line and near Patrol Bases Clifton and Bridzar. Initially, C Coy also had to detach a platoon to Forward Operating Base Budwan, a job which was soon to be taken over by B Coy.

Photo taken by OC B Coy, Danish Battle Group: No 3 Platoon in Camp Viking, January 2011.

Because there were so many patrol bases in the Patrol Base Line much effort had to be put into force protection, leaving limited resources for genuine patrolling. As a result, the British D Coy got the Upper Gereshk Valley East near Khar Nikah and down to the Helmand River. In reality, however, the company merely controlled an area around Forward Operating Base Khar Nikah and Patrol Bases Bahadur and Oqob. In order to remain in control, the company established an often-used post in an abandoned compound between Forward Operating Base Khar Nikah and Patrol Base Bahadur. E Coy, the US Light Reconnaissance Company, patrolled Highway 1 in a zone stretching about two kilometres to each side and including Artillery Hill and a hill to the east of Gereshk. The British F Coy's area of operations was to the north-east of C Coy's and covered the densely inhabited Rahim and Adinzai locale. F Coy, comprising only two infantry platoons and a fire support group with heavy machine guns, mortars and Javelin rocket launchers, had for some time had one platoon detached to service in Patrol Base Line, reducing its patrolling capability. By the end of August, the company was re-united, and intensified patrolling around Rahim, stressing the insurgents there considerably and continuously. On Route 611 a private construction company was busily occupied with carrying out repair works and was, from 19 August onwards, secured by a British Viking company conducting mobile operations in the area between this road and the Green Zone.

Photos taken by OC B Coy, Danish Battle Group: B Coy Infantry Fighting Vehicle in Green Zone, September 2010.

The combined Force's headquarters was in Main Operating Base Price, the Danish Kandak Advisory Team (DKAT) was in Gereshk, and the Police Operational Mentoring and Liaison Team (P-OMLT) co-operated with British police advisors stationed at the Afghan National Police headquarters in Gereshk.

Not surprisingly, changes to the force laydown occurred throughout the tour. Having liaised with his Afghan opposite number, Lieutenant-Colonel Mohammad Hamayun, commanding officer of 3rd *Kandak,* Colonel Fredskov realised that in the long run more troops would be needed, and could be employed more effectively, around Gereshk. Conversely, bases such as Budwan and Zumbalay, which Hamayun had declared to be useless, might have to be abandoned. Moreover, giving up these bases fitted with the Afghan plans on which bases to take over and which not.

Upon arrival in early August the new Danish Battle Group began their initial all-round assessment of the military situation that they would have to address. However, two specific issues were given priority. The first was that the insurgents – i.e. the Taliban – who had both a physical and psychological sway that extended from Gereshk in the south to Qaleh ye Gaz in the North, had several key nodes of command and control in locations that they used for projection of both soft influence and violent activity. The second was that the insurgents had apparently put much effort into perfecting a multi-layered system of improvised explosive devices based on caches, as well as large production and storage facilities. Their objective appeared to be tying down ISAF forces in their bases or limit their freedom of movement, hampering their influence operations in support of the Afghan Government's outreach. This situation had been developing over the previous year when units had prioritised force protection of their bases at the expense of patrolling. Should this situation remain unchanged it would severely curb the Combined Force's Human Terrain Mapping, which required units to get out amongst the population. These early indications called for agile intelligence driven operations to disrupt and degrade the improvised explosive device networks' capabilities, creating a more permissive environment for influence operations to take place.

As we have already seen, the *raison d'être* of the Combined Force NES (N) was to prepare the Afghans themselves to become able to take care of security in the area. But it was not everyone who was convinced that this would ever happen because, as an officer who had been training the Afghans for almost six months told a Danish colleague: 'The Afghan police and military loath each other, and as soon as the ISAF leaves, the army will kill the police and introduce a military dictatorship.' Nevertheless, the realities of the Danish ISAF team arriving in August 2010 were that the companies had to practice the aforementioned 'rule of three', which required co-operation with Afghan army and Afghan police sub-units. While B Coy did so from its base and at Check Point 5 at the Bandi Barq Road, C Coy executed its task from positions in Patrol Base Line. Conversely, the 'rule of four,' which had been tried by previous British contingents and had included also the local elders, was never implemented by the Danish Battle Group, who did not believe in its usefulness. At their levels, the two Danish officers commanding – B Coy's Major Michael Toft and C Coy's Major Eskil Berger – experienced reasonable successes with their 'rule of three' collaboration.

The common planning process invariably involved either Task Force Helmand's J3 branch – i.e. the operations branch – or the Brigade Advisory Group. Whenever the Combined Force expressed the desire to plan and launch an operation jointly with their Afghan collaborators, Task Force Helmand would talk to the Afghan 3rd Brigade, which, when agreement was reached, would order the Afghan *Kandak*s [battalions] and *tolay*s [companies] to co-operate with their ISAF counterparts. This required that almost all operations were conducted in a partnering fashion, and it was put into practice with the assistance of Task Force Helmand's Brigade Advisory Group and the Police Development and Advisory Team (PDAT), who would, in each case, detail the necessary advisors. Although, co-ordinating the actions of not only his own company but Afghan police and military sub-units as well, a company commander might have sensed some overstretch of his capabilities, due to the advisors' active rôle as communications links passing orders and information amongst the various elements this never materialised as a problem.

Photos taken by OC B Coy, Danish Battle Group: OC B Coy, Major Michael Toft in his Coy HQ, January 2011.

Major Berger saw the partnering this way:

> We were ordered to co-operate in a sort of one-Dane-to-one-Afghan-fashion on all matters, and no exceptions were allowed. The partnership was to take place at all levels and in order to make this function smoothly the Danish Battle Group had at its disposal a so-called DKAT or *Danish Kandak Advisory Team* whose task it was to advice the staff of the Afghan *Kandak*, or battalion, on how best to perform the military trade. We had trained intensively before deployment and used linguist officers as stand-ins for the true Afghans.

Since the Combined Force took over about the beginning of the Islamic festival season of Ramadan – which, in 2010, was running from 11 August to 10 September – it conducted operations with concentration of effort around the town of Gereshk, from whence it was supposed to help expand the Afghan government's influence.²³ Initially the security situation in and around Gereshk was improved through intensified co-operation with the Afghan National Police, the ANP, in the town and by increased presence of ISAF, and the Afghan National Army, ANA, on the outskirts. Moreover, during the following weeks, partnered activities of the Afghan Heavy Weapons *Tolay* and B Coy were initiated. Even though the intensity of these actions was limited due to the Ramadan, E Coy, the Light Armoured Reconnaissance Company, still carried out patrols with the Afghan Army on Highway 1 and in the Yakchal area ca. 12 km south-east of Gereshk. Moreover, the Afghan Army contributed to security in Patrol Base Line and Forward Operating Base Budwan.

Having arrived at the beginning of August, the Danish Battle Group found the first few days of their presence tumultuous. On 7 August, a Danish infantry fighting vehicle of B Coy of the outgoing ISAF Team 9 was hit by an extraordinarily powerful improvised explosive device that flipped over the thirty-five tonnes heavy vehicle. Two Danish soldiers were killed, and two British as well as two Danes were wounded. Since the two Danes killed were from the Royal Lifeguards, as was most of the incoming team, they were well known by many of the new arrivals; this incident was indeed a note of caution to those setting their feet on the ground. C Coy was detailed to clear up the remainder of the vehicle's radio equipment, weapons and personal belongings and to make sure that nothing was taken by the insurgents. This turned out to be a complicated task, not least because the Taliban insurgents sent in children to steal grenades and other valuables from the vehicle. They knew that Danish soldiers would not fire on kids, and their *modus operandi,* thus, was carefully adapted to this reality.

Then on 10 August, a British Chinook helicopter hit a *sangar* close to Patrol Base Bahadur, its rotor almost cutting through the watch tower and sadly resulting in one British fatality. The airframe crashed, and it had to be recovered before the Taliban got the chance to cannibalise it. At 6 pm on the 11th, staff officers from Task Force Helmand, the Danish Battle Group and US Marine Corps sat down to plan, and at 4 am on the following morning the recovery operation went ahead. Three Chinook helicopters carried the covering force from Task Force Helmand's reconnaissance unit, and in the early hours of 12 August the damaged chopper was successfully recovered.

On 14 August, Colonel Fredskov and Major Byrholt, the Chief S3, joined a party conducting reconnaissance of the ground from Main

---

[23] The official designation was the Government of the Islamic Republic of Afghanistan (or GIRoA in coalition new-speak).

Operating Base Price to the bases in Patrol Base Line and further east, where C and F Coys were establishing themselves. The drive was quiet and the four bases – Clifton, Bridzar, Sponden and Malvern – seemed to be well set up. In Clifton an enthusiastic platoon commander had already struck an excellent relationship with his Afghan opposite number. Having been given an impressive briefing by the platoon commander on the situation and his present activities, S3 drove on to Bridzar, the headquarters of C Coy. Here he met the officer commanding,[24] Major Eskil Berger, and since it was soon getting dark they moved on immediately to Patrol Base Sponden to be put into the picture by the sergeant in charge. After twenty minutes they went back to Bridzar to eat and rest for the night. As Major Byrholt crept into his sleeping bag, the company's mortars and the sangar posts opened up with an overabundance of pyrotechnics in support of a patrol leaving the base. It was, he recalls, like being part of shooting an action film. The next day, the party carried on visiting Forward Operating Base Budwan and from there retraced their steps to Patrol Base Rahim, where Byrholt went on a foot patrol with F Coy, who was manning the base. On Monday 16 August they went back to their headquarters in Price, settling en route an operations plan for C Coy with the officer commanding.

C and F Coys patrolled the high ground where they interacted with local Afghans and conducted minor offensive operations against insurgents in the Green Zone, the arable trunk of land along the Helmand River. Nonetheless, on 19 August, the Taliban carried out a major attack against construction works at the eastern edge of the combined force's area of responsibility and managed to run over a number of checkpoints manned by a local militia. Based on information provided by the private construction company the Combined Force estimated that there were about 100 well armed insurgents involved in the clash. They had managed to kill a number of constructions workers and security employees and capture personnel as well as weapons, ammunition and vehicles, all – it was assessed – in order to prohibit the road construction, which would have considerably limited the insurgents' freedom of movement upon completion.

When the attackers were about five kilometres from the junction of Highway 1 and Route 611, the Combined Force's tank troop and the escort and security platoon were launched together with a US Marine Corps platoon. Colonel Fredskov moved with his forward command post to the road junction where he met with Lieutenant-Colonel Eid Mohammad, the commanding officer of the Afghan Highway *Kandak*, and with the leader

---

[24] A 'commanding officer' is a Lt-Col or Col commanding a Battalion or a Regiment, whereas a company commander is a major called 'the officer commanding.'

of the construction team. While the marines and the escort and security platoon were unleashed in a northerly direction, A Troop was ordered to take up positions on the high ground west of the road and cover the advancing infantry's movement. As the commanding officer moved north, he saw the devastation wreaked by the insurgents – burnt down checkpoints and corpses strewn along the road – but there was no longer any opposition. The insurgents had vanished into thin air.[25] Within a matter of a few hours the combined force had managed to clear the road allowing the construction workers to continue their toil.

The escort and security platoon that had continued north ran into an improvised explosive device. Calmly the platoon commander, Lieutenant Martin T. Andersen, whose Eagle IV armoured patrol car was actually hit, informed the commanding officer that his platoon had made contact with the enemy. As Colonel Fredskov hurried ahead with his forward command post he beheld a vehicle damaged beyond recognition, but, miraculously, Lieutenant Andersen was completely unscathed. His driver and one more soldier were only slightly injured and were evacuated by helicopter shortly after. Returning to camp in the early hours of the evening the party came under renewed attack – this time by rocket propelled grenades (RPGs) – but still they were lucky. Subsequently, while D Coy maintained its ability to patrol and liaise with the locals, the British 'Viking Coy' was tasked with protecting logistics columns to the north and with dissuading insurgents from further harassment of construction workers on Route 611.

On 23 August, close to Patrol Base Bridzar in the northern part of the Patrol Base Line a new bridge was completed replacing an older one, which had been sabotaged on the previous Saturday – probably because the Taliban wished to isolate the troops in Patrol Bases Sponden, Malvern East and Malvern West, all of which lay close to the Helmand River. This was intolerable to the combined force as well as to their Afghan partners, and in the autumn the Patrol Base Line was extended and improved in order to prohibit, or at the very least reduce, the insurgents' emplacement of improvised explosive devises on the roads and near bases. During this period of strengthening the line, the bases were also prepared for the upcoming winter season. The following days, a dust storm rose and on 25 August, under cover of the bad weather, the Taliban chose to attack C Coy. All the company's bases came under assault and after a while the officer commanding requested close air support. An F/A-18 Hornet fighter aircraft was on patrol and the pilot accepted the risk of dropping a GPS-guided

---

[25] Presumably because of the tanks being present.

bomb despite the target being as close as 185 metres to own troops[26] in contact. However, as the GPS was unaffected by the dust he was instructed to go ahead and seconds later the pilot reported, 'Bomb dropped, 30 seconds to impact.' There was some anxiety because of the proximity, but eventually the signal came through that the target had been hit, and C Coy was henceforth left in peace.

Saturday 28 August there was another forceful attack, this time against a British sub-unit. Since small arms fire was far from enough, 468 mortar bombs and two air launched bombs were fired. Moreover, the Guided Multiple Launch Rocket System battery at Forward Operating Base Budwan (GMLRS) fired three rockets.

However, at the same time Forward Operating Base Budwan – one of the two bases which the Afghans found useless – was amidst the process of being reduced to a fire base. It hosted one of B Coy's platoons having relieved the one from C Coy, which had rejoined the bulk of the company in Patrol Base Line, a Danish home guard platoon for force protection, a British Gunline battery and the British 'Guided Multiple Launch Rocket System' unit. Consequently, patrolling in this area ceased and contact to locals was reduced to occasional walk-ins.

On the last day of August, the Combined Force initiated planning for two operations aimed at finding the compounds where improvised explosive devices were being manufactured. Chief S3 sensed huge enthusiasm on the part of his British opposite number, who had apparently been looking forward to increased activity in this field.

In early September, the military situation seemed stable. There was relative quiet within the Combined Force's area of responsibility, although with some minor incidents of small arms fire. Unfortunately, the improvised explosive device threat continued to affect operations and restrict movements. It appeared that the Battle Group's determined efforts to overwatch frequently used areas did not seriously reduce the threat, but rather inspired the insurgents to even more daring innovation of their methods. The Patrol Base Line was particularly vulnerable and continued to sustain high casualty rates due to improvised explosive devices. However, the successful removal of a key insurgent commander contributed to maintaining the military status quo prior to the general elections. Removing this individual, who reportedly had been ready to direct his violence against participants of the pre-election campaign, had probably greatly increased security up to Election Day.

The combined operational planning, begun on 31 August, continued

---

[26] 'Own troops' is a proper military term for friendly troops in such a situation.

over the first days of September including Danes, British, Americans and Afghans. Amongst the external experts working with the Combined Force was a Scot major who became a person most fondly appreciated by the Danes not only due to his impressive courage, but also because of his ability to come up with the right questions at the right time during staff work. However, this officer also provided a classic example of the language gap which is bound to materialise in coalition warfare. The Danish officers and soldiers of ISAF Team 10, whose English, generally, was acquired at school, were often at a loss to understand his highland accent and had to ask for repetition over and over again.

Initial orders for the action, Operation Tufaan Qimaat, were issued on Friday 3 September, and the final order group was held on Saturday 4th.[27] Under the appalling conditions of a failing air-con system, Lieutenant-Colonel Funch Pedersen and Major Byrholt gave out the orders followed at 3 pm by the so-called roc (rehearsal of concept) drill – a sand table exercise where the participating commanders moved little vehicles and other markers around to illustrate their perception of how to solve their missions. Moreover, during this drill all pertinent queries were solved. Then the operation went ahead. The companies involved did their utmost to find and remove the insurgents' homemade explosives and close down their production sites in order to prevent further manufacture of road side bombs. Nothing was found, however, but there was a sense of immense pleasure gained from the smooth staff work, which had overcome all conceivable cross-national difficulties. While the mission objectives had not been achieved, important breakthroughs in multinational collaboration had been made.

Colonel Fredskov assumed that the insurgents had managed to retain their capability of laying bombs along one of the important thoroughfares, the Bandi Barq Road, but this was a nuisance that could be dealt with later again. What was worse was the possible effect the action might have caused on the Combined Force's reputation amongst civilians in the area. While the positive outcome of Operation Tufaan Qimaat was the realisation that the co-operation with the Afghan Army had worked efficiently, the downside of this enterprise was that local roads and sewage canals as well as crops were damaged. Quite reasonably, therefore, the Combined Force assumed that because of the obvious lack of operational success as well as the damage done during the action, Operation Tufaan Qimaat might affect negatively the relationship with the local civilians.

Two operations had aimed at opening and securing the road to Patrol

---

[27] 'Tufaan' was the nickname for Danish led operations.

Bases Malvern East and Malvern West, achieving freedom of movement on the route called 'Sephton,' the road which connected all the bases in the line. While the Combined Force had finished the first, Operation Tufaan Ghumbesa, by the end of August, the second, Operation Tufaan Kala 1, was begun on 1 September. These two operations had become necessary because of the need to operate safely from the line and to be able to resupply its patrol bases. As Route Sephton was cleared, this goal was achieved. During the following week, supported by the Engineer Platoon C Coy effectively razed their most essential arcs of fire and emplaced barbed wire in order to avoid insurgent surprise attacks and prevent reseeding of improvised explosive devices. The company commander, Major Berger, recalled that:

> We entered a relatively calm period [of time] and were able to move around in Patrol Base Line. The biggest problem was that surveillance was hampered by the many masks in the surrounding terrain and we had to keep up a constant readiness in order to monitor changes. Under these circumstances our establishment strength was simply too modest. It was obvious that we had to do something about security of the camps and not least the road connecting them. In order to improve visibility we had to cut down entire corn fields manually, thus greatly annoying the local farmers whose subsistence basis we destroyed. We needed to blow up trees, fill ditches and roll out about twenty kilometres of razor wire. And all this had to be done in temperatures about 45° Celsius and humidity close to maximum. On top of that the fine dust found its way into all creases and cavities and the security situation did in no way encourage discarding either helmets or body armour.

Amongst the masks limiting C Coy's free vision and fields of fire there were a wall running along the road as well as an Afghan 'compound' obscuring a considerable piece of ground. Frequently, the insurgents hid behind these masks from which they ambushed the company's patrols and supply convoys. This area was where Combined Force Nahr-e Saraj North had previously had one killed and one wounded. Therefore, though after much procrastination and further attempts to clear the place causing several more wounded in the process, Major Berger managed to ascertain authority to demolish the compound, which was subsequently done by a number of rounds from A Troop's three tanks followed up by British specialists finding six unexploded improvised explosive devices in the process. The commanding officer was away on leave, but he had been consulted on the matter before he left and agreed wholeheartedly to what his deputy, Lieutenant-Colonel Thomas Funch Pedersen had suggested. The deputy commander recalled that:

At long last, having met with headquarters Task Force Helmand on the matter, one day queuing up together with J35 before cook house, I managed to covey to him the imperative that this compound be obliterated using tanks in lieu of air strikes as he had suggested.[28] Having got final approval, I went up with A Troop's three tanks and, spending 63 HEAT (high explosive anti-tank) grenades, we levelled the compound, thereby improving security around Patrol Base Bridzar significantly.

More generally, early September saw security deteriorating through a swell in the number of small arms fire incidents. Over the period 24 August to 4 September there had been a significant increase in reports on gunfire and explosions of which the majority had taken place between 6 a.m. and 6 p.m. In spite of Ramadan the daytime activity was high, which was probably due to the fact that having conducted a slightly more modest activity for some time, the Combined Force was now returning to normal patrol patterns. It was, however, impossible to gauge whether this was the reason or the increased shooting activity at some distance from Combined Force bases might rather have been connected with rivalries between insurgent groupings seeing a need to assert themselves in the wake of ISAF's successful targeting of a key insurgent leader. Moreover, one might speculate that the increased activity might have been due to a belief amongst insurgents that, if sufficient pressure was applied on the junior partner in Task Force Helmand, the entire coalition project would collapse.

Although early September was dominated by the Afghan Army's preparations for the general elections, partnering Anglo-Danish units with Afghan forces proceeded according to plan. Simultaneously, the Combined Force's efforts to keep up the flow of supplies to Forward Operating Bases and Patrol Base Line proved successful. During the subsequent weeks, joint patrolling by Heavy Weapons *Tolay* and the Danish B Coy, as well as initial preparations for joint operations in the beginning of October, came into focus. Partnering was improving daily with astonishingly positive attitudes on the part of Danish soldiers. B Coy's partnering with Heavy Weapons *Tolay* engendered excellent inter-personnel relationships and, in the following period, these two sub-units were able to patrol jointly.

In mid-September there was an unexpected ebb of the insurgents' activity level, which was lower than it had been for a long time. This was particularly surprising because a high number of confrontational activities had been expected in connection with the general elections coming up on the eighteenth. The Combined Force believed that one likely explanation for this might be found in the insurgents' lack of funding and supplies, which

---

[28] J35 was the UK officer heading the two branches of J3 (operations) and J5 (plans).

would also explain the pervasive use of outdated ammunition for their improvised explosive devices. The Combined Force saw extensive use of obsolete munitions and less small arms fire than what had been the case hitherto, although with increased accuracy, probably brought about by the need for saving ammunition. Another reason for low activity might have been the execution of Operation Highwayman that had taken place in the 'brigade battle space' – a vast empty just north of the Combined Force's area of operations where no single battle group was responsible. During the operation, US Special Operations Forces and Afghan Army units had taken out a number of high value targets and disrupted insurgent operations as well as countered local narcotics production. The Combined Force assessed that these activities had been successful in reducing the insurgents' ability to target the force.

Planning to support security provision during the Afghan general elections was designated Operation Wolesi Jirga. This brought the Combined Force and the Afghan security forces even closer together. The operation aimed at propping up local authorities' during their preparations for and execution of the elections. It was apparent, at least to the Combined Force, that these elections would pose a challenge to the Afghan security forces. During the weeks prior to 18 September combined planning sessions involving Afghan army and police officers as well as the Combined Force staff took place in order to help the Afghans to structure their deliberations. The Afghan central authorities' wish was that Afghan security forces should take sole responsibility for planning and execution of this operation, but the realities of the Afghan staff work routines soon made it clear that this would not work. On request of the Afghan forces, the Combined Force changed its plans three times because of increasing demands for more ISAF involvement and, eventually, security provision during the election became a truly combined effort. In the beginning of the planning process the Afghan officers appeared somewhat uncertain as to what to ask, but as time went by they grasped the purpose and braced themselves for asking questions and requesting support from ISAF. On the day the elections went ahead, 18 September 2011, the chief of police, Mr Ezmerai, the *kandak* commanders Lieutenant-Colonels Mohammad Hamayun and Eid Muhammad as well as the brigade commander General Sheren Shah co-operated efficiently with the Combined Force's staff. Prior to the opening of the polling stations, the ballot boxes and the voting papers were collected in Lashkar Gah protected by Afghan police and army units, and they were safely returned there in due course.

In the Combined Force's area of responsibility the day was generally peaceful, although the Afghan police in Bandi Barq were attacked eight times with rocket propelled grenades and small arms fire. Here and there minor incidents occurred but all was dutifully taken care of by the Afghan security forces with only modest involvement of ISAF. The aims of high election turnout, peaceful completion and local avoidance of ballot rigging were generally achieved, and the enterprise taught many useful lessons and established ties which would turn out to become useful at a later stage.[29] It became particularly apparent at the general elections that the dedicated campaign by higher authorities against the Taliban commanders had greatly reduced planning and co-ordination of Taliban action intended to disrupt the voting process.

With the elections completed ambushes and improvised explosive devices again became the primary concerns. On Sunday 19 September, the cameras on board the blimp hovering 300 metres above Camp Price discovered four persons engaged in digging on the road near an Afghan police check point, probably trying to emplace an explosive charge. The British police advisers and an Apache attack helicopter were alerted, but when they arrived the insurgents had disappeared. The next day, however, a team consisting of a Danish infantry platoon and British ordnance disposal experts was send out to dispose of the charge. They found two improvised explosive devices and a house with weapons and plenty of ammunition. Following this event Major Byrholt chose to employ an infantry section with a team of snipers to pick off insurgents in future before too much harm could be done, a decision met with Colonel Fredskov's silent approval.

During the days of 20-21 September the British F Coy conducted Operation Tufaan Sheen Azda 10 to disrupt the insurgents' ability to use local compounds in the area west of Forward Operating Base Rahim. While successful, this operation did not put an effective kibosh on the ongoing emplacement of explosives as, in late September, the number of improvised explosive device incidents increased. On 21 September, such a strike destroyed one vehicle in a track leading to Fire Base Budwan, and on the 22nd in the Patrol Base Line, another improvised explosive device struck a dismounted patrol, killing a twenty-one year old Danish soldier of C Coy, while another private was seriously wounded. The soldiers had dismounted their Piranha – a wheeled lightly armoured personnel carrier, or APC – to secure their comrades as the explosion happened. For the moment, operations were temporarily halted as the replenishment groups (or 'log trains') had to sweep their way in to avoid hitting anti-personnel mines or improvised explosive devices. Explosive charges placed at important nodes on frequently used routes remained a constant hazard.

---

[29] Although at the national level accusations were raised concerning irregularities during the elections, similar allegations were not made in the Combined Force's area of responsibility.

Private Claus of the Mortar Section, C Coy, was in the patrol hit on 22 September. He wrote in his diary:

> We had driven but briefly when a powerful explosion reverberated. We hoped that it might be the engineers who were minding their own business, but it wasn't. The engineers' activity was to be covered by some of our blokes, who – as had been done on earlier occasions – were to take up position on the roof of an Afghan compound close by, but either the floor or the roof was booby trapped with an improvised explosive device detonating as soon as the boys came up there. One died in the explosion, while the other was severely wounded. He is now stabilised and seems to be ok.

The bomb exploded at 7:23 am and the message reached Combined Force headquarters immediately after. A chopper was called in and it was only 8:18 when the casualties were brought into the field hospital at Camp Bastion. At 9 am one was declared dead.

On the 23rd Operation Tufaan Tor was launched. This operation aimed at improving security on the Bandi Barq Road for Afghan National Security Forces as well as for the ISAF by annihilating the insurgents, who were caught in the process of placing improvised explosive devises on the road. This, however, was not successful. The insurgents laying bombs were detected, but they were chased out of the area by snipers from B Coy who did not, however, manage to eradicate the menace. The Combined Force realised that they had to try again.

It was obvious that insurgent actions were becoming more daring, although their success was indeed limited. On 23 September a small gang ambushed C Coy. Private Claus of the Mortar Section noted:

> I was standing securing with two of the blokes as we heard the supersonic cracks as projectiles flew over our heads. One second elapsed and then the fire was returned in full measure. I personally fired about half a magazine, but the blokes went absolutely berserk and emptied up to two and a half magazines... There were a lot others in the position now, and so the three of us ran to the mortars. Bombs were quickly charged and shortly afterwards three salvoes were fired. We corrected and four more salvoes went. Now, also several GPMG[30] teams and at least one vehicle mounted heavy machine gun opened up – it was a pleasure! After additional salvoes from our mortars, an Apache helicopter circling overhead joined with a row of 36 mm shots.

Securing the Afghan civilians and supporting the peaceful development of

---

[30] The GPMG is the colloquial term for the standard issue light (7.62 mm) machine gun (general purpose machine gun).

the district was the Battle Group's most important task, but there were other duties to care for. VIPs came visiting one after the other, and depending on the level of their importance security measures could be more or less demanding. On 26 September Operation Banjani Fil was launched in order to provide security during a visit by the Danish prime minister who had desired to have a stroll in the streets on the outskirts of Gereshk. It was a very time and resource consuming security enterprise, but it was carried out successfully.

As suggested earlier by the Combined Force, the bases Budwan and Zumbalay ought to be abandoned, and as Task Force Helmand's approval of closing down the latter was obtained the Combined Force launched Operation Tufaan Woraan, aiming at doing this. Chief S3, Major Christian Bach Byrholt, had spent a lot of his time preparing this, which proved to be a major operation. The 'roc drill' for Operation Tufaan Woraan was held with British, US and Danish participants on the 27th. As he had appointed himself officer in charge of the operation, this 'roc drill' was especially important to him, and he was delighted to see that it went all right.

On 30 September Colonel Fredskov launched the operation. The task force put together for the operation consisted of about 55 vehicles and more than 270 officers and soldiers, viz. four Lithuanians, twenty-four US Marines, ninety British and 150 Danes. There were three infantry platoons: one Ghurkha, one Danish Life Guards and one from the US Marines. There were three tanks, three infantry fighting vehicles, a lot of APCs, seventeen large trucks, engineer equipment, small UAVs (unmanned aerial vehicles) and scout cars of the Eagle type. A British mine clearing team was to spearhead the advance followed by the American platoon. H-hour was 0500 hours, but for some unknown reason the US Marines were twenty-five minutes late. At 7:45 am the main force left the base, Major Byrholt amongst them. The day was used clearing the route to Zumbalay bringing the combat sub-units into positions and setting up a Forward Supply Area called Bibi. Byrholt had established his observation post with the tank troop, which appeared to be a lucky choice as the troopers were service-minded, the radio equipment adequate and the British Bowman Operator a huge asset – circumspect and helpful. Once the track had been secured, the plan was to let the infantry guard it while the sappers went into Zumbalay to dismantle the base. On Sunday morning 3 October they reported that the task was satisfactorily executed and they were ready to leave. It had been planned that a UK Apache combat helicopter should cover the disengagement and, as this chopper appeared about noon, the retreat commenced, Ghurkhas first followed by Life Guards. The whole operation had been carried out according to plan and a lot quicker than had been foreseen.

At the end of September, all companies of the Combined Force were partnering with their counterpart *tolay*s of 3rd *Kandak*. Its staff and Heavy Weapons Tolay were in Camp Gereshk, 1st *Tolay* in Forward Operating Base Rahim, 2nd *Tolay* in Patrol Base Line with platoons in Patrol Bases Sponden, Clifton and Bridzar, and 3rd *Tolay* was in Forward Operating Base Kah Nikar although one platoon was detached to operate from Patrol Base Bahadur. Moreover, in partnering of staffs considerable progress in staff work procedures had materialised, signifying an important step forward towards the goal of smoothly running routines with daily meetings between branches, timely planning and central co-ordination by the chief of staff.

Similarly, the development of the Afghan National Police continued to meet plans and expectations, and the police force were deservedly credited for providing security during the Eid religious festival and the general elections. As to security provisions, there were several lessons learned during these events, which would prove useful in future operations led by Afghan National Security Forces.

At the beginning of October, the military situation appeared to remain stable. There had been a sustained reduction of complex attacks on the Combined Force, a point which was also reflected in reduced radio chatter throughout the area of responsibility. It appeared that the insurgents were having a period of indecision, as a *shura* they held in the eastern part of the Combined Force's area of operations, which was attended by around twenty persons, ended with an agreement not to make any attacks. The cause of this hesitation and reflection was probably to be found in the recent losses in the insurgents' leadership tier and the attrition of their infrastructure used for manufacturing bombs and communication. The targeting of Taliban commanders was assessed to have been reasonably accurate, and the disruption of the improvised explosive device organisation was obvious. The Combined Force's patrols now began to get out amongst the population, allowing the conduct of human terrain mapping: finding out who knew whom, who influenced what, who were the bandits and who were the ones menaced by them.

Colonel Fredskov, too, assessed that the improved security in the Combined Force's area was caused mainly by the losses amongst the insurgents' leadership that had hampered their co-ordination as well as their efficiency and determination. The frequent and successful strikes against the rebel leaders had had a perceptible impact on their activities, and it was assumed that the disruptive effects would probably reverberate for several months. Moreover, there was another, and much more mundane, reason for the insurgents' sudden lack of zeal: the poppy planting season was coming on and they were doubtless now more interested in securing the crops than challenging the Afghan government's authority and ISAF's military superiority. In Gereshk, too, the political and security situations remained stable, but a risk of suicide bombers remained. Therefore, in order to not provoke new ambush attempts, patrolling in this town was temporarily adjusted, though, as this risk later declined, the patrol activity reverted to the usual pattern.

The Combined Force's partnership with the Afghan Army was improving by the day and, in particular as far as operations were concerned, disagreements were solved and issues of importance were co-ordinated concurrently. The Afghan National Police headquarters observed that the security situation had seen a significant development for the better due to the numerous check points which had been established amongst the citizens of Gereshk. However, improvements depended on thorough training and commitment, and there was one snag that remained obstructive to this – namely some Afghans' lack of precision, dedication to tasks, and sense of duty. Many attempts at training the Afghan officers foundered on their lack of enthusiasm, and particularly in the field of intelligence, which is of key importance to security building, progress was not as quick as was hoped.

As mentioned above, the need to create a secure environment for the workers repairing the hydro-electric power plant and laying hard surfaces on the road leading to it required increased attention to the Bandi Barq and Deh-e Adam Khan areas. Police checkpoints in the area had been attacked repeatedly over the last few months, and the district governor had stated he hoped that the battle group would increase its presence there. To a large extent, it was the governor's priorities that had been pivotal in creating the battle group's campaign plan, which included the recommendations that the bases Budwan and Zumbalay be closed in order to free troops for the tasks in the Bandi Barq area. As these two bases were huge establishments, and as equipment could not be allowed to fall into enemy hands, the closures required considerable preparations and meticulous tidying up. Nonetheless, while for various reasons the closure of Budwan would have to wait almost to the end of the tour, Patrol Base Zumbalay had already been closed on 3 October, as detailed above. While to the Danes the closure seemed the only sensible thing to do, it was feared that the insurgents might sooner or later try to re-occupy the remainder of the base. This would be ominous to the local citizens, because the Taliban was notorious for not treating kindly those who had collaborated with NATO forces once these forces were no longer present. For the same reason, the Afghan security forces' personnel living locally were in danger should it be known that they had co-operated too closely. To make sure that, in the eyes of the local citizens, there would be no tainting of the Afghan National Security Forces' reputation – and no suspicion of too close relations with NATO – they were left out of the process, which was carried out as a co-operative effort by Danish, British and US forces.

Another way to ensure the availability of as many combat soldiers as possible for surprise action against the insurgents was finding other troops for guard duties in fixed the positions. Planning for insertion of Danish Home Guard personnel and Bosnian guards units proceeded with a view to discharging combat units from force protection duties in i.a. Budwan, which, as covered above, had recently been reduced to a fire base. This novel development would free one of B Coy's platoons for other employment, which would better exploit the high cross-country capability of their infantry fighting vehicles. Moreover, preparations for relief of the Light Armoured Reconnaissance Company by D Squadron of the Household Cavalry Regiment, the HCR, were initiated.

All in all, in early October the Combined Force was happy with its results and confident as to the outlook for the immediate future. Planning of the ISAF Team 11 'commanders' recce' – the pre-tour reconnaissance by the commanding officer and sub-unit commanders – proceeded as planned and was finished by 16 October – leaving Chief Operations with the feeling that now, eventually he could see the end of his tour.

Also on 16 October, Major Byrholt, being senior duty officer that evening, was told that four persons had been observed placing bombs on Bandi Barq Road. As he had no snipers in the vicinity other measures had to be taken, and he decided that a missile strike would be appropriate. However, there were two caveats: civilians must not be harmed and the digging persons should be clearly identified as engaged in a hostile act. Having ascertained himself on these aspects, he decided to request a missile strike, an Apache combat helicopter to follow up should some of the insurgents not be hit, and a reconnaissance aircraft to monitor those who might escape. After more than twenty minutes of nervous waiting the missile struck and only one person was seen running away. The helicopter did not manage to find him, but the monitor tracked him to an Afghan compound, which had been under suspicion for some time for harbouring insurgents. Although S3 would have liked to send in Special Forces to finish the job, none was available and his own reserves were insufficient for the task, so the mission was brought to its conclusion at this juncture. Although total success was not achieved, the result was not bad and Task Force Helmand was most appreciative. The missile strike was probably the reason for a temporary improvement of security at Bandi Barq Road. The strike had been successful because prior mapping of the insurgents' pattern of life had made it easier to spot in time those who were engaged in placing improvised explosive devices and to target them without unreasonable risks to innocent civilians or jeopardy to their property.

It was important that not only the Afghans were made aware of the ongoing progress. The world at large – and in particular the tax payers at home – should be left with no doubt that the efforts were worth every penny paid and that soldiers did not die in vain. In October, a Danish media visit to Gereshk was successfully used to promote the achievements of the Afghan National Security Forces, the outreach of the Afghan government's authority and more specifically the development in Gereshk. The most important part of the visit turned out to be a press conference held with participation of the district governor, the sector chief of police and the district community council's members. The Danish media were also shown a school with 1,100 girls and a women's affairs centre.

In late October, the Combined Force started initial planning for a human rights seminar scheduled for 3 November in the Blue Shura Hall in Gereskh, during which ISAF would support security with its quick reaction force. Similarly, the formation of a trade committee was seen as an important step towards strengthening Gereshk as the commercial hub of the Helmand Province.

From the Combined Force Headquarters' point of view mid-October had proved a generally stable period, although armed clashes did indeed happen within the designated boxes of operations. Moreover, it was noticed that, remarkably, there were no attempts at placing improvised explosive devices on Bandi Barq Road, which used to be one of the most troubled areas. The Combined Force believed that this was due to the successful strike on 16 October against a team laying such an infernal device. Nevertheless, the insurgents were continuously probing for weaknesses as well as places and possibility for reseeding their bombs, an activity which seemed likely to continue as an unwelcome extra hazard to operations. This was a major concern during Operation Omid Char, which aimed at establishing a new patrol base called Hazrat. It was an enterprise planned and carried out entirely by the Afghans, though under strong supervision by the Brigade Advisory Group. However, much to the chagrin of Colonel Fredskov this operation was planned and launched without proper involvement of the Danish Battle Group, in whose area of responsibility it went ahead. Moreover, it happened in a neighbourhood where the Battle Group had sustained severe casualties in mid-October. Consequently, he found reason to take up discussion on the pertinence of this procedure – infringing on his perception of 'mission command' – with Brigadier Chiswell, who eventually agreed that future operational planning would happen in full collaboration amongst all units with potential interests.

Nonetheless, Operation Omid Char was concluded successfully, and the end of October it appeared evident that it had had a positive influence on the local Afghans' perception of the ways in which security had been improved and insurgency activities had decreased across the Combined

Force's area of operations. Moreover, October's operations had created further improvement of the security at Bandi Barq Road, which was of immense value to the ongoing reconstruction works at the power plant.

While the officer commanding B Coy, Major Michael Toft, had begun his initial operations as early as 4 September, the company's first major challenge was its move into the town of Deh-e Adam Khan at the end of October. This action, called Operation Tufaan Gulaabi, happened after the company's and its Afghan partners' meticulous preparations over the autumn, including intensive intelligence collection and *shuras* – meetings with the local population. While this operation was being executed by B Coy, D & F Coys exploited the insurgents' relative passivity expanding their areas of patrol.

Political and commercial activity now sprang up, and the district governor held his first, very successful, *shura* in Hyderabad. Forty Afghan local civilians attended, and the atmosphere was good. He also held a *shura* where all attendees agreed to provide better access to markets in the towns of Lashkar Gah and Gereskh, and the Business Development Association's 'Interim Committee of Four' started reviewing the local draft bylaws designed to regulate commercial life. A 'Committee for the Livestock Association' was soon established, a 'goat programme' was expected to start within the following month and the Food Zone Programme in Khar Nikah, aiming at selling seed cheaply so to allow the farmers to shift their production away from poppy cultivation, was conducted successfully and all seed was distributed. However, especially in F Coy's area of responsibility there were continuing clashes with insurgents, and the company commander later told Major Byrholt of his surprise by the change of quality amongst the opposition: 'These are no longer the usual peasants who have just recently got rifles in their hands.' Whether this can be ascribed to improved training standards, new recruits fresh from training camps in Pakistan or simply a changed *modus operandi* is hard to say.

The military situation in the Combined Force's area of responsibility at the beginning of November was stable but dynamic with four concurrent operations being conducted. After an initial increase in violence due to the new operations, it appeared that senior key insurgent commanders had been given guidance to disengage from sustained engagements and return to a programme of placing improvised explosive devices. Buttressed by improved information sharing, over the previous two weeks the situation had changed from a long static period to becoming much livelier, favouring Combined Force's sustained initiatives to frustrate and stress insurgent networks. Seen from the vantage point of the insurgents this must have appeared as a completely new Combined Force *modus operandi*. There were clear indications that the increasingly agile actions were stressing the insurgent networks in terms of weapons, forces and ammunition.

However, success had not come easily. Major Byrholt had had a hard time during most of October; the early days of November, too, boded ill as far as his workload was concerned. Moreover, he was well aware that quite a few had died and many others had been maimed on operations that he had planned and ordered. A risk of mental and physical stress was certainly present, and this was not eased by the arrival of judge advocates who had been flown out to investigate alleged excessive use of violence by C Coy. Their working methods were time-consuming, and Byrholt, already hard pressed for time, was forced to spend hours with this team trying to explain issues that he sensed they knew everything about.

On 7 November, the British D Coy launched an operation aimed at finding and annihilating a band of Taliban insurgents and destroying a weapon with which they had caused much mischief. Early on the operation a soldier was struck by an improvised explosive device and was badly mauled, but since he wore Kevlar reinforced underwear the rest of his body was undamaged. He was evacuated by helicopter, and the operation went on. At a later stage an Afghan soldier tripped on yet another device, was blasted several metres away and knocked unconscious. A British female medic immediately came to his rescue and tore off his garments to start searching his body for injuries. He was unscathed, it appeared, but was somewhat bewildered as he came to and realised he was almost naked in front of the Amazon Samaritan. On the same day F Coy ventured deep into 'Taliban territory' to stress the insurgents by their presence. They did not encounter any roadside bombs but three NCOs and one Guardsman were shot. One NCO remained in theatre and the remainder were sent back to the UK for further treatment.

At this moment, US attacks from armed unmanned aerial vehicles against Taliban bases in Pakistan produced a wave of insurgents surging back into Afghanistan in early November, and it was speculated that this might increase violent activity.

The Combined Force's tank troop, A Troop, took over responsibility for Route 611 during the Viking Coy's hand-over to the incoming Warthog Group. From mid-November onwards, A Troop acted both as quick reaction force and as escort to the two Irish Guards companies D and F Coys. From November to January, B Coy, whose Operation Tufaan Gulaabi had been successfully concluded, established themselves in Deh-e Adam Khan and Bandi Barq, opening Patrol Bases Viking and Shir Agha, and started conducting operations under Afghan Army leadership. Henceforth, B Coy patrolled jointly with Heavy Weapons *Tolay* and was, for some time, engaged in operation Tufaan Laram. C Coy prepared their tents in Patrol Base Line for the upcoming winter while simultaneously patrolling jointly with their local Afghan Army partners. In the same period, D Coy steadily expanded its area of influence and, on 10 November, sent one platoon to man the newly established Patrol Base Hazrat. Although since then this tied down the a large part of the company in a fixed installation, the officer commanding D Coy managed to detail soldiers for continuous and very active patrolling. The Household Cavalry began patrolling Highway 1, and F Coy steadily expanded their area of influence and had a number of contacts which, however, resulted in some casualties.

As the Taliban wanted to protect the seeding of their poppy fields, November turned into a peak season for placing improvised explosive devices, and the various Taliban groups appeared to have received considerable quantities of explosives for this purpose. On 14 November, while patrolling north-east of Gereshk a soldier of B Coy was killed by a roadside bomb. His platoon was securing British sappers working in the area, and as he leant against a wall, the device placed on the opposite side of the partition was set off.

On the 17th D Coy spotted a Taliban doing what they had reason to suspect as emplacement of an improvised explosive device and asked for instructions. The Combined Force had an observation aeroplane airborne close by and directed it onto the objective. While Task Force Helmand was initially reluctant to authorise killing this individual, Major Byrholt asked the officer commanding to send a personal request to his British superior that annihilation be sanctioned. This helped, and Byrholt was authorised to launch all necessary means available. He asked his Air Liaison Officer to call in the waiting Apache helicopters, and the ensuing conversation with the pilots was made brief but effective: 'Target hot! Hellfire and follow up with 30 mm cannon.' A few seconds later, a missile struck. The Taliban warrior was later found thirty metres from the remnants of his motorbike.

Later that day, a patrol from Patrol Base Hazrat was ambushed and sustained five casualties. As we have seen, the Combined Force had been ambivalent about that base right from its establishment in late October, as they felt that it unnecessarily tied up forces in a fixed position that would have been more useful had they been mobile. One dead, three severely and one slightly wounded were picked up by a Chinook helicopter less than thirty minutes later. Although an Apache was covering the rescue team, the Taliban managed to send a few bullets in the direction of the Chinook. The day after, Brigadier Chiswell went out to Hazrat to see for himself, and he took Byrholt along with him. At the base Chiswell gave a pep-talk to the only 14 persons remaining of the platoon, which had once been 28 strong. Byrholt took the opportunity to ask Chiswell into a sangar from whence he was able to illustrate the difficulties of the position, pointing out the dangerous spots on the surrounding ground. Returning to Camp Price they met Lieutenant-Colonel Funch Pedersen and talked the situation over again. Christian Byrholt presented a number of suggestions on how to improve matters at Patrol Base Hazrat. He was a bit unsure, he later told me, of which kind of impression his less than deferential attitude might have left with the brigadier, but the conversation ended amicably with the latter's question, 'Anything else you want from me?' He was satisfied that, apparently, all was well.

About the midst of November, the Taliban increased their level of aggressive rhetoric. They issued a statement concluding that the Lisbon summit declaration, calling for NATO's withdrawal of combat troops, was an admission of defeat on the battlefield.[31] They claimed that pulling out was a cowardly act and a consequence of Washington's failure to persuade allies to provide more troops. Moreover, they threatened to execute children turning up for instruction in schools built by or supported by the ISAF. Nonetheless, the Combined Force saw the propaganda campaign previously launched by the Taliban in the vicinity of Patrol Base Rahim as an indication that they might have felt a loss of legitimacy due to the widespread local appreciation of ISAF's presence.

In the last days of November, Task Force Helmand had indications that the Taliban had now moved to Pakistan for the winter, recommending those staying behind in Helmand to limit their activity to emplacing improvised explosive devices, abstaining from regular attacks on western and Afghan forces. Nevertheless, so-called Close Quarter Attack Teams consisting of four persons mounted on two motorbikes were observed cruising around in Gereshk, where they were assumed to go for local officials in order to intimidate the population. The Taliban needed to assert themselves following a row of setbacks, as apparently, among other things, the local Taliban leader was annoyed by the reconstruction of the bridge at Rahim. The district community council member Haji Abdul Khaliq had been murdered in mid-November, and the investigations carried out so far pointed to his influential status as a member of the council as well as chairman of the Gereshk bazaar as the reason for the liquidation.

For the Combined Force November was concluded by a visit by the Danish Crown Prince and two ministers. It appeared a tremendous success, and even the Battle Group's leftist officer – there would hardly be more than one of these – was enthused by HRH's humour and casual attitude.

As December opened the military situation remained stable. Since,

---

[31] The actual wording of the part of the summit declaration referred to by the Taliban was "Looking to the end of 2014, Afghan forces will be assuming full responsibility for security across the whole of Afghanistan." In NATO's *Lisbon Summit Declaration* of 20 Nov. 2010.

during the Advent season, the Danes normally go out of their way to make their surroundings cosy and Christmas-like, loads of parcels with Christmas decorations had arrived, as had a container with two tonnes of sweets donated by a factory in Denmark.

Reporting suggested the insurgents were slowing down their operations, although sending more requests for personnel, ammunition and explosives. This was possibly due to a combination of factors including the Combined Force's high operational tempo, in particular with regard to targeting. The poppy season was coming on, which would normally entail reduction in Taliban operations or at least shift their focus squarely onto improvised explosive device production and emplacement. At this juncture, the Combined Force's operations were moving into new areas in the north, forcing the insurgents to shift men and equipment around. This provided a rich source of indications of the insurgency's routes and facilities – a Godsend for intelligence.

On 4 December, Major Byrholt with a few engineer officers and some British colleagues went out to reconnoitre for a new patrol base at the Bandi Barq Road. There had been so many bomb incidents there that an extra base seemed to be necessary. At 4 am they left Camp Price and arrived in the area before dawn. A platoon of B Coy had swept for mines and placed infra-red lights along the secure path and at a crossing of a small brook.[32] As the sun rose they were at the desired spot. They did their reconnaissance and withdrew without anything happening. The new patrol base was not established at this moment, but more reconnaissance parties were sent out during December to keep an eye on developments and gain basic knowledge to be used in future planning.

In early December, the work on Route 611, a prestige project of the central government connecting the Helmand and the Sangin Provinces, was interrupted by insurgents attacking two check points, one of which was situated on a dominant terrain feature. Since this was of importance to the entire road construction project, the Warthog Group and the private security company responsible for the area braced themselves for countering the attacks. Likewise, the bridge at Rahim, called crossing point Locket, was once again destroyed by insurgents and a school in Rargay came under pressure. As a result, the confidence in security measures among the citizens in F Coy's area of operation declined, but it was speculated that intensive information activity might repair some of the damage done in this respect.

On 6 December, President Karzai's animosity against the private

---

[32] Infra-red lights can be seen through night vision goggles, only.

security companies – frequently described as PSCs – resulted in a central government decision to step down the activities of the PSCs operating in the country. The reason behind this was that there had been frequent and indiscriminate shootings at innocent civilians by American members of such companies. President Karzai considered the PSCs amiss in accountability and adequate supervisory controls. Henceforth, the Afghan National Police would become solely responsible for the upkeep of security. A few of these companies, which had been contracted by embassies, construction companies and international forces, were allowed to continue their activities until expiration of their contracts, after which they would have to cease working in Afghanistan. Many of them subsequently moved their bases to Pakistan, from which they carried on offering protection of convoys.

On 9 December, the local police chief, Mr Ezmerai, arrested seven Afghan policemen of the crew at Check Point 2 on grounds of illicitly having collected excise from local Afghans. Other irregularities appeared to have taken place, and the episode confirmed the suspicion that some rogue elements within the police routinely got as much extra out of their jobs as they possibly could.

On 10 December, the Combined Force managed to spot and follow two cars containing suicide bombers. While one blew up, the other bomber was successfully apprehended, which was indeed lucky as much useful information on similarly inclined individuals could normally be expected to be forthcoming from such a prisoner-of-war.

On 22 December, B Coy launched two platoons in a reconnaissance mission along the River Helmand towards Patrol Base Malvern in order to gauge passability and sound out the opposition's reaction – there was none. Later in December, B Coy took part in Operation Omid Panch, whose aim was improving security at the Bandi Barq Road. Moreover, from December onwards the Afghan army started patrolling, themselves being up front. The Combined Force increased its manœuvrability by using British IED-disposal assets and good sweeping by their own sub-units. They focused on improving the sub-units' mobility, and the Quick Reaction Force was launched on several occasions with good results. In January 2011, there was a major attack on Fire Base Budwan in the north-east of the Combined Force's area of responsibility, where one of B Coy's platoons was positioned.

Around the New Year, the military situation remained stable with very limited insurgent violence in response to the Combined Force's operations. The Battle Group retained a clear upper hand, causing frustration amongst its opponents, who had limited ability to conduct offensive operations. Despite some reports indicating that the insurgency's higher leadership had given directions ordering increased focus on improvised explosive devices and high-profile attacks, this did not materialise.

On the threshold of 2011, it appeared that a high level delegation led by Mr Rabbani of the Afghan Peace Council was on its way to Pakistan to try to establish preliminary contacts with Afghan Taliban with a view towards including them in the peace process. Allegedly it was expected that this endeavour might lead to further travels abroad in order to persuade various other countries to support the effort. At the same time, it seemed a promising sign that in Helmand the provincial governor, Mr Mangal, established a 'High Peace Council' with representation not only by his own acolytes but with representatives of the opposition and of the tribes living in the province.

Early in the New Year, insurgents attempted intimidation of the Afghan Local Police by issuing liquidation orders. A Taliban chief was reported to have ordered the killing of twenty-five tribal leaders plus the district governor. Moreover, there was skirmishing with insurgents and attempts at emplacing improvised explosive devices on Route 611, probably intended as an attack on the road construction company working there.

In January, Fire Base Budwan, whose tasks had for some time been limited, was ordered to be decommissioned. Shortly after Christmas it had been ridded of its Guided Multiple Launch Rocket System, which was moved to Main Operating Base Price, where it was operative as of 5 January. From early January until beginning their leave, the Bosnia-Herzegovinian sub-unit detached fifteen men to provide force protection at the base. Similarly, B Coy detached a minor force for guard duties at Budwan and a platoon for the security of Patrol Base Shir Azha, which would stay until the end of their stint in Afghanistan.

Regardless of setbacks and casualties sustained by the Combined Force, its work over the autumn had been largely successful. As a clear indication of the improved state of security, in mid-January 2011 an Afghan company commander was able to arrange and conduct a *shura* comprising about seventy local citizens. While B Coy conducted Operation Omid Kher with Heavy Weapons *Tolay*, D Sqn co-operated with 3 *Kandak* doing joint patrolling of Route 611, the responsibility for which they had taken over from the Warthog Group. At the beginning of January, also, there were indications that the Deh-e Adam Khan area would become a focus of the insurgent's attention as they would probably want to live up to their earlier 'success' at Patrol Base Hazrat, where they had managed to place improvised explosive devices followed up by small arms fire, inflicting considerable casualties on the British platoon there.

In early February 2011, in spite of some Al Qaeda recovery in eastern Afghanistan, there were indications that the sustained American employment of unmanned aerial vehicles was stressing the terrorists to an extent engendering panic and internal distrust. In the area of operations of the Combined Force, enemy activity had calmed down as well, and the insurgents seemed to be hard pressed. However, now and then armed clashes did occur, and these, it was feared, might eventually benefit the insurgency which appeared to have started to infiltrate the area around the Bandi Barq Road, possibly with the aim of hampering construction works.

By the end of ISAF Team 10's deployment to Afghanistan, the commanding officer had got the impression that the Taliban had been considerably weakened in the Danish Battle Group's area of responsibility, the Nahr-e Saraj North. Where previously large scale attacks were not uncommon, now only minor disturbances by five to ten men at a time occurred, there were very few incidents caused by high trajectory fire, and the few that did happen were normally satisfactorily countered. The perception among the Afghan members of the district council was that security had greatly improved. It was obvious that a dedicated campaign by higher authorities against Taliban commanders had reduced planning and coordination of Taliban actions. This became particularly apparent at the general elections on 18 September 2010.

The Afghan National Police's development in Nahr-e Saraj North had met expectations, and with respect to the army partnering, the battle group remained confident as to the successful outcome of activities scheduled for the future. A highly capable Afghan unit – whose business it was to create confident relations between the foreign and national armed forces and the local citizens – had conducted eleven *shuras* in the assigned areas. The Combined Force had continuously supported the UK Military Stabilisation Support Teams deployed within the Battle Group's area of operations, primarily in the town of Gereshk, as well as the 'Food Zone Programme' – Operation Tufaan Koshala Bazgar – which had been initiated by the District Governor, Mr Mangal, in order to allow 65% discount for farmers buying seeds and fertilisers as substitutes for poppy. This aid programme was conducted through distribution from three outlets within the Battle Group Area, viz. Gereshk Mail Warehouse, and Patrol Bases Rahim and Khar Nikah, and was promoted through the Combined Force's information and provision of security. The Food Zone Programme appeared to have been well received among farmers around Gereshk and it provided positive effects as far as the overall participation of the Afghan governmental authorities were concerned. The programme saw 1,100 farmers through the distribution sites.

Although development on the security front had proved positive in Gereshk, allowing freedom of movement for Afghan National Security Forces and ISAF as well as for local Afghans, the Bandi Barq Road remained a security problem. Development of the combined security layout for the Bandi Barq and Deh-e Adam Khan areas was ongoing; and the Combined Force was still awaiting the final decision concerning timing.

Apparently, it came as an unpleasant surprise to the insurgents that the Afghans themselves were taking over counter insurgency operations, as they appeared to be considerably better than the ISAF in pin-pointing and identifying the insurgents' hide outs and activities.

Afghan National Army Build-Up

One of the key aims of being in Afghanistan is, and has been for years, to bring about a situation where the Afghans can take care of their own security and ensure by themselves that terror organisations would no longer be given the opportunity of operating from bases in their country. This change would be served primarily by training of the Afghan National Security Forces.

Photos taken by OC B Coy, Danish Battle Group: Afghan soldier, trucks and equipment during Operation Omid Panj 10 Dec 2010.

Danes and Afghans patrolling together (photo by Major K.V.F. Ahlefeldt-Laurvig who was Force Protection mentor to Garrison Support Unit of 3rd Brigade, 215th Corps).

From the beginning of the tour, Colonel Lennie Fredskov had set up aims for the development within the Afghan National Army units within his area of responsibility, i.e. 3rd *Kandak*, and much to his delight, throughout the early months, he could observe almost daily improvements. Focussing on leadership, operations, infrastructure and institutional processes, Colonel Fredskov anticipated a steady improvement month by month. As far as leadership was concerned, daily meetings between the staffs of the Danish Battle Group and 3rd *Kandak* were to be introduced, exchange of intelligence was to happen regularly, officers commanding the companies were to liaise with their Afghan counterparts – the *tolay* commanders – concerning planning and co-ordination at the various bases, and the *Kandak* should be persuaded to have representatives at the meetings of the District Security and the Gereshk Security Groups. As to operations, 3rd *Kandak* had NCOs and signallers permanently present in the Operational Co-ordination Centre District (OCCD), future courses in disposal of explosive ordnance, recovery of vehicles and first aid were decided upon, and regular intelligence exchange with the local police was commenced. However, in spite of his optimism, the commanding officer had to realise that not all was well, and there were many hurdles to overcome. Concerning infrastructure and institutional processes, the lack of computers was an abiding issue and plans for logistic support were inadequate. Moreover, the 3rd *Kandak* did little to plan for the training of individuals and sub-units, which was possibly due to a belief that this would be the business of higher formations.

Nonetheless, right from the beginning, partnered activities between companies of the Danish Battle Group and the *tolays* of 3rd *Kandak* were carried out daily. The combat sub-units, except Heavy Weapons *Tolay*, were all collocated with British or Danish companies and patrols and guard duties – and occasionally genuine operations – were carried out in common. Heavy Weapons *Tolay* was based in Camp Gereshk where it worked with the Danish B Coy and the Headquarters and Logistics Company. However, while collaboration generally went smoothly, training was a different matter, and although 3rd *Kandak* seemed to be kindly disposed to partnering, it remained a problem that training was not taken too seriously.

The Danish Kandak Advisory Team, in daily parlance DKAT, was under the command of the commanding officer of the Danish Battle Group in his capacity as Senior Danish Representative, but the day to day business lay in the hands of Major Kim Kristensen. The team was at a par with the British Kandak Advisory Team, which was part of the Brigade Advisory Group, and they were both working in an advisory capacity to 3rd *Kandak*. However, while the British had a complete organisation matching the staff of 3rd *Kandak*, the Danish team's staffing was limited. From the start, a number of private soldiers were detached to work with the Police Operational Mentoring and Liaison Team, the P-OMLT. Thereupon, the daily leader of the Danish DKAT had only himself, eight officers and one NCO. The latter worked efficiently as an advisor to the Afghan adjutant's branch (S1), while the remainder (one reserve officer (S2), six regular majors and a linguist (Dari)) covered the areas of operations (S3), fire support, engineer, logistics (S4), communications (S6), training (S7) and interpretation. Linguistically, collaboration did have some hitches. Only the interpreter spoke Dari, and the Afghans had a very limited command of English. However, from interpreting from English to Dari and from Dari to English, the Afghans' linguistic skills seemed to undergo certain improvement. Major Kristensen held the opinion that the difficulties in the task of advising were generally solved, though the modest size of the team from time to time necessitated British assistance. Moreover, he believed that Danish military authorities were mistaken in sending senior majors to do this job, because most of the Afghan co-operation partners were merely junior captains. However, in practical matters he did not encounter serious trouble from the disparate levels of seniority.

Amongst the DKAT's most important challenges was the establishment of personal relations. The team had thought about this prior to deployment, but in the end it was the simple needs of the day that shaped the actual approach leading to positive results. Though the Danes had foreseen sports activities as the prime mover of collaboration, regular meetings turned out to be the motor of cross-national friendships, leading to an informal co-operation based on mutual trust and facilitating instruction without loss of dignity on the part of the Afghans.

The Ramadan period from 11 September onwards was an operational challenge because patrolling had been restricted in order to avoid offending Muslim believers. However, this restraint on the part of the Task Force Helmand was balanced by a similar reduction in activity by its opponents. The DKAT did not stop their activity completely but adjusted it to the cultural and religious constraints of the actual situation, maintaining a reasonable continuity in advising the *Kandak* Staff.

The general elections of 18 September were a security challenge demanding foresight and planning. On the one hand, the DKAT managed to convince the *Kandak* Staff of the adequacy of thorough analyses and farsighted plans for employment of forces. On the other hand, the Afghans' traditional faculty of improvisation contributed to success and facilitated smooth running of the electoral process.

Major Kristensen found that cultural differences, not least as far as women were concerned, had little importance. The DKAT dealt solely with the Afghan National Army, which had neither female officers nor soldiers. However, it appeared that the general view on women was somewhat more relaxed than initially believed.

The tasks at hand were solved amicably in co-operation with the British, because the latter were very forthcoming and willing to help. Nevertheless, the DKAT sensed some dissatisfaction on the part of the British because the Danish organisation was so small that its inherent limitations necessitated frequent drains on British resources. Therefore, upon returning to Denmark in February 2011 Major Kim Kristensen suggested to the Army Operational Command that in future, Denmark should either mount a full Advisory Team like those of the British or simply attach individuals to British teams. Eventually, the latter solution was accepted.

The co-operation with the Afghans was respectful, though burdened by their limited literacy and limited traditions for planning, but it was also furthered by their pragmatism and their skill at improvising. The Afghans showed a willingness to learn, but also a professional pride clearly curbing Danish inquisitiveness (e.g. in personnel and administrative matters). The DKAT got the impression that although the Afghan army was differently organised and decision-making happened on an ad hoc basis, it was an organisation which had its own professional ethos and a will to get things done. Long term planning, which to the Afghans meant more than a few days ahead, was alien to them, but they compensated through an extraordinary capacity for improvisation, and they did not like the Danes supervising their methods. Similarly, they had their own mysterious way of finding personnel for the various posts. The Danes suspected that nepotism was an abiding part of this, and the Afghans most certainly declined to reveal anything in such matters.

A few Afghan officers spoke reasonable English, although not quite good enough for all of them to make official briefings. Their ethical standards were different from those of Danes and British, but they were – contrary to a widespread belief amongst westerners – not cavalier in their attitudes to collateral damage or casualties. They had a philosophical approach to losses, but they were not trigger happy. Moreover, it was an interesting observation to make that religion did indeed have some importance, but did not influence activities negatively.

Camp Shorabak was collocated with Camps Bastion and Leatherneck (photo by Major K.V.F. Ahlefeldt-Laurvig who was Force Protection mentor to Garrison Support Unit of 3rd Brigade, 215th Corps).

Apart from the DKAT there was a Garrison Support Unit (GSU) mentoring team of five officers, five NCOs, two corporals and one civilian. This team's mission was mentoring the Garrison Support Unit of 3rd Brigade of the 215th Corps in Camp Shorabak, which was collocated with

Camps Bastion and Leatherneck. The S3 Force Protection mentor, Major K.V.F. Ahlefeldt-Laurvig, observed that one of the primary challenges was acquiring the confidence and acceptance of the mentees.

Much time and effort were required for creating mutual understanding and a common concept. Moreover, the time issue was an important factor when implementing the solutions, which had been accepted and agreed upon. Implementing new routines and methods took time – even if the mentee and the mentor agreed on improvements, these would have to be understood by the whole organisation. Effective use of the mentoring system at all levels helped create a common acceptance of the new procedures and sped-up their implementation. The educational level of the mentees was not the best, and there was an obvious penchant for allowing personal relations to matter more than merit in issues concerning promotion. Moreover, it was a difficult process getting an incompetent officer replaced, if he had good contacts.

Dano-Afghan collaboration (photo by Major K.V.F. Ahlefeldt-Laurvig who was Force Protection mentor to Garrison Support Unit of 3rd Brigade, 215th Corps).

Ahlefeldt found that personal relations and rapport mattered more than cultural knowledge of the area. Linguistically, things were not too difficult, as the team had a good establishment of interpreters. As far as local customs and the attitude towards women were concerned, he was adamant that this was neither the place nor the time to try introducing western norms. Asked about the five most inopportune occurrences, Ahlefeldt pointed out that the team was too focused on doing everything at once, that the Afghan National Army did not approach work the way this was done in northern Europe, that the mentees in the Security Company was not properly understood right from the beginning, that the team was not fully aware of the pressure the mentees were under and, thus, did not schedule their meetings with proper respect for their conditions and, finally, that they did not know how to motivate the interpreters properly.

Collaboration with the British partner team went smoothly inasmuch as both parties understood the challenges and saw the possible solutions, and in particular Ahlefeldt found that the Royal Air Force Regiment excelled. Danes and British shared the same understanding of the Afghan National Army, collaborating well together towards workable resolutions. Conversely, joint patrols with the US Marine Corps (USMC) presented some challenges. The US Marine Corps had less patience in understanding and employing the Afghan unit properly and the latter, consequently, felt some apprehension. The Danish team therefore undertook to help ascertain decent conditions for the Afghans, finding adequate and workable solutions to the conundrum.

Religion was an important and integrated part of the Afghan soldiers' daily life, and it was remarkable to find a function in the Garrison Support Unit dedicated specifically to this matter, the 'S3/Religious and Moral Affairs.'

In the effort to influence society to move in a peaceful and confident direction and to boost the Afghan National Army's profile in the area, joint patrols were carried out routinely, normally three times a week. The Afghans were the only ones allowed to perform searches of inhabited compounds, which they frequently did. In these cases while some contact was established with the male inhabitants, the women were ushered from room to room as the search progressed. During patrolling small radio sets were distributed to allow the population to listen to programmes broadcast in Pashto from the local radio station, where leading Afghan National Army officers frequently spoke in order to encourage mutual understanding, knowledge of the government's programmes and co-operation with authorities.

The local Afghans were obviously intimidated by the Taliban – a term mostly used synonymously with insurgents – and if there were no Taliban there would be no need for ISAF to be in Afghanistan. The citizens needed to tread carefully so long as the Taliban threat was as manifest as it was, and for this reason they were keen on not collaborating too openly with Afghan military forces or the Combined Force Nahr-e Saraj North.

Only very few of the private soldiers could write and calculate, and they were taught as part of their army service. They were willing and easily motivated, although the concept of motivation was not too well established within the Afghan armed forces. Nor were there many social offers available to the soldiery, and the three year contracts must have been a bit of a dreary task. NCOs were hardly ever employed in a leadership rôle, but rather used as competent privates. With the Garrison Support Unit they were, though, occasionally used as leaders of small patrols of the Quick Reaction Force. They could read and write but rarely use a computer. Many of the officers had been in the army for many years and some had even started their careers in the army before the Russians came to Afghanistan. Commanders wished to control every individual personally, though younger officers were prepared to thread new paths and try new methods. However, frequently they were checked by older officers, who did not like changes. Ahlefeldt observed that it was hard to get a career in the Afghan National Army with no contacts – good performance was definitely not enough. The build-up of Afghan forces was massively supported by the USA, without which everything would probably have ground to a halt. Equipment was generally adequate and it seemed likely that with better motivation and premium for high performance the necessary talent would probably be available for developing a well-oiled military machinery.

By the end of 2010, Afghanistan was still mired in internal disagreement between the Pashtuns on the one hand and almost all the other tribes on the other. This conundrum added to the difficulties of handing over responsibility to the Afghan National Security Forces, which all members of the coalition did their utmost to train as effectively as possible. The Afghan army was then at about 134,000 (all ranks) and the police ca. 109,000. However, these figures belied their effectiveness. Eighty-six percent of the soldiers were illiterate and drug abuse was widespread, and this was believed to be even worse with the police. Moreover, while army units and sub-units operated successfully in partnership with ISAF troops; few were able to take on the leadership.[33]

Nonetheless, within the entire Combined Force Nahr-e Saraj North's area of responsibility, the training of and partnering with the Afghan Army proceeded better than it might have been feared. In December, for instance, the Afghan army carried out a major operation during which they established a patrol base, and they grew more efficient by the hour. One

---

[33] Ahmed Rashid, "Vejen ud af kaos [The way out of Chaos]" in *Weekendavisen* #52 31 December 2010.

night two Afghan *tolay*s commanded by the commanding officer of 3rd Kandak, Lieutenant-Colonel Hamayun searched a village in order to find weapons, ammunition and explosives. The village was known to be a hide-out of the Taliban and the operation aimed at demonstrating that the Afghan Army was present and able to dominate. ISAF sealed off the area with tanks and infantry fighting vehicles, but the search operation was done exclusively by the Afghans themselves. While the enterprise was completed in a quiet and amicable fashion, Hamayun and Colonel Fredskov sitting around the camp fire with local citizens had a casual chat. Three men were arrested on suspicion of having links to the Taliban and on having previously been seen placing improvised explosive devices, but little else was found in the village. The Afghan officers and soldiers had carried out their task in a quiet, disciplined and professional manner, signifying that the Afghan army was developing satisfactorily.

Afghan National Police Build-Up

In parallel with developing the Afghan army units within their area, Danish ISAF Team 10 was tasked with supporting the training and professionalisation of the Afghan National Police (ANP). This assignment was completed by the Danish Police Operational Mentoring and Liaison Team – the P- OMLT – and through intense co-operation with partners from the UK Ministry of Defence Police and its Police Advisory Team (PAT). Under this programme offices were set up, along with means of communication in Main Operating Base Price and at the District Operations Co-ordination Cell. Moreover, the activities laid down in the Afghan National Police development plan were concluded, and detailed plans for mentoring and training were finalised. The desired end state as of September 2011 was that the ANP should be qualified to work on their own securing the town of Gereshk, gaining the trust of the local population and inspiring confidence in the Afghan governmental authorities. The plan seemed to work satisfactorily and, as early as September 2010, the police appeared to behave with enhanced decency, inspiring trust amongst merchants and customers in the bazaar, providing tranquillity during the Ramadan and Eid festival season, as well as contributing to security at the general elections on 18 September 2010. Moreover, it merits attention that the screening and selection of candidates for police jobs were done efficiently by the Helmand Police Training Centre.

Public Debate

Immediately prior to deployment of ISAF Team 10, Professor Poul Villaume of the University of Copenhagen claimed that in spite of the fact that the Taliban's strength was ascending, public debate seemed oblivious to 'the failure of the coalition-of-the-willing fighting them.' Referring to opinion polls over the last couple of years he claimed that between two thirds and three quarters of the Afghans were against ISAF's presence in their country and that a similar majority favoured top-level negotiations with the Taliban and other insurgent groupings on power sharing, compromise and reconciliation. However, at the same time less than ten percent of the population wanted the Taliban regime back the way it was before 2001. Villaume argued that the more troops NATO and other foreign powers poured into Afghanistan the stronger the Taliban would get; at the same time western intelligence sources maintained that, in spite of the exorbitant amounts that NATO had spent supporting this campaign, the terror threat had not lessened – rather the opposite. Thus, Villaume asserted, the coalition's strategy had failed. His conclusions were that an international peace conference should be convened including all major players; and that the Danes should 'start discussing the pertinence of 'Denmark's militarised foreign policy' of the last ten to fifteen years.'[34]

Three weeks later, in an article in the same paper James Ferguson, the author of three books on Afghanistan, reminded the readers of the atrocities that were still being committed by the Taliban against women whose behaviour did not fit with Islamist norms. However distasteful the acts, he expressed the conviction that our forces were not in Afghanistan primarily to change society but to assure that the country would not in future again become a base for Al-Qaeda's worldwide terror, and in this respect he was in full concordance with the NATO Commander-in-Chief for Afghanistan, American General David Petraeus, writing that: 'Women's rights and the protection against violence are important goals, but they are of no direct relevance to combating Al-Qaeda.'[35]

Although only a few agreed that such a discussion would serve any purpose in 2010, the Danish political parties got increasingly anxious to fix a date for pulling out of Afghanistan. The UK and Danish Prime Ministers Cameron and Rasmussen agreed, as did later the remaining NATO

---

[34] Poul Villaume, "Er selvbedraget om Afghanistan under afvikling? [Is the self-delusion over Afghanistan being wound up?]" in *Information* 7 August 2010.

[35] James Ferguson, "Et stolt patriarkalsk samfund [A Proud Patriarchal Society]" in *Information* 19 August 2010.

countries, that all combat troops would pull out of Afghanistan by the end of 2014. However, as early as August 2010, the Danish Social Democrats (Labour) demanded that withdrawal be commenced as early as 2011, and at the same time the Social Liberals suggested that all 800 Danish troops should be home by 1 July 2011.[36]

On 24 August 2010, the *Information* interviewed four high profile public figures, all opposing Danish participation in the coalition-of-the-willing. All four recommended withdrawal from Afghanistan regardless of the consequences. If the Taliban would re-emerge as power-brokers this would have to be accepted. The left wing politician Frank Aaen claimed that the longer the procrastination, the stronger the Taliban would become. He was convinced that sooner or later the Taliban would assume power and wield country-wide dominance, though not necessarily as totally as before. The social critic Carsten Jensen, who had travelled extensively in Afghanistan, did not believe that anything positive would result from military withdrawal, but that disengagement was necessary because there was no probability that the effort would succeed within a reasonable number of years. He believed that, since the coalition was not willing to spend what was necessary in terms of personnel, equipment and funds, the endeavour was doomed. The leftist politician Holger Nielsen opined that a negotiated settlement would have to include representatives of the Taliban, although this boded ill for the Afghans. He suggested that what was needed was to get the Afghan economy and social fabric up and running and then get out as quickly as possible. Finally, Professor Poul Villaume suggested – naïvely, perhaps – that Denmark should wield her influence with the United States to ascertain that negotiations be commenced immediately rather than in five years. The Taliban should be forced to promise that never again would al-Qaeda be allowed to seek refuge in Afghanistan, and UN troops should be deployed to make sure that the promise be fulfilled.[37] The day after, these views were echoed in an editorial concluding that, cynically speaking, the USA and Europe, and thus Denmark, could live with Afghanistan being cast back into the darkness of the eighth century as long as it posed no serious threat to peace in the rest of the world – though this,

---

[36] "S afviser borgfred om Afghanistan [Social Democrats dismiss agreement on Afghanistan]" in *Berlingske Tidende* 14 August 2010.

[37] Charlotte Aagaard, "War opposition is prepared to let the Taliban return to power" in *Information* 24 August 2010.

of course, would be bad for our conscience.[38]

In *Berlingske Tidende* on 25 September 2010, Senior Researcher Lars Erslev Andersen of the Danish Institute for International Studies asserted that Afghanistan had already been made an al-Qaeda free area, and there was no longer any indication that the Taliban was co-operating or had taken part in any kind of terror against targets outside Afghanistan. Concentrating on Afghanistan, he did not realise that Taliban had been quite active in neighbouring countries. However, he suggested that what the Taliban craved for was an Islamic emirate in Afghanistan though, probably, most Afghans did not want this. Nevertheless, the Afghans themselves would be responsible for negotiating with the Taliban if, indeed, negotiations were the proper way ahead. Andersen concluded by emphasising that the primary aim of having Danish troops in Afghanistan was to prohibit al-Qaeda from operating out of that country. This had already been achieved and the second aim was to establish a viable democracy, which, he claimed, would not happen for many years to come, so why on earth remain there?[39]

In October 2010, the sneaking acceptance that negotiations with the Taliban were unavoidable led to the Minister for Development Søren Pind's recognition that a certain collaboration was already taking place inasmuch as funds allocated to companies carrying out Danish development programmes had been spent paying for protection by the Taliban. This was new, as, so far the government's policy had been clear avoidance of collaboration with bandits and terrorists.[40]

The aim with any given war is a political matter for which the military machine is harnessed, with a view to achieving it as quickly and as cheaply as possible. In former times the aim might have been conquest or remuneration, but today this has changed. In an interview in the weekly *Weekendavisen*, Senior Lecturer at University of Copenhagen Peter Viggo Jakobsen claimed that while great powers like the USA or the UK go to war to win, small states like Denmark do so for the sole purpose of joining: "If Britain and the USA are happy when we go home, we have won. We are not in Afghanistan to win but to score goodwill with our major allies in NATO.

---

[38] "Afghanistan – ingen lette løsninger [Afghanistan: no Easy Solutions]" in *Information* 25 August 2010.

[39] Lars Erslev Andersen, "Hvorfor er vi (stadig) i Afghanistan? [Why are we (still) in Afghanistan?]" in *Berlingske Tidende* 25 September 2010.

[40] Anna von Sperling, "Bistandsstrategi lægger op til samarbejde med 'terrorister' [Development Strategy Suggests Co-operation with 'Terrorists']" in *Information* 7 October 2010.

In particular it is important to keep good relations with USA."[41]

For this reason, he claimed, even the Danish participation in the War in Iraq from 2003-7 was a splendid endeavour and a huge success. However, he also expressed worry for the public debate. He maintained that the debate was uninformed, because unlike Britain and the US, Denmark still had not got, to any significant extent, think tanks and university departments staffed with ex-professionals capable of criticising government policy and the use of military forces. Most Danish experts were found within the services, and it was against their professional ethos to comment on political decisions as well as on the dispositions of their military superiors. In Jakobsen's opinion, this dearth of independent scrutiny was a severe shortcoming, which might lead to faulty decisions taken on the basis of inadequate and insufficient advice.

The Downsides

Apart from the hitches already mentioned, a number of disparities existed amongst the coalition partners in the Helmand Province from 2010-11, making co-operation slightly difficult in certain areas. While many of these had to do with dissimilar weapons and equipment, some pertained to tradition, outlook and educational aspects. This reality did not make Task Force Helmand's component elements completely incompatible, but constituted constraints that could not and should not be ignored.

Operational Matters

Regarding equipment, cross-country mobility was where Danes and British differed. While the British companies of the Combined Force Nahr-e Saraj North used small, agile, but un-armoured or lightly armoured vehicles, the Danish infantry sub-units drove rather heavy, armoured personnel carriers and infantry fighting vehicles which, in most circumstances, were excellent cross country vehicles. As long as they did not sink into the quagmire, they provided good, protected transport. On the other hand, because of their weight and size, now and then they had to be left behind when operating on boggy ground, forcing the soldiers to walk while carrying all their paraphernalia on their backs. On these occasions one major draw-back

---

[41] "I krig med eliten [Waging War with the Èlite]" in *Weekendavisen* 29 October – 4 November 2010.

materialised: the excessive load of the equipment that the soldiers had to carry: body armour, water, ammunition, radios etc. Moreover, some soldiers, and in particular those who were on their first mission, tended to bring along on patrols a lot of unnecessary stuff – just to be sure.

Disparate standards of dedication to mission was seen off and on in collaboration between Danes and Afghans. Operation Omid Char [Hope Four] was mentioned in the section on operations above. In late October, the partnering activity with the Afghan National Army was dominated by the launch of this operation, which was conducted by 3rd and 6th *Kandak*s from the 3rd Brigade of the 215th Maiwand Corps. The brigade worked closely with the chief of police in Nahr-e Saraj, ensuring that all authorities were co-ordinated in their response to the operation. The operation aimed at establishing a new patrol base called Hazrat in order to secure key terrain near Gereshk, restrict Taliban freedom of movement, and display a higher profile Afghan National Army presence in what was currently an area suffering severely from insurgent influence. 3rd Brigade's British mentors, from the 1st Battalion Irish Guards, were on hand to observe and support throughout the operation when required. At *kandak* level, the Danish *Kandak* Advisory Team's tasks were mentoring the rear tactical operations centre (TOC) in Camp Gereshk in order to evaluate the level of situational awareness related to the operation, training of trainers related to medics, as well as providing continuous intelligence and force protection training. However, neither the 3rd nor the 6th *Kandak* bothered to report either for the medic 'train the trainer' or for intelligence tutorials. As the officers eligible for participation in these programmes appeared to have set their priorities differently, the Danish DKAT had only slight success with their training effort.

Moreover, the Danish advisory team observed that the operational effectiveness of 3rd *Kandak* suffered from the fact that headquarters staff officers worked as individuals rather than as a coherent body. Sub-units differed in quality as to leadership as well as to the number of officers missing. Generally, the *tolays* were seen as being able to conduct no higher level of activities than those of platoons. As far as the Afghan National Police was concerned, the Danish Police Operational Mentoring and Liaison Team were aligned with their UK opposite numbers and proceeded in accordance with the police development plan, which called for setting up five sub-stations and five super check points. Weekly meetings with Chief of Police and Coy Commanders were institutionalised, and the P-OMLT officers were requested to participate, which they mostly did.

Communication

While enhanced communication is one of the post-Cold War era's huge advantages, the Danish Battle Group suffered from various shortcomings in this field because with no encryptable equipment of their own they had to rely on external support. Communication – both voice and data transmission – was of immense importance to operations, but coalition partners did not always enjoy the same advantages in equipment. Bandwidth connectivity was poor because Danish troops did not have their own classified radio net, and were therefore dependent on the availability of British Bowman radio operators. However, some of the deficiencies were offset as British Bowman operators were attached to the tanks – A Troop – all the companies and patrols as well as to forward headquarters, facilitating unfettered classified voice communication. Encrypted VHF communication was being introduced but was not completed during ISAF Team 10's tour in Afghanistan.

Photos taken by OC B Coy, Danish Battle Group: B Coy 10 August 2010.

The ability of sub-units to pass data around the battlefield was limited. The Officer Commanding B Coy, Major Michael Toft, saw it as a major disadvantage that Mission Secret, a valuable communication system for classified data, was not available until late during his company's tour. Not only was this system indispensable for transmitting classified data, it also possessed a good geo-viewer.⁴² This left data transmission a bit thorny, and although all sub-units should be able to receive and transmit through Mission Secret stations, there were frequent problems for C Coy units in Patrol Base Line. Therefore, this company relied on the Danish national *Fiin* (Danish Defence Integrated Information Network), which was an administrative net not vetted for classified and operational matters. Lots of messages, not least intelligence, were lost between this company and the Combined Force's headquarters in Camp Price. Nonetheless, Mission Secret was used by C, D and F Coys and eventually by B Coy as well. Because of the limited availability of this system the US Marine Corps Long Range Reconnaissance Company and later on its successor unit, D Squadron of the Household Cavalry Regiment could communicate only in 'voice', which was deplorably inadequate as it created considerable communications constraints. Personal Role Radios (PRR) were excellent means of communications amongst individuals, but unfortunately – unlike the British – Danish troops did not possess these in an encryptable version.

Communication matters in modern information warfare, and in many respects Taliban was far more efficient than their opponents. Whenever an ambush had been carried out text messages popped up on the mobiles of journalists in the area almost immediately. And not only did Taliban use text messages, they also managed to send footage of the incidents. For some years NATO had been aware that there was a considerable information deficit on their part that had to be filled up if the aim of 'winning hearts and minds of the Afghans' should ever be fulfilled. However, so far NATO had not succeeded, and more often than not it turned out that the Taliban information was correct.⁴³

Cultural Differences

Recently it has become fashionable to put 'winning hearts and minds' of the locals high on the list of priority tasks for coalitions-of-the-willing. However, not every officer agrees that such subtlety is a matter for the military. Major Michael Toft found that persuading the local citizens of the merit of supporting the Afghan government's agencies and representatives

---

⁴² An interactive mapping tool designed to let users preview geospatial data.

⁴³ "Taleban vinder informationskrigen [Taliban wins the War of Information]" in *Berlingske Tidende* 17 November 2010.

was none of his business.⁴⁴ He was there to fulfil his mission, and provided he succeeded this would support the authorities and his achievements would speak for themselves. However, as many of the patrols mounted by his company included CIMIC personnel, and because these officers frequently distributed radio sets to local civilians, on which they might listen to the messages of the official radio stations, rapport was indeed established and some positive influence was exercised.

There are many reasons why coalition partners' cultural background, edification and working habits affect the efficacy of the coalition. The degree to which national commanders and staffs understand each other and are able to participate in joint planning impacts the time required to plan and the sharing of knowledge of every component of operations. Language is a prime concern and working through interpreters is a cumbersome affair. For this reason officers of a coalition must have command of a common working language. In this particular coalition English was the operational language and the means for getting notions across, and it soon became obvious that for many Afghans this was not easy. Although the linguistic affinity of Danes and Brits was a lot better, it was not perfect; learning a language at school does not provide the same versatility in grammar and command of vocabulary as does an upbringing with one's native tongue. Not only did the Ghurkhas conduct their radio conversations in Nepalese, the Scot ordnance disposal expert's dialect was almost incomprehensible to Danish officers. Moreover, the limited foreign language familiarity was an additional trouble when working under duress. During a period of intense activity in mid-September the Chief S3, Major Byrholt, observed that Danish staff officers frequently switched to Danish, even though a Bosnian and three British officers were collaborating with them in their Tactical Operations Centre. This was not ideal, but it is hard to see how in a multinational environment such stress reactions can be completely avoided.

Each nation has peculiarities of its own, and while Danes tend to believe in the merit of extensive informality, they might also be seen as a grumbling and pugnaciousness lot believing themselves to be the champions of 'mission command' and widespread independent decision-making by commanders even at the lowest levels.⁴⁵ It is, therefore, relevant to consider whether or not the cultural backgrounds of two separate

---

⁴⁴ In military reports the government was invariably designated GIRoA (Government of the Islamic Republic of Afghanistan).

⁴⁵ 'Mission Command' [in German *Auftragstaktik*] is a notion originating with the Prussian Great General Staff. It was introduced in the second half of the 19th century by the Field-Marshal Helmuth von Moltke (the elder) on the basis of the ponderings of Clausewitz.

militaries, like the British and the Danish, allow common planning. Brigadier Chiswell saw this conundrum as being a matter of perspective, because, basically, the UK and Denmark shared a common 'North European viewpoint' as opposed to – or at least different from – an American one. However, in spite of possible cultural differences between Americans and Europeans, the relationship with the American Marines of Regional Command South-West ran smoothly and satisfactorily. A commander works with various sub-commanders, some of whom are stronger and more easily adaptable than others, but fundamentally they all share the same values. The concept of 'mission command' is as highly appreciated by Britain as it is by Denmark, and UK forces regard themselves as being strong in this respect. While each of the battle group areas within Task Force Helmand's area of responsibility posed different challenges and required different approaches, the concept of 'mission command' remained central.

Nonetheless, there were situations where adequate fulfilment of Task Force Helmand's overall mission necessitated certain dispositions which might be seen as interference with subordinate commanders' freedom of decision. While – probably for political reasons – the Danish 'campaign plan' foresaw reduction in the north-east and concentration as close as possible to Gereshk, the commander of Task Force Helmand perceived the situation differently. Brigadier Chiswell realised a need to 'deepen security' in the north comprising the area of the Upper Gereshk Valley delineated by Gereshk-Rahim-KharNikah-Gereshk including the eastern bank of River Helmand. He believed that withdrawal as foreseen in the Danish campaign plan would have undermined the confidence of local inhabitants. Moreover, in the cases of platoon patrol bases situated relatively remotely from the rest of Combined Force Nahr-e Saraj North and in rather dangerous surroundings, from the task force commander's point of view it was inadequate to leave sole responsibility to the platoon commander on the spot. What was needed was, rather, the experience and clout of the officer commanding the parent company. However, pressing this point might be seen as undue interference with a subordinate commander's decisions and therefore a violation of the notion of 'mission command' – so he abstained. Conversely, from a Danish point of view, Task Force Helmand was too keen on regulating details rather than leaving the method of execution of any given mission to the responsible subordinate commander. Prescribing precise ways of executing a mission was an approach that did not go down well with the Danish commanders, who were used to being allowed greater freedom of decision than their British fellow officers – or, at least, so they believed. From the Danish perspective a liberal approach was preferable, and the Danish staff believed to have recognised indications that the two

Irish Guards companies attached to the Combined Force were rather happy with being allowed more freedom of action than what they had usually experienced with the Irish Guards.

The proliferation of Task Force Helmand's ISTAR assets (Intelligence, Surveillance, Target Acquisition & Reconnaissance) made it possible for the task force and battle group commanders to engage more fully and more precisely at certain points, as long as sub-units were adequately deployed. For this reason, at one particular point the 'non-interference' notion of mission command was at loggerheads with sound operational thinking. As in early February 2011 the Danish ISAF Team 10 was being relieved in place, large areas were stripped of forces without prior agreement by Task Force Helmand. This caused considerable consternation as the task force commander was obliged to move in the task force's tactical reserve, which might be needed elsewhere at any time.

Over recent years, Danes have developed a rather liberal attitude towards commenting and criticising the wisdom, decisions and prerogatives of superiors. As a result, today it is fairly common to raise doubt about the ingenuity of a commander's plans – although only until they have been transformed into orders. From a non-Danish point of view, this might – and did – cause occasional apprehension. Moreover, Danish soldiers and NCOs tend to have a rather casual relationship with many of their officers, which frequently surprised coalition partners. In Brigadier Chiswell's eyes, however, like the Danes, UK forces too have close bonds among those fighting together and, although there are no first names used between rank groups, a manifest camaraderie does exist. Moreover, the generous presence of facial hair amongst Danish private soldiers was a source of much curiosity with the British, who had chosen to stick with traditional notions of battlefield hygiene, but in the end no trouble arose from these normative disparities.

Co-operation between countries of particularly incommensurable cultures, such as the Germans and the Turks or the British and the Chinese during World War I, is frequently hampered by a mutual lack of understanding for each other's cultural preconditions. Lack of knowledge of what lies beyond one's own borders, linguistic inability, and cultural narrow-mindedness are the primary reasons for this predicament. Therefore, the level of exposure of military organisations to other cultures in the pre-coalition stage determines their ability to minimise cross-cultural tensions with fellow coalition partners. This, however, according to Brigadier Chiswell, turned out to be no problem at all during his tenure of the post as commander of Task Force Helmand.

Most of the time, Americans, British and Danes – and even the Afghan professional soldiery – worked smoothly together, and Major Michael Toft, officer commanding B Coy, found that his soldiers were sufficiently prepared culturally to co-operate well with their Afghan partners. Nonetheless, there were rumours in the press that Afghan soldiers were undisciplined, frequently went absent without leave, had homosexual relationships with local boys, and were on drugs most of the time. The commander of the Danish *Kandak* Advisory Team, Major Kim Kristensen, who worked closely with the Afghans, would not totally dismiss these accusations, but his impression was that, by and large, the negative consequences did not percolate through to operational matters. He found that in general discipline was reasonably good and the enthusiasm of the officers made the Afghan National Army an effective part of the security architecture. Likewise, the then-Danish Minister for Defence, Ms Gitte Lillelund Bech, commented on the rumours, stating that, whatever the trouble, Danish forces were in Afghanistan to help establish security forces good enough to protect the Afghans. She believed that, in spite of all their shortcomings, the Afghan society and security forces were developing in the right direction. Apparently, the rumours were based on stories spread by Danish soldiers returning from Afghanistan, and Ms Bech commented that although the troops were there to support development, they were not expected to do so by trying to change the cultural norms to fit those of their own. The Afghans would have to make changes in their own way and at their own pace, because, eventually, Afghan issues would require Afghans solutions.[46]

The officer commanding C Coy, Major Eskil Berger, had had a strange experience of cultural clash in the early days of his stint. Suddenly, an Afghan platoon commander had marched up to a Danish colleague and asked him to remove the Danish flag from his position, as this 'banner of the infidels' was an open insult to the Afghans signifying that they were under foreign occupation. The request was refused, but, according to Major Berger, the incident was a bad omen indicating on the part of the Afghans a general lack of understanding for the seriousness of the task at hand. However, it merits notice that similar incidents did not re-occur and that, over the months of his stint, Major Berger got a favourable impression of and a good relationship with his Afghan opposite number.

---

[46] "Minister: Afghansk hær er på rette vej [Minister: Afghan Army is on the right track]" in *Berlingske Tidende* 11 October 2010.

Discipline, Spirituality and Mental Health

In the British system of military promotion Annual Confidential Reports are written on every officer, the commanding officer being the first reviewing officer, the brigadier the second. While, generally speaking, the first reviewing officer writes to enlighten the officer being reviewed, the second writes to inform authorities of his or her promotional perspectives. Since all reporting is now open, the officer will always be aware of both these positions. The British brigade commander's rôle as a second reviewer, however, does not extend beyond the realm of the British officer corps. Therefore, although the officers of the Danish Battle Group were subordinated to Brigadier James Chiswell, he had no say in their promotional evaluations. While theoretically, the different evaluation conditions for British and Danish officers of the same battle group might have caused some uneasiness and perhaps disgruntlement, this never materialised. On the other hand, seen from a Danish perspective, the officers commanding the D and F Coys (detached by the Irish Guards Battalion) seemed a little too keen on contacting their British seniors to enquire about the reasonableness of orders given by their Danish commanding officer or his staff. They were, however, rare occurrences and quite logical reactions to doubt, which Colonel Fredskov assumed might have happened under any circumstance to any coalition partner. In the end, no harm was done and no offence was taken, but it was striking that to make a good impression in order to further their chances of promotion, the British company commanders went out of their way to ascertain that their national superiors remained conscious of their existence and, again, Colonel Fredskov found this a self-evidently understandable inclination.

It may sound like a quotation from Robert Graves' memorable World War I account in *Goodbye to all that*, but it is, nonetheless, what the Danish weekly *Weekendavisen* chose to tell its readers in July 2010: 'In the heated Afghan desert the army chaplain Thomas Aallmann blessed the soldiers before they went into battle. Then he made the sign of the Cross and the V-sign exclaiming: "V for victory."'[47] The article informed that this chaplain, who had taken leave from his parish obligations to join the army, did not go 'merely to be a bore of a pacifist or war sightseer.' He lived with the soldiers at the front, walked with them on patrols – armed with the standard

---

[47] Robert Graves, *Goodbye to all that* (London: The Folio Society, 1981) p. 170: "A Roman Catholic padre, on the other hand, had given his blessing and told them that if they died fighting for a good cause they would go straight to Heaven or, at any rate, be excused a great many years in Purgatory."

issue carbine – and was generally there whenever they needed him. However, not all clergymen shared Rev Aallmann's gung-ho enthusiasm. A professor of divinity at the University of Aarhus told the Christian news daily *Kristeligt Dagblad* that 'it is deeply troubling that we have an institution called the 'army chaplain.' This is a bad confusion of soldiery and clergy.' Another parish priest opined that if a clergyman wished to wear weapons he should join the army as a soldier. And a senior lecturer at the Theological Seminary 'Løgumkloster' went even further, stating that 'we cannot have priests who glorify war.' On the other hand Rev Aallmann was not alone. Many army chaplains sympathised with his views, which were basically that if the soldiers were prepared to give their lives in the service of a peaceful world, his obligation was to assure them that God was near them: 'God has no part in the war, but He is the safe haven of all those who are weary and seek refuge.'[48] There are differences in how the spiritual service is performed with British and Danish units – let alone Afghan ones – but there is no doubt that during their tour officers and soldiers of ISAF Team 10 appreciated what was done. On 5 September, mortar man Claus of C Coy wrote in his diary:

> It was Sunday after all and there was a service in the 'church.' I was ten minutes late, but it was nice participating. First, there were quite a lot there – some, too, whom you would not have expected to see – and it gave me a feeling of a kind of unity, and also such a pleasant peace. It was so informal and the Reverend had a remarkable talent for getting matters right down to earth, where one could feel comfy and secure.

Regardless of the efforts of the army chaplains, there will always be soldiers to whom the impressions during active service and the mental stress of repeated overseas tours suddenly become too much. Many soldiers derive huge social and mental benefits from being overseas, but some are pushed over the edge, brake down mentally or become violent. Because of the modest size of the Danish armed forces, most soldiers are deployed overseas up to six times during their careers, some even considerably more. From February 2008 to August 2009, the Danish Institute for Military Psychology surveyed soldiers coming home with respect to their increase of either benefit or strain resulting from their recent deployments. It was observed that after the first two deployments the benefit increase shrank, possibly because their motivation fell and new deployments no longer yielded the same level of fresh experiences. The level of strain, however,

---

[48] Pernille Stensgaard, "Fjenden er også Guds skabninger [The Enemy are God's creatures, too]" in *Weekendavisen* No 30, 30 July 2010.

appeared to be constant, probably because the risks were more or less unchanged. While, remarkably, the Danish survey did not encounter any significant difference as to the occurrence of Post-Traumatic Stress Disorder caused by an increasing number of deployments, a similar American research finished in 2009 pointed to an increase of psychological post-operational reactions with each extra deployment.

Whatever the strain on the individual due to repeated deployments, the discipline within an army sub-unit holds it together and lessens the burden caused by danger and operational stress. Conversely, the consequences to the sub-unit's combat efficiency from lack of discipline amongst coalition partners might well materialise as despair and drop of enthusiasm. The feeling of being alone in shouldering an important task leads to questions about the meaning or futility of the whole enterprise. As a result, the apparent indifference to training schemes set up to improve the Afghans' professional skills as well as their unreliability when on parade was demoralising to instructors and collaborators alike. Major Eskil Berger noticed with apprehension that upon his arrival in Patrol Base Line in August 2010, the Afghan policemen had already left without ascertaining that someone would take over their responsibility. Although the line had not been unmanned for very long, this brief interregnum had been enough for the Taliban to place a considerable number of improvised explosive devices.

Partners and Dominance

Among the difficulties of being one of a number of subordinate commanders in a coalition force is the balance that must be struck between national policy and coalition operational imperatives. While the Ministry of Defence in London as well as its counterpart in Copenhagen agreed that the ultimate goal of the Afghan enterprise was leaving security provision to the Afghan Security Forces, there was no prior agreement on the sequence of actions leading to that objective. Inasmuch as the Danish commanding officer of ISAF Team 10, Colonel Lennie Fredskov – so Brigadier Chiswell sensed – was tasked with thinning out in remoter parts of his area of responsibility and concentrating troops closer to the town of Gereshk to establish a larger manœuvre force, this hampered the task force commander's freedom in deciding tactical issues concerning, among other things, the deepening of security in the Gereshk Valley area. Conversely, Colonel Fredskov strongly emphasised that his decisions were not results of political pressure but entirely based on his own estimates of what was tactically adequate. He wished to reduce presence where the Afghans did not want to take over, and at the same time enhance flexibility of his military muscle. But although national priorities seemed to be slightly different, the brigadier did not feel that there were serious caveats from any quarter decisively limiting his operational choices.

Moreover, although excellent means of communication might have tempted the commander of Task Force Helmand to attempt some level of micro-management, both Chiswell and Fredskov claimed that they had avoided that trap. The commander Task Force Helmand realised that, at all levels of command, invariably from time to time there would be wishes for specific actions which would make interference in subordinate commanders' dispositions unavoidable. Although this was not desirable, it was a logical consequence of tactical imperatives, which should be ignored by no one.

Rivalries amongst coalition partners may materialise in various ways, and that of criticising a partner's willingness contribute or to share in risk taking is not uncommon. The American criticism of Montgomery's cautious and thoroughly rehearsed operations in Europe in 1944-45 is only too well known, and it has a parallel in that of Anglo-Danish counter insurgency in Afghanistan's Helmand Province disclosed by WikiLeaks 2010. In December 2010, American as well as Afghan decision-makers were quoted for severe criticism of the British strategy, resource allocation, risk avoidance and allegedly poor ability to create security in the Helmand Region.[49] Although such denigration behind one's back is rarely appreciated, Brigadier Chiswell told me that he could easily see a rationale behind it, because he realised that at the heart of this censure lay disapproval of the limitation of resources rather than of lack of good will. Nine thousand troops were obviously insufficient for the task of protecting the Helmand Province in its entirety through full spectrum counter insurgency. Because of insufficient resources Task Force Helmand had let itself to be driven up and down the Gereshk Valley of the Helmand River depending on where trouble happened to occur. This was utterly inadequate, and, especially concerning the Sangin area, the criticism, he believed, was justly severe. There was, however, willingness to secure deeper involvement in spite of difficulties, and there were indeed advances as to governance in the Helmand Province. Some criticism was fair enough as Britain had not always been able to resource all her ambitions, and because there were frequent disagreements between Foreign and Commonwealth Office and Ministry of Defence concerning resource allocation.

---

[49] The Guardian, 3 December 2010, "WikiLeaks cables expose Afghan contempt for British Military."

Not only resources but also safety regulations might be dissimilar and thus hamper operations differently. While previous British commanders stationed in Helmand have complained that peacetime safety measures and limitations had influenced operations negatively, this was not the case at Task Force Helmand level 2010-11, although the commander speculated that it might have been to some of his battle groups. However, Colonel Fredskov also found the precautions adequate and reasonable to handle, and since he was allowed, under certain circumstances, to make exceptions, there was not any complication seen from his point of vantage.

Disunity and Dissolution

When the immediate threat seems to vanish and partners, for political or economic reasons, cannot afford participation any longer, coalitions tend to fall apart. In the spring of 2010, Britain and Denmark decided to start withdrawing combat troops from Afghanistan in 2011 with the intention of completing this process by the end of 2014. Later that summer, a NATO summit agreed similarly that it was about time that the Afghans took over responsibility for their own country – the disintegration of the coalition-of-the-willing had begun. Then in the autumn, surprisingly to some, the Afghan President Hamid Karzai announced that Afghanistan as well as the US would start negotiations with the Taliban. In October, *The Washington Post* claimed that secret top level negotiations with the so-called Quetta Shura had been initiated. Although this was initially denied by Afghan authorities, it was later confirmed by the president. According to *The Guardian*, the NATO Commander-in-Chief General David Petraeus noticed that high-level Taliban leaders had made an opening to the Afghan government, "and this is how wars are brought to a conclusion." *The Guardian* further claimed that the negotiations had come about because both sides to the conflict were war weary and had realised that their economies could hardly support protracted warfare. The Taliban hideouts in the so-called Federally Administered Tribal Areas in Pakistan had for long been under NATO's keen observation and frequent attacks by unmanned aerial vehicles targeting leaders and infrastructure, and NATO, too, had sustained enormous costs economically and in terms of casualties.[50]

---

[50] "Forhandlinger med Taleban er en realitet [Negotiations with Taliban is now reality]" in *Information* 12 October 2010.

Experiences

The Danish commanding officer saw his UK colleagues as particularly good at extracting experiences and sharing them with him. Similarly, Brigadier Chiswell believed he had received many useful analyses and pieces of information from the Danes based on their experiences in action. Best practises were taken over and continued and, in particular, the Danish Intelligence Branch's (S2) and the police mentors' procedures had been appreciated. It was clear that both parties learned from the other, and to the British one particularly useful Danish experience was the use of armoured vehicles in counter insurgency combat.

By ensuring continuity, providing personal teaching and support of future contingents, Chiswell and his brigade would make sure that subsequent Herrick teams would benefit from their experiences. Moreover, there would be a lessons learned compilation work with the Field Training Unit and the preparation for deployment of future contingents.

The British Land Warfare Centre has a team permanently deployed with Task Force Helmand tasked with feeding back the lessons, equipment needs, etc. Getting this feedback loop tight is an area in which the UK Army has invested sizeable efforts to make sure that they got the continuity bit right. Because of the presence of this team, future equipment and weapons procurements could be relied on to happen on the basis of the experiences made by Herrick 13.

Although in the past there had been some problems with unity of command because of the American Operation Enduring Freedom and the simultaneous running of the NATO ISAF in adjacent areas, at the tactical level unity of command was now secured. At the operational level, Operation Enduring Freedom was still on, but, since the rise to the post of Commander-in-Chief of American General Petraeus, things had been moving in a more 'joint' direction.

Coming home, the Danish Battle Group believed they had solved more or less all the tasks that had been foreseen to solve or which had materialised while in the area. During their stay in Helmand the Danes saw positive developments in many fields – not least security. Towards the end of their stint, the Battle Group demonstrated the improved level of security by various events: the commanding officer went shopping in Gereshk without body armour, his deputy went fishing at Check Point 5 and the entire C Coy walked the so-called Dancon March, a completely leisurely activity devoid of any combat or security tasks. Thanks to a decidedly manœuvrist approach Colonel Fredskov believed that he had managed to maintain the initiative *vis-à-vis* the Taliban most of the time. However, manœuvre requires a certain redundancy, which in turn requires adequate numbers of personnel. Therefore, it was a considerable problem that there had been no personnel replacement when fatalities or repatriations occurred. In C Coy one infantry section was completely missing because of casualties.

Post-Redeployment

In Denmark, for some years an acclimatisation programme has been in place, concerning various measures to let the troops wind down and adjust to either new military jobs or re-entering civilian life. For ISAF Team 10 a special arrangement had been agreed with the Royal Arsenal Museum. During the first nine months of 2011, the museum was busily occupied with setting up an Afghanistan Exhibition, and soldiers wishing to have a quiet job allowing them to process memories of their tour were offered work under the aegis of the museum for about six months subsequent to coming home.

Thus, a quite reasonable task was solved to ease the individual soldier's way back to peacetime conditions, but with respect to the battle group as an entity it was different. With regret, the officers noticed that coming home sub-units would be practically broken up and the personnel dispersed ubiquitously to take up various other jobs. While some of the private soldiers would have to leave the army as their contracts could not be extended, of those staying only few would remain with their original companies to prepare them for their new mission as ISAF Team 15 in 2014.

While the dismissal of a considerable number of soldiers with combat experience was a loss to the army and a waste in terms of funds spent to train them, the further splitting up of companies showed an even more destructive trend. As from psychological studies of previous wars we know that soldiers do not fight for high ideals, but for their small fighting communities and for their buddies whom they know they can rely on to protect their lives to the utmost of their abilities. The combined effect of these two consequences of braking up companies will naturally be that training and bonding must start anew, while much money, combat efficiency and experience are lost.

Chapter III – Vital Partnership

> *"In working with allies it sometimes happen that they develop opinions of their own."*[1]

TODAY'S COALITION WARFARE unfolds in a manner dictated by the political problems of our time and on conditions characterised by the technological and economic standards of the twenty-first century. The costs are enormous, the number of soldiers involved is huge in comparison with coalitions of Antiquity, but modest compared with the world wars, the reasons given are not always consistent, and partners to the various coalitions join with disparate motives. But if this is so today, how was it in the past? Is the concept the same, or is it merely the term that has not changed? Are we more liable to bandwagoning now than humankind used to be?

Across history, sometimes the results of coalition battles have been complete disaster, such as the Greeks at Thermopylae in 480 BC, while at other junctures the fighting coalitions have been unequivocally successful, as for instance when Wellington defeated Napoleonic France at Vitoria in 1813, or when the Prusso-German states eliminated French resistance during the Franco-Prussian War of 1870-71. Recently, the post-Cold War era's interdependence amongst many developed and developing countries has contributed decisively to making coalitions the modern vehicle *par*

---

[1] Churchill, 1942.

excellence for successful warfare: the Balkans in the 1990s and Kuwait in 1991 are but two obvious examples. Henceforth, the successful outcome depends on careful mixture and adjustment of key ingredients. These include mutual trust, command relationships, doctrine, technology, organisation, training, personnel and equipment strength and cultural relationships.[2]

There is not today, and there never was, a *vade mecum* (a book for ready reference) for coalition warfare – no commonly accepted and everlasting doctrine laying down the general principles for this kind of contest. Every coalition in history has been unique in purpose, character, composition, aims and scope, but there are some basic commonalities that will invariably confront the commander. Part of the basis on which we may form at least some ideas of a template doctrine is historical experience, and this is what this chapter is about. History plays such a prominent rôle in understanding the enduring aspects of warfare that modern commanders continue to study the battles of our forebears for understanding the challenges of their present.[3]

To get a reasonable grasp of these challenges and in order to answer other pertinent queries we are best served by starting with Antiquity and meandering our way through the paths of the history of war onto the present, picking illustrative examples of the key features of coalition warfare. However, we shall not endeavour to describe entire wars or campaigns, nor shall we make a full chronological survey of the history of coalitions. We will merely endeavour to get a firm comprehension of key characteristics, along with what has changed and what has not.

Ancient Greece

War is as old as mankind, though written sources of its manifestations are not. Some of the oldest writings about war were made by Thucydides, telling in great and eloquent detail about military combat amongst the Greek city states of Antiquity. Thucydides was an historian and an author from Alimos, living c. 460 BC – c. 395 BC. In spite of his standing as an historian, modern scholars know relatively little about his life. However, some most reliable information may be gleaned from his own *History of the*

---

[2] Willie J. Brown, USMC, *The Keys To Successful Coalition Warfare: 1990 And Beyond*. From internet, URL http://www.globalsecurity.org/military/library/report/1991/BWJ.htm accessed 11 August 2010.

[3] Mitchell, Network Centric Warfare and Coalition Operations, p. 3.

Peloponnesian War, which goes some way to illustrate his nationality, paternity and native locality. Thucydides has informed us that he fought in the war, contracted the plague and was exiled by the Athenian democracy. He has been called the father of scholarly history because of his strict emphasis on evidence and analysis in terms of causes and effects and, as outlined in his introduction to his work, he accorded no rôle to divine intervention. He has also been dubbed the father of the school of political realism, which views the relations between nations as based on might rather than right. Moreover, he took a keen interest in developing our understanding of human nature to explain behaviour in such crises as war, plague, massacres and civil war.

Like warfare in general, coalition warfare is no novelty and the work of Thucydides' *History of the Peloponnesian War* bears witness to one of the earliest known examples. For this reason the Peloponnesian War's start and close are well-established, but since coalition warfare has been practiced for thousands of years it will hardly be possible to point out an exact origin or time of its first appearance. However, thanks to the written testimonies still extant, it can without any difficulty be traced at least as far back as the Archaic and Classical eras of the Greco-Roman world, i.e. starting with the 8th century BC. However, the sources at our disposal do not allow us to conclude that coalition warfare was the only way the Greeks fought; bilateral battles were seen as well, albeit more rarely, and the written sources are sparser.

There are reasons to believe that in most conflicts political negotiations have preceded war in order to sound out the possibilities to forge strong alliances. One example is provided by a battle fought in 510 BC, in which the big and powerful city-state of Sybaris in southern Italy was decisively defeated by its neighbour Kroton.[4] No contemporary source describes this battle, which quickly became the object of legendary tales, so we have to rely on later sources.[5] Herodotos, writing two generations after the battle, ascribes the defeat of Sybaris to Kroton alone, although he also records a rumour that Kroton might have received support from Sparta. There are equally good indications that Sybaris received support from allies in this battle. Sybaris is, in fact, known to have been the leader of a coalition of allies: an inscription announcing the entrance of a new member into the

---

[4] T.J.. Dunbabin, *The Western Greeks: The History of Sicily and South Italy from the Foundation of the Greek Colonies to 480 B.C.* (Oxford University Press: Oxford 1948), p. 359.

[5] *Thukydides's Historie* [Thucydides' History] translated by M. Cl. Gertz, 1st and 2nd Books. (Kjøbenhavn: Karl Schønberg, 1897-98) pp. v-xxxvi.

league has been found at Olympia, where major Greek states often published important acts of state. It seems unlikely that Sybaris would not have drawn on the resources of its own alliance, in particular since Kroton was a formidable enemy and the battle was the end product of a conflict which had begun some years before the final clash. So this battle, one of the great battles of archaic Greek history, was in all probability fought between two coalitions. [6]

In many instances, it is amazing how rich the written sources are when it comes to testimonies of coalition warfare in the Ancient world. Dons Adam Schwartz and Thomas Heine Nielsen of the University of Copenhagen tell us that, as they searched for relevant information on battles of the Archaic and Classical ages, they compiled from the Ancient Greek historians' battle descriptions and other sources a sample list of forty-one battles chosen more or less at random for their particular wealth of information on how the Greek heavy infantry, the hoplites, actually fought. Through this compilation they became aware that all of these battles except one were, in fact, coalition battles.

We began to ask ourselves whether the perceived standard battle was, in fact, not so average after all. So far, we have examined some 200 battles, and the sprawling source material keeps yielding relevant data. These have only confirmed our suspicion that alliance and coalition battles were the norm, rather than the exception. That this should have gone unnoticed is all the more surprising as our principal sources for Greek interstate wars, the great historians of the classical period — Herodotos, Thucydides and Xenophon in particular — made war the very subject of their works. Their point of view is always that of the great city-states and their wars, which were almost without exception fought by coalitions; they did not take any particular interest in the martial history of the hundreds and hundreds of small or even very small city-states which may often have been involved in minor battles against each other.[7]

Thucydides' *History of the Peloponnesian War*, fought between 431 and 404 BC, recounts the conflict between Sparta and Athens up to the year 411 BC, and for reasons of its clarity epitomising coalition battle this work is still being studied at universities and advanced military colleges

---

[6] In Adam Schwartz's and Thomas Heine Nielsen's joint paper presented at the coalition warfare conference in Copenhagen May 2011.

[7] Adam Schwartz's and Thomas Heine Nielsen's joint paper presented at the coalition warfare conference in Copenhagen May 2011.

worldwide.[8] *The Peloponnesian War* remains a classic example of the interactions between politics and war, though Thucydides is not always quite clear about the coalition aspect, because:

> Our sources, then, often focus on the famous and powerful city-states to the exclusion of their often minor allies: Thucydides, for instance, says in his introduction that the Peloponnesian War of the later fifth century BC was fought between "Athens and the Peloponnesians", though in fact the Athenians were supported by numerous allies. In this case we have plenty of evidence that the Peloponnesian War was fought between two great coalitions; but when sources refer in a compressed way to wars of the Archaic period (that is, from 750 to 480 BC), we sometimes run into difficulties in determining whether a given battle was a coalition battle.[9]

Generally, Sparta fought her wars as the leader of a coalition. All the battles that were attributed to Sparta were actually coalition battles fought by the Lakedaimonian armies.[10] Lakedaimon was the region in which Sparta was situated, and its inhabitants were called Lakedaimonians. The Lakedaimonians were settled in several city-states, of which Sparta was the unquestioned political leader. The other city-states of the region were obliged to follow the Spartans in war and had to supply hoplites for coalition campaigns. Accordingly, the famous Spartan army, which was invincible for long periods of time, was not strictly speaking Sparta's own army, but a Lakedaimonian one, or, in other words, a coalition force. Similarly, and different from the conditions of today's coalitions-of-the-willing, the 'allies of allies' were taken for granted as coalition members with no right to opt out. Thus, as a remarkable aspect of 'the four-city alliance' of 420 BC, which is carved in stone and set up on the Athenian Acropolis, we may note that all four powers referred to in the treaty were each in themselves coalition powers: each was the head of a military league of some size, and the co-operation and obedience in turn of these allies, or rather vassal states, was taken for granted — at any rate, their compliance was not required for the treaty to be valid. Certainly, Athens in the second

---

[8] *Thukydides's Historie* [Thucydides' History] translated by M. Cl. Gertz, 1st and 2nd Books. (Kjøbenhavn: Karl Schønberg, 1897-98) passim.

[9] Adam Schwartz's and Thomas Heine Nielsen's joint paper presented at the coalition warfare conference in Copenhagen May 2011.

[10] Following the current trend amongst scholars of Ancient Greece, the modern 'Greek' transliteration has been used throughout rather than the 'Latin' version traditionally used in English texts.

half of the fifth century ruled an empire of at least 250 lesser city-states and even more vassals of other kinds (peoples, tribes etc.). These states may nominally have been allies, but disobedience would not be tolerated, as several of these subservient city-states found out at their peril.[11]

Apart from her leadership of the Lakedaimonians, over the years Sparta engaged in larger coalition arrangements, and in a number of great battles in 480-479 BC, The Hellenic League – a coalition of some thirty-one states under the leadership of Sparta – managed to defeat the invading Persians.[12] Similarly, the Peloponnesian War was not simply a war between the cities of Sparta and Athens. It was a contest between two coalitions both wanting to prevail as the dominant power of Greece, viz. the Peloponnesian League led by Sparta and the Delian League under the aegis of Athens.[13] Moreover, the Corinthian War of the early fourth century BC was fought between Sparta and the Peloponnesian League on one side and a strong coalition between Argos, Athens, Corinth and Thebes on the other. Eventually, in the renowned battle of Leuctra in 371 BC, the Lakedaimonian army was decisively defeated, and it never regaining its former strength. This battle was fought by Sparta with its Peloponnesian associates against the forces of the Boeotian Confederacy, led by Thebes. Finally, the battle of Chaeronea in 338 BC was fought by the Macedonians and their allies against a Greek coalition led by Athens and Thebes. This was Phillip II's Macedon's – an emerging great power's – eventually successful attempt to establish peace under its domination in a country consisting of competing and often warring small states, and the similarity to what happened in Germany during the closing decades of the 19th century and in the European Union during those of the 20th is for everyone personally to compare.[14]

Therefore, though many of us may not very often think of it, from Ancient times coalition warfare has been an integral part of warfare *per se* and an abiding companion in power politics – at least since European Antiquity. From the wars of the Archaic and Classical ages to the most recent anti-terror operations in Afghanistan, coalition warfare has made

---

[11] This treaty included Athens and a triad of important Peloponnesian city-states: Argos, Elis and Mantinea.

[12] Fuller, J.C.F., *The Decisive Battles of the Western World 480BC-1757* (London: Paladin Books, 1954), pp. 54-73.

[13] Keegan, John, *A History of Warfare* (New York: Vintage Books, 1993), p. 256.

[14] For further details on the Macedonian ascension see: Keegan, John, *A History of Warfare* (New York: Vintage Books, 1993), pp. 257-63.

significant contributions to the defence of values and security cherished by coalesced groups of states as well as to aggressive entities striving for hegemony and aggrandizement.

Coalition building was a convenient and effective way for a major city-state to increase its power on the battlefield. But the system of coalition fighting was probably also an advantage seen from the point of view of the minor city-states. In many cases there was probably no truly viable alternative to joining a coalition, and in these cases it must have been highly desirable to strike a deal by negotiating a treaty rather than risk it all on the field of battle.

Medieval Italy

Alliances and coalitions serve the states they are made up of. They are not necessarily friendships, but they are, invariably, prudent, egotistical security arrangements serving the self-interests of each individual participant state. Recent coalitions like the Austro-German alliance in World War I as well as the Anglo-American one in World War II were based on assumptions of common ideal values, such as cultural pre-eminence and democracy respectively, common languages and shared security needs, which were genuinely believed to be of decisive importance at the time these coalitions existed. Therefore, they carry historic weight as explanations of the emergence of these alliances, but generally coalitions are formed to protect interests rather than ideals. It is fair to say that coalitions are entered into for motives of national endurance, and usually for economical or self-protection reasons.

This was as true in Antiquity, in the Middle Ages and in the Early Modern Era, as it is today, and employing 15th century Italy as an example the American historian Paul Dover very aptly points to the need for unity among states with matching interests. He tells us that:

> In chapter 11 of *The Prince*, Machiavelli remarks that in the years before the French king Charles VIII's invasion of Italy in 1494, the five major powers of Italy (the Pope, the Republic of Venice, the Republic of Florence, the Duchy of Milan and the Kingdom of Naples) had "two principal cares: the one being that an outsider should not enter into Italy with arms; the other being that none of them should increase his state."[15] Machiavelli looked back at this fifteenth-century history, inescapably seen through the lens of the subsequent twenty

---

[15] Nicolo Machiavelli, *The Prince, with Related Documents*, ed. and trans. by W. Connell (Boston: Bedford/St. Martins, 2005), p. 74.

years of war involving the major European powers, with a view to demonstrating the dangers of Italian disunity. For Machiavelli, the root of the calamities that had descended on Italy with the so-called Wars of Italy after 1494 was the inability of Italians to band together against the ultramontane barbarians. This is why he ends *The Prince* with an almost chiliastic call for an Italian redeemer who can bind together the states of Italy to face down the invaders.[16]

Niccolò di Bernardo dei Machiavelli, 1469-1527, was an Italian historian, philosopher, humanist, and writer based in Florence during the Renaissance. A diplomat and a political philosopher, he was among the main founders of modern political science. Born into a family which had already produced thirteen *Gonfaloniere* of Justice to Florence, he was himself to assume high office as a civil servant. His position in the government of Florence as a secretary to the Second Chancery of the Florentine Republic lasted from 1498 to 1512, a period in which the Medici were not in power. However, Machiavelli's most well-known writings appeared after this period, during the time when the Medici recovered control and Machiavelli was removed from all positions of responsibility.[17] When he wrote *The Prince* in the twilight years between the Middle Ages and the Early Modern Era, the Italian League, the defensive alliance which had included the five major powers of Italy, was long gone. This coalition had been agreed upon in March 1455, at the end of an extended period of political instability and exhausting warfare that had spanned several decades.

The League had been designed to last for twenty-five years, and it had called for the signatories to maintain standing armed forces that could be used should any member of the coalition be threatened by extra-Italian powers, or, in other words, an obligation not too different from that among NATO countries. It had been blessed by Pope Nicholas V as a most holy (*sanctissima*) league, and the language of the treaty suggested that it was in part directed outwards, in particular against France, who entertained aspirations of conquest in Italy, and against the Ottoman Empire, which had captured Constantinople a few years earlier.[18] The Italian League was agreed upon as a common sense result of the Italian states' limited military

---

[16] Paul Dover in a paper presented at the coalition warfare conference in Copenhagen May 2011. This paper has also provided inspiration for a large part of the remainder of this paragraph.

[17] Machiavelli, N. *The Art of War*. Cambridge: Da Capo Press, 2001), pp. ix-xii.

[18] Keegan, John, *A History of Warfare* (New York: Vintage Books, 1993), pp. 321-2.

muscle and their pervasive desire for enduring peace. Its immediate precursor was the bilateral agreement in the Peace of Lodi of April 1454 between the Duchy of Milan and the Republic of Venice, which had agreed to end their mutual hostilities. The two states were convinced that the ongoing efforts of Pope Nicholas V to bring about universal peace in Italy were doomed, but that their own peace agreement deliberately left some hope, as it was open to other Italian states to join, creating a wider perspective of a pan-Italian reconciliation.

As already intimated, alliances are neither friendships nor are they set up in order to uphold moral ascendancy. It is reasonable to downplay the pre-eminence of idealistic aims behind the creation of such defensive arrangements. In a Medieval Italian context it is essential not to perceive the balance of power as a general remedy for all political deadlocks, a universal acceptance of the need for peace and harmony, or a result of the endeavours of far-sighted politicians. It was rather the logical consequence of economic exhaustion and awareness that there was no such thing as easy conquests, even though the inclination to promote national interests remained alive. Nonetheless, the Italian states persisted in seeking to seize what opportunities might appear for political and territorial gain, while at the same time bracing themselves for countering similar aspirations by foreign states. Therefore, in spite of their wish for peace and territorial status quo, the Italian states regarded the art of war and the martial and diplomatic professions as guarantors of successful political activity and economic growth.

The Italian League was a coalition of states which shared security interests, but it was not an obligatory supra-state organisation with a well-defined set of rules for inter-state political and economic relations.[19] However, this coalition was the manifestation of a concept for maintenance of the territorial integrity of partners and the preservation of Italy as an entity shielded against trans-Alpine and Ottoman aggression. Thus it aimed at achieving peace and prosperity through preservation of a geographic status quo in Italy and by providing Christianity with a bulwark against the menace from the Sublime Porte. However, the existence of the Italian League did not prevent the Italian states from forming ad hoc alliances against each other. It would be alliances of two or three of the primary powers, nearly always resulting in a counter-league formed of the remaining peninsular powers. Such a smaller-scale coalition was called a partial league, as opposed to the large scale universal coalition of the Italian

---

[19] M. Mallett, "Diplomacy and War in Later Fifteenth-century Italy" in *Proceedings of the British Academy* LXVII, 1981, p. 268.

League. It was mostly to such alliances rather than the universal Italian league that states looked for protection, and it was in such opposing political constellations that wars would be conducted in Italy for several decades.[20] These partial leagues, while serving to achieve a rough balance, did little to reduce political instability and mutual suspicion, especially because few of them lasted particularly long. By the end of the century, alienation and suspicion among Italian states made it all the more likely that Italian states would be willing to call upon non-Italian states to ameliorate their own positions vis-à-vis their Italian rivals. In this way, the Italian League lost importance and faded silently away.

During subsequent years, Nicolo Machiavelli looked back at the fifteenth-century invasions and the Italian internal disunity. To him, the root of these calamities was the inability of Italians to band together against the common foe. Thus, he ends *The Prince* with a call for an Italian statesman who can form a coalition amongst the Italian principalities and republics.[21]

Despite the fact that the Italian League failed to operate in the fashion for which it was designed, it remained a point of reference in negotiations and in diplomatic discourse. The spirit of the League, and its pretensions to maintain an Italian peace, were repeatedly invoked as states competed for advantage.[22]

Early Modern Europe

The early Modern Era was no less brutal than the Middle Ages; on the contrary, power was concentrated in three rather large entities: France, the Holy Roman Empire, and the informal union of Protestant princes, and the clashes between the latter two of these were disastrous to European citizens, agriculture and economy.[23]

On 25 September 1555, in the imperial city of Augsburg, now in

---

[20] Rubinstein, N. "Das politische System Italiens in der zweiten Hälfte des 15. Jahrhunderts", in *'Bündnissysteme' und Außpolitik im späteren Mittelalter*, Ed. P. Moraw (*Zeitschrift der. Historische Forschung. Beihefte*, 5, 1988) pp. 106-17.

[21] N. Machiavelli, *The Prince, with Related Documents*, ed. and trans. by W. Connell (Boston: Bedford/St. Martins, 2005), p. 74.

[22] V. Ilardi, "The Italian league, Francesco Sforza and Charles VII (1454-`461)" in *Studies in the Renaissance* VI, 1959, pp. 142-3.

[23] J.C.F. Fuller, *The Decisive Battles of the Western World 480BC-1757* (London: Paladin Books, 1954), pp. 467-93.

present-day Bavaria, the Peace of Augsburg, or the Augsburg Settlement, had been signed, temporarily settling the religious dispute between the Emperor on the one side and the Protestant princes on the other. It was a treaty agreed between Emperor Charles V and the forces of the Schmalkaldic League, which was an alliance of Lutheran princes. This officially ended the religious struggle between the two groups, and it was intended to make the legal division of Christendom permanent within the Holy Roman Empire. The Peace established the principle *Cuius regio, eius religio* [he who rules decides the religion], which allowed princes of the Holy Roman Empire's states to select either Lutheranism or Catholicism within the domains they controlled, ultimately reaffirming the independence they claimed for their states.

Half a century later, the Protestant (or Evangelical) Union was formed in 1608 as a coalition of Protestant German rulers wishing to defend the rights, lands and persons against the demands by the Holy Roman Emperor. It was formed as, in 1607, the Emperor and Duke Maximilian I of Bavaria had attempted to re-establish Roman Catholicism, and because, in early 1608, a majority of the Reichstag had decided that the renewal of the Peace of Augsburg of 1555 should be conditional upon the restoration of all church land appropriated since the Reformation. The Protestant princes then met near the town of Nördlingen on 14 May 1608 and formed a military coalition under the leadership of the Elector of the Palatinate, Frederick IV. In response, the Catholic League was formed in the following year, headed by Maximilian, duke of Bavaria. These two coalitions, both of a magnitude hitherto unseen in Europe, fought each other for thirty years with devastating results not only to the militaries involved but to civilians, property and state finances.

The Thirty Years War ended in 1648 by the peace settlement of Westphalia, but soon new power struggles emerged. "The Habsburgs turned towards the east, and when a generation later the Ottoman Empire began to crumble, they sought on the Danube compensation for their losses on the Rhine."[24] King Louis XIV's France was not only an absolute monarchy, himself being a determined ruler of his own country, but one with far-reaching aspirations of territorial aggrandizement at the expense of neighbouring countries – in particular the United Provinces and the principalities along the Rhine. Moreover, he aimed at setting up political alliances with all Catholic states, creating a threat to the rest of non-Catholic Europe. He had close ties with the English King James II, who was in the

---

[24] J.C.F. Fuller, *The Decisive Battles of the Western World 480BC-1757* (London: Paladin Books, 1954), p. 496.

process of re-introducing not only absolute rule but also Catholic dominance in Britain – an activity for which, in 1688, he was deposed by a group of parliamentarians and replaced by his son in law, William III.

In response to the looming danger from Louis XIV's grand designs, a number of coalitions were formed in order to prevent him from taking what he wanted. One of these was the coalition between Britain and Denmark of 1689, which aimed at providing Danish troops for the multi-national coalition being assembled with a view to ridding the British crown of an imminent Hiberno-French menace.[25]

To the Danes three strategic aims made joining the British-led coalition-of-the-willing a tempting notion. First, there was the common European aspiration of eliminating the threat of French aggression. In this respect, Danish war aims were identical with those of Britain, the Netherlands and Brandenburg. Secondly, King Christian V of Denmark needed funding for the upkeep of his armed forces, which were relatively larger than those of any other European country measured by the soldiers-to-population ratio, and he required training opportunities for officers and men in order to maintain and sharpen the military instrument for later action. Finally, having defeated Sweden in the Scanian War of 1675-79 and then, due to French intervention, been deprived of her territorial gains, Denmark had to retain a credible force to deter her old enemy from renewed attempts of aggression.[26]

National defence is not merely a matter of territorial assertion – it is equally concerned with discovering potential hostile military threats, wherever these might materialise, and combating them where they appear before they turn into genuine threats to the homeland. This was as true in the case of King Louis XIV's quest for European hegemony in the seventeenth century as it is with respect to the modern dictators with ambitions to acquire weapons of mass destruction.[27] It is, and was, also a matter of never letting the armed forces stand idle, but keeping the military instrument sharp and prepared for action whenever the need might arise. Thus, logically, indirect security was as crucial a need in the 1690s as it is today and the anti-French coalitions of the wars in Ireland and the Netherlands, as well as the War of Spanish Succession, where coalition forces ventured far afield, are some of the cases in point.

---

[25] Kjeld Hald Galster, *Danish Troops with the Williamite Army in Ireland, 1689-91: For King and Coffers* (Dublin: Four Courts Press, 2012), p. 84.

[26] Ibid., p. 11.

[27] Ibid., pp. 19, 23.

Today, the enormous costs involved in having troops employed far away from home as well as the highly contentious issue of casualties give rise to popular and political demands for cost limitation, end-states and exit strategies.[28] In this respect very little has changed since the Williamite coalition's expedition to Ireland in 1689. Then, the Danish Treasury was determined to make the war a low budget one, the troops were left to themselves as far as the daily needs were concerned, and no extra personnel were dispatched to complement the establishment during the years abroad. Moreover, in order to define an end-state, at which coalition obligations would expire, the coalition treaty of 1689 stated that the Danish expeditionary contingent would remain with the British colours only until the conclusion of a peace treaty. This was signed in Ryswick in 1697. By then sickness and battle casualties had depleted the Danish contingent considerably, but those troops, which did actually return home, were battle hardened and a good tool in the king's inventory of foreign policy instruments.

Like war in general, coalition warfare has its organisation and its rules. The Danish contingent in Ireland was subordinated to the lead-nation, Britain, and to the British commander-in-chief, initially King William III himself, with the Danish contingent commander, Lieutenant-General Ferdinand Wilhelm, duke of Württemberg-Neustadt, as his immediate subordinate.[29] Prior to assembling the Danish contingent, the aforementioned treaty was set up and ratified by the two kings, and within this a number of rules, amounting to what we would now call a Status of Forces Agreement – a SOFA – set out the rights and obligations of the Danish commanders, fixed personnel numbers, and determined the order of battle, as well as the much-coveted recompense and various other administrative details.[30]

The friction in war was as abiding a companion of the coalition fighting in Ireland and on the Continent, as it is in any war. Friction may be caused by the weather, shortage of food and ammunition, sloppy commanders or misguided actions. Throughout history we hear about so-called friendly fire over and over again. In modern times we have electronic gadgets which are

---

[28] Ibid., p. 215.

[29] Jahn, J.H.F. *Det danske Auxilliaircorps i engelsk Tjeneste fra 1689 til 1697* [The Danish Auxilliary Corps on English Service, 1689–1697]. Kjøbenhavn: Udgiverens Forlag, 1840, pp 6-25.

[30] John Childs, *The Williamite Wars in Ireland 1688-91* (London: Hambledon Continuum 2007), p. 194.

supposed to distinguish between friend and foe, but nonetheless troops and aeroplanes continue to shoot at the wrong targets. On the ground visual recognition is difficult because uniforms and equipment are much the same the whole world over, and this predicament was indeed extant in the 1690s, too. Therefore, since in the heat of battle recognition is never easy, and as the uniforms of the two sides were not too different, it was a wise precaution that the Williamite troops were ordered to fasten green sprigs in their hats to distinguish them from the opponents, who wore pieces of white paper in their headgear – an old-time friend-or-foe recognition.

As stated above, in joining the coalition-of-the-willing under British lead, the Danish king had three strategic aims in mind: eliminating the French menace to 'global' security and checking King Louis XIV's territorial ambitions; funding the upkeep of his armed forces and providing training opportunities for officers and men; and ensuring that Denmark retained credible military and naval forces to deter Swedish aggression.

It was the first of these aims that had been the coalition's common purpose. By eliminating resistance in Ireland thus prohibiting the deposed King James II's attempt at a comeback as British sovereign, French wishes to use that country as a stepping stone to Britain had obviously been foiled. Because of this feat, coalition forces were freed to take up the contest on the continent of Europe.

Moreover, although the Danish contingent had not been paid and replenished by the British as foreseen in the treaty, the expedition had been largely without the need of any disbursements on the part of the Danish Treasury.[31] Personnel numbers had sunk but, generally speaking, coming home in 1697 the contingent organisation was intact and its officers and men were in possession of much valuable combat experience.

Late Modern Era

The French Revolution of 1789 heralded the start of what was to become known as the Late Modern Era. It sent shockwaves through Europe, as monarchs and aristocrats feared ripple effects of the French revolutionary ideas in their own societies, and this apprehension only increased with the arrest in 1792 of King Louis XVI of France and his execution in January

---

[31] Account of Cha. Fox & Thom Coningsby Esqrs Receivers & Paymasters Gen. of their Maties Revenue and Forces in Ireland.

1793.³² The anxiety amongst neighbouring countries was so deep that several attempts were made to crush the French Republic. The wars that followed began as coalition warfare against a single country – France – but they developed into two coalesced blocs fighting each other for more than two decades. While, over the years from the beginning of these wars and right through to the end, the opposition formed seven consecutive coalitions of great powers supplemented by lesser contributors, France developed her own, steadily expanding, coalition under unquestioned French leadership, and in this respect the Revolutionary and Napoleonic wars were and remained asymmetric throughout. The anti-French First Coalition was formed as early as in 1793 when the Holy Roman Empire, the Kingdom of Sardinia, the Kingdom of Naples, Prussia, Spain and the Kingdom of Great Britain joined forces in order to defeat the revolution, or at least prevent it from spreading its dangerous intellectual seeds. Issuing the *levée en masse* ordinance (introducing conscription) and implementing the military reforms devised by the Secretary for War Lazare Carnot – including the *ordre mixte* in which conscript units were interspersed amongst regulars, the operational use of divisions allowing the army to move fast along different routes and then unite before the battle as well as improvements of the supply system – France managed to defeat the First Coalition.³³ In this context the war was asymmetric, though this type of asymmetry faded away as, after the battles of Ulm and Austerlitz in 1805, most continental powers saw the writing on the wall and began to develop similar systems. The war of the First Coalition ended on 18 October 1797, when General Napoleon Bonaparte forced the Empire to accept his peace terms under the Treaty of Campo Formio.³⁴

The Second Coalition sprang up in 1798, including the Holy Roman Empire, Great Britain, Naples, the Ottoman Empire, the Papal States, Portugal, Russia and Sweden. At the time, France suffered from corruption and internal division under the Directory. France also lacked funds, and no longer enjoyed the services of Lazare Carnot, the war minister who had guided it to successive victories following his extensive reforms during the

---

³² J.C.F. Fuller, *The Decisive Battles of the Western World 1792-1944* (London: Paladin Books, 1954), pp. 29-34.

³³ *Levée en masse*: ordinance issued by the French National Convention 23 August 1793. Lazare Carnot "The Organiser of Victory" was an engineer officer turned politician during the revolution. He managed to devise the *ordre mixte* allowing the remainder of the old Royal Army to operate in unison with the new revolutionary forces.

³⁴ Lindquist, Herman, trans. Henrik Eriksen, *Napoleon* (Oslo: Schibsted, 2005), pp. 94-124.

early 1790s.[35] Bonaparte, the main architect of victory in the last years of the First Coalition, had gone to campaign in Egypt. Missing two of its most important military figures from the previous conflict, the Republic suffered successive defeats against revitalised enemies whom British financial support brought back into the war. However, Bonaparte returned to France on 23 August 1799 and with the coup of 18 Brumaire (9 November 1799), replaced the Directory with the Consulate. He reorganised the French military anew and created a reserve army positioned so it might support campaigns on the Rhine as well as in Italy. From that moment onwards, French advances invariably caught the Austrians off guard and succeeded in knocking Russia out of the war. In 1800, Bonaparte won decisive victories over the Empire at Marengo in Italy and at Hohenlinden on the Rhine. The Treaties of Lunéville and Amiens signed in 1801 ended the War of the Second Coalition.

Nonetheless, Britain retained important influence over the continental powers, encouraging continued resistance to France, and she was willing to spend considerable sums on subsidies to those who would join the tussle. As General Bonaparte, from 1804 onwards the Emperor Napoleon, realised that without either defeating the British or signing a treaty with them he could not achieve complete peace, he planned an invasion of Great Britain and amassed 180,000 troops at Boulogne. However, in order to mount the invasion, Napoleon needed to achieve naval supremacy – or at least superiority in the English Channel. Over the summer of 1805 this looked increasingly unlikely to happen, and as on 21 October the Franco-Spanish fleet under Admiral Villeneuve was caught and defeated by a Royal Naval squadron under Admiral Lord Nelson in the Battle of Trafalgar, invasion was finally given up. France never got another opportunity to challenge the British at sea, and the plans to invade England were cancelled. Now, as Britain was busy organising the Third Coalition against France, Napoleon again turned his attention to enemies on the Continent and set the French army in motion towards Austria.

In April 1805, Britain and Russia had signed a treaty with the aim of removing the French from the Batavian Republic, roughly today's Netherlands, and the Helvetian Confederation, the modern Switzerland. The Holy Roman Empire and Sweden, who had agreed to lease Swedish Pomerania as a base for British troops, also entered the coalition. The Austrians began the war by invading Bavaria with an army of about 70,000 troops under *Feldmarschall-Leutnant* [Major-General] Karl Mack *Freiherr* [Baron] von Leiberich, and the French army marched out from Boulogne

---

[35] Lindquist, Herman, trans. Henrik Eriksen, *Napoleon* (Oslo: Schibsted, 2005), pp. 95, 98.

in late July 1805 to confront them. Arriving from a variety of approaches the French army corps united at Ulm in late October 1805, where Napoleon outmanœuvred Mack's army, forcing its surrender with only insignificant losses sustained by his own army.[36] The French then moved on to occupy Vienna. Far from his supply bases, Napoleon now faced a larger Austro-Russian army under the command of Field-Marshal Mikhail Illarionovich Golenishchev-Kutuzov. The French doctrine again demonstrated its superior flexibility, allowing Napoleon's armies to march along different routes and unite at exactly the right moment and, on 2 December 1805, Napoleon crushed the joint Austro-Russian army at Austerlitz in Moravia. He inflicted a total of 25,000 casualties on a numerically superior enemy army while sustaining fewer than 7,000 himself.[37] On 26 December 1805, France and the Third Coalition signed the peace Treaty of Pressburg – today's Bratislava, capital of the Slovak Republic – and the coalition was dissolved.

Shortly after the collapse of the Third Coalition, Prussia, Russia, Saxony, Sweden, and the United Kingdom formed the Fourth Coalition against France.

Now the French too espoused the concept of coalition warfare forming, in July 1806, *La Confédération du Rhin* [the Confederation of the Rhine] out of a number of minor states of the western and southern parts of what is today Germany. Consequently, the Holy Roman Empire of German Nation ceased to exist as Napoleon elevated Baden, Hesse, Clèves and Berg to grand duchies and Württemberg, Bavaria and later Saxony to kingdoms, all of them allied with France. His coalition had one single lead-nation – France – as opposed to the various anti-French coalitions with two or three contenders for leadership, viz. Austria, Britain and Russia.

In August 1806, the Prussian king, Friedrich Wilhelm III decided to go to war independently of all other Fourth Coalition partners apart from Russia. In September, Napoleon unleashed all the French forces east of the Rhine and while, on 14 October 1806, he defeated a Prussian force at the Prussian city of Jena, Marshal Davout beat another one at Auerstädt. With an army of approximately 160,000 France managed to put the Prussian military machine out of action before it could reach Berlin. Napoleon entered the Prussian capital on 27 October 1806 after merely 19 days of campaigning. He then went on to drive the Russians out of Poland, creating

---

[36] J.C.F. Fuller, *The Decisive Battles of the Western World 1792-1944* (London: Paladin Books, 1954), pp. 98-99.

[37] J.C.F. Fuller, *The Decisive Battles of the Western World 1792-1944* (London: Paladin Books, 1954), pp. 99-100.

the Duchy of Warsaw. Subsequently, Napoleon turned north to confront the remainder of the Russian army and to try to capture the temporary Prussian capital at Königsberg [Kaliningrad]. His manœuvre at Eylau on 7/8 February 1807 forced the Russians to withdraw further north, routing them on 14 June at Friedland. Following this defeat, on 7 July 1807 the Russian Tsar Alexander I signed a peace treaty with Napoleon on a raft – which was considered neutral ground – tethered off Tilsit.

While in 1809 the United Kingdom and Austria formed the Fifth Coalition against France, Britain, on her own, engaged in the Peninsular War. Although during the time of the Fifth Coalition, the Royal Navy won a succession of victories in the French colonies; on land this coalition attempted only few larger military endeavours. One, the Walcheren Expedition of 1809, involved a dual effort by the British Army and the Royal Navy to relieve Austrian forces under intense French pressure. It ended in disaster after the Army commander, John Pitt, 2nd Earl of Chatham, failed to capture the objective, the naval base of French-controlled Antwerp. British military operations on land apart from those in the Iberian Peninsula remained restricted to hit-and-run operations. These rapid-attack operations were aimed mostly at destroying blockaded French naval and mercantile shipping and the disruption of French supplies, communications and military units stationed near the coasts. Often, when British allies attempted military actions, the Royal Navy would arrive, land troops and supplies, and aid the Coalition's land forces in a concerted operation. Royal Naval ships even provided naval gunfire support against French units fighting sufficiently close to the coastline.

A new element in coalition warfare was the economic warfare pursued by both sides. However, breaches of the Continental System occurred continuously as French-dominated states engaged in trade with British smugglers.[38] When Spain failed to maintain the continental system, the uneasy Spanish alliance with France ended in all but name, and the government of Napoleons brother, Joseph Bonaparte, could do little when popular resistance developed into guerrilla warfare supporting the British in the Peninsular War.

Austria, established in 1806 on the remnants of the Holy Roman Empire, had briefly been allied with France, but took the opportunity to attempt to restore its imperial territories in Germany as held prior to Austerlitz. Austria achieved a number of initial victories against the thinly dispersed army of Marshal Louis Alexandre Berthier, 1st Prince de Wagram, 1st Duc de Valangin, 1st Sovereign Prince de Neuchâtel. Napoleon had left

---

[38] Lindquist, Herman, trans. Henrik Eriksen, *Napoleon* (Oslo: Schibsted, 2005), p. 269.

Berthier with only 170,000 men to defend France's eastern borders. On 16 January 1809, Napoleon had enjoyed easy success in the battle of Corunna, forcing a temporary withdrawal of the out-numbered British troops from the Iberian Peninsula. However, when he left, the guerrilla war continued to tie down great numbers of French troops. Austria's attacks on Berthier's troops in the East, necessitating Napoleon's departure from Spain, prevented him from wiping out the British resistance in the peninsular theatre.[39] In his absence and that of his best marshals – Davout remained in the east throughout the war – the French situation in Spain deteriorated, and then became dire when Lieutenant-General Sir Arthur Wellesley arrived to take charge of Anglo-Portuguese forces.[40]

French coalition forces defeated the Austrians at Wagram, on 5/6 July. It was during that battle that Marshal Bernadotte was stripped of his command after retreating contrary to Napoleon's orders. Humiliated, Bernadotte accepted the offer from the Swedish court to fill the vacant position as Crown Prince of Sweden. On 14 October 1809, the War of the Fifth Coalition ended with the signature of the Treaty of Schönbrunn. In 1810, the French Empire reached its greatest extent, and, hoping to achieve a stable alliance with Austria, Napoleon married Archduchess Marie-Louise of Austria.

In 1812, at the apex of the Napoleonic coalition system, Napoleon invaded Russia with a pan-European *Grande Armée*, consisting of 650,000 men. For this expedition the Polish born *Maréchal de France* [Marshal of France] Prince Józef Antoni Poniatowski Poniatowski became commander of the nearly 100,000 strong Polish forces, the 5th Corps of the *Grande Armée*. Napoleon's army marched through Russia, winning a number of relatively minor engagements, along with the major Battle of Smolensk on 16-18 August. However, the fate of the invasion was to be decided in Moscow, where Napoleon himself led his forces. The Russians used a scorched-earth doctrine, and retreated for almost three months. Finally, on 7 September the two armies engaged in the Battle of Borodino on the outskirts of Moscow.[41] The battle was the largest and bloodiest single-day

---

[39] Richard Holmes, *Wellington: The Iron Duke* (London: Harper Collins Publishers, 2002), pp. 127-94.

[40] Wellesley was promoted to full general in 1811 and field-marshal in 1813. Following Napoleon's exile he was made ambassador to France and elevated within the English peerage as 1st Duke of Wellington.

[41] J.C.F. Fuller, *The Decisive Battles of the Western World 1792-1944* (London: Paladin Books, 1954), p. 117.

action of the Napoleonic Wars, involving more than 250,000 men and resulting in at least 70,000 casualties. The French captured the main positions on the battlefield, but failed to destroy the Russian army. Napoleon entered Moscow on 14 September, after the Russian Army had retreated yet again. But, by then the Russians had largely evacuated the city and even released criminals from the prisons to inconvenience the French. Furthermore, the governor, Count Fyodor Rostopchin, ordered the city to be burnt. Tsar Alexander I refused to capitulate, and the peace talks, attempted by Napoleon, failed. In October, with no sign of a clear victory in sight, Napoleon began the disastrous Great Retreat from Moscow. In the following weeks, the *Grande Armée* was dealt a catastrophic blow by the onset of the Russian Winter, the lack of supplies and constant guerrilla warfare by Russian peasants, Cossacks and irregular troops. When the remnants of Napoleon's army crossed the River Berezina in November, only 27,000 fit soldiers remained, with some 380,000 men dead or missing and 100,000 captured. Napoleon then left his men and return to Paris, to prepare the defence against the advancing Russians, and the campaign effectively ended on 14 December 1812, when the last enemy troops left Russia. The Russians had lost around 210,000 men, but with their shorter supply lines, they soon replenished their armies.

On 21 June 1813, at Vitoria, the combined Anglo-Portuguese and Spanish armies defeated Joseph Bonaparte, finally breaking French power in Spain. The belligerents declared an armistice continuing until 13 August.[42] During this time negotiations on the Sixth Coalition brought Austria out in open opposition to France. Two Austrian armies took the field, adding an additional 300,000 men to the Coalition armies in Germany. In total the Allies now had around 800,000 front-line soldiers in the German theatre, with a strategic reserve of 350,000 formed to support the frontline operations.

Nonetheless, Napoleon succeeded in bringing the total imperial forces in the region to around 650,000 – although only 250,000 came under his direct command, with another 120,000 under Nicolas Charles Oudinot and 30,000 under Davout. The remainder of imperial forces came mostly from the Confederation of the Rhine – and from Denmark contributing an expeditionary corps under Davout. In addition, to the south Murat's Kingdom of Naples and Eugène de Beauharnais's Kingdom of Italy added another 100,000 armed men. In Spain, another 150,000 to 200,000 French troops retreated before Anglo-Portuguese forces numbering around

---

[42] Richard Holmes, *Wellington: The Iron Duke* (London: Harper Collins Publishers, 2002), pp. 170, 186-87 and 200.

100,000. In total, around 900,000 Frenchmen in all theatres faced around 1,800,000 Sixth Coalition soldiers.

Following the end of the armistice, Napoleon seemed to have regained the initiative at Dresden in August 1813, where he once again defeated a numerically superior Sixth Coalition army and inflicted enormous casualties, while sustaining relatively few himself. However, the failures of his marshals and a slow resumption of the offensive on his part cost him any advantage that this victory might have secured. At the Battle of the Nations at Leipzig, 16-19 October 1813, 191,000 French fought in vain against more than 300,000 Allies, and Napoleon's defeated troops had to retreat into France.[43] His remaining ally, Denmark-Norway, became isolated and fell to the coalition.

Napoleon determined to fight on, incapable of fathoming his massive fall from power. During the campaign he had issued a decree for 900,000 fresh conscripts, but only a fraction of these ever materialised, and Napoleon's schemes for victory eventually gave way to the reality of the hopeless situation. The Emperor of the French abdicated on 6 April 1814.

The Seventh Coalition of 1815 pitted the United Kingdom, Russia, Prussia, Sweden, Austria, the Netherlands, Denmark and a number of German states against France. While the period known as the Hundred Days began after Napoleon's escape from the Mediterranean island principality of Elba, subsequent to the defeat at Waterloo it ended in Paris on 22 June 1815, when Napoleon was forced to abdicate for the second and final time.[44]

Although the initial asymmetry lessened over the twenty-two years of war, it is fair to say that its development hinged not only on the differences of French military doctrine from those of her opponents, but also on the novel freshness of French popular and national political views vis-à-vis those of the old empires mired as they were in tradition and fear of social unrest.

Having acquainted ourselves with the asymmetric warfare of the opposing coalitions of the Napoleonic era, we will now move some thirty-eight years ahead to look at a rather different war of coalition – the Crimean War. On the surface, the contested issue was Russia's demand for status as the protector of the orthodox Christians in Constantinople, but in reality it was instead the Tsar's threat against the declining Ottoman Empire in general, and the right of passage through the Strait of Bosporus in

---

[43] J.C.F. Fuller, *The Decisive Battles of the Western World 1792-1944* (London: Paladin Books, 1954), pp. 119-55.

[44] Richard Holmes, *Wellington: The Iron Duke* (London: Harper Collins Publishers, 2002), pp. 236-54.

particular. Britain was committed to the upkeep of Turkey and the France of Napoleon III was on the lookout for glorious engagements in far-flung parts of the world. As the Sublime Porte turned down a Russian ultimatum concerning the protection issue, Russia invaded the Danubian Principalities, and on 30 November 1853 destroyed a Turkish naval squadron at Sinope in the Black Sea, Britain and France began preparing for a war which was duly declared in March 1854.

British naval historian Andrew Lambert observed that:

> The Anglo-French coalition that fought against Russia during the Crimean War was an unlikely combination. The two powers had been at war, off and on, for close on five hundred years, and well within living memory had been the principal opponents in a twenty-two year long total conflict, the Revolutionary and Napoleonic wars, 1793-1815. In the years since Waterloo their relations had rarely been other than hostile. Yet they managed to agree on policy, strategy, and operate in apparent harmony to invade and defeat Imperial Russia – the continental superpower of the age. Beneath the surface, however, things were far from smooth. The clash of strategic cultures, and continuing long-term rivalries manifested in a major arms race, made the coalition a fragile instrument. Long before the war terminated the two parties anticipated an early end to the relationship, and apart from the latter stages of the Second Opium War (1856–1860), the two powers would not take the field again as partners until 1914.[45]

Britain sent an expeditionary force of 26,000 men under Field-Marshal Fitzroy James Henry Somerset, 1st Baron Raglan (1788-1855), containing five infantry and one cavalry divisions. There were no reserves, logistic arrangements were inadequate, and the administration was chaotically decentralised. Under the command of *Maréchal de France* Armand Jacques Achille Leroy de Saint-Arnaud (1798-1854) France initially dispatched 40,000 men in four infantry divisions and two cavalry brigades, but later supplemented them by another seven infantry divisions. Many officers, both British and French, had experience from either the Napoleonic Wars or – as far as the French were concerned – from the conquest of Algeria. The Russian army of about 1,400,000 men was inadequately prepared for modern warfare and poorly trained, and it had no experienced general staff.

After a brief employment of Anglo-French troops on the western shores of the Black Sea, the Russians withdrew from the Danubian Principalities. The Western Allies had to look for new objectives and the British

---

[45] Andrew Lambert in paper presented at the coalition warfare conference in Copenhagen May 2011.

government decided that the Russian naval base at Sevastopol was worth taking. The coalition army sailed for the Crimea, where they landed on 14 September 1854. 35,000 Russian troops under General-Admiral Prince Aleksandr Sergeyevich Menshikov (1787-1869) waited for them in well prepared positions on the River Alma. On 19 September, the allies advanced with St Arnaud's French on the right and Ragland's British troops on the left. Despite resistance the river was crossed quickly and the positions taken. At the end of the day the Russians had lost around 5,000 casualties, the allies about 3,000. On 29 September, St. Arnaud died of natural causes and his post was temporarily taken over by General François Certain de Canrobert (1809-95).

Canrobert agreed with Lord Raglan on softening up Sevastopol before attacking, and on 17 October naval and military bombardments of the city were commenced. However, Sevastopol held out longer than expected, and on 25 October a Russian force of 25,000 approached from the east, apparently planning to take Balaclava, the harbour where the allies got their supplies. This threat was repelled by the 93rd Highlanders' Thin Red Line and a flanking attack by the British heavy brigade. The Russians made several sorties from Sevastopol and on 5 November 20,000 troops from that city, 16,000 from the east and another 22,000 from the south attacked the allies at Inkerman. In the end the Russian attack was warded off with 2,500 British and 1,700 French casualties, while the Russians lost about 12,000. After this failure Prince Menshikov was replaced by Prince Mikhail Dmitrievich Gorchakov (1793-1861). It was not only the Russian high command that experienced change. Amongst the allies, frustrations on account of confusing orders and counter-orders from Paris led to the resignation of Canrobert, who went back to command his old division and was replaced by General Aimable (sic!) Jean Jacques Pélissier, 1er Duc de Malakoff (1794-1864). This changed the war's progress. Pélissier commanded by relentlessly pressuring the enemy, and he sustained an unalterable determination to conduct the campaign without interference from Paris. His perseverance was crowned with success in the storming of the Tower of Malakoff on 8 September 1855, which ended the Siege of Sevastopol and crowned the Anglo-French Crimean War against Russia with victory. On the 12th September he was promoted to *Maréchal de France.*

The Anglo-French coalition had indeed been one of diverging interest between the partners. Because Britain's main interests were her empire and the upkeep of the balance of power on the Continent, including the Ottoman Empire, British policy could be conducted without the need to enter into binding alliances with other major powers. Britain sought coalitions with other powers only when needed to secure her political and military aims in

Europe. This approach was epitomised by Lord Palmerston noticing in March 1848 that: "We have no eternal allies, and we have no perpetual enemies. Our interests are eternal, and those interests it is our duty to follow."[46]

In contrast, during the realm of Napoleon III, France led an activist foreign policy trying to make her great power influence felt more or less globally. While apparently Napoleon had no other aim with the Crimean conflict than adding to the country's glory, the coalition in itself was more important than doctrinal and operational details. The key issues of the Anglo-French strategy were offensive action, and concentration and economy of effort. From a British point of view the primary aim with the war would be persuading the Tsar to abandon his designs on Turkey through destruction of the Russian Black Sea Fleet and its base at the Crimean town of Sevastopol. An early success would release resources for the Baltic theatre, where another, larger Russian fleet would need to be dealt with later through operations against Russian naval bases in the Gulf of Finland. However, as soon as Britain suggested this kind of escalation, France terminated the war. Napoleon III was not interested in spreading the conflict or in procrastinating its conclusion, and a compromise peace was therefore quickly patched up in Paris.

Towards the end of the 19th century, the world saw anew a coalition of a great power lead-nation buttressed by contributing partners of lesser punch. As the 2nd Boer War loomed on the horizon, Great Britain encouraged her dependencies to rally to the preservation of their common Empire. This happened to a large extent, but not without reservations by some Dominion countries and colonies. Thus, this war provided examples of political disunity and doubts over coalition participation, the merit of supporting the imperialist intentions of the lead-nation and the way a troop contingent might be assembled. In this context we may regard the 2nd Boer War as different from the seven coalitions fighting hegemonic France or the Anglo-French effort against Russia in the Crimea. Conversely, in the 2nd Boer War we see a form of coalition which is rather like that of Napoleonic France *contra mundum* (against the world), as Paul Kennedy wrote:

> It was those Britons who feared for the future of the empire and who saw world tendencies through Darwinian spectacles, who sought alliances, and in two forms. The first was to attempt an imperial defence coalition, to attach the largely self-governing Dominions to the common cause, a notion perceived with scant enthusiasm by Canada as well as South Africa. The second form

---

[46] Kenneth Bourne, *Palmerston: The Early Years 1784–1841* (London: Alien Lane, 1982), p. 627.

was a willingness to contemplate alliance with another great power – with Japan in 1902 and, less formally, with France and Russia a few years later. Thus British statesmen, too, sought the dual security which they believed a coalition offered: aid from partners in the event of hostilities; and a pledge to assist partners, whose defeat by a mutual foe might be disastrous to one's own interests.[47]

The Second Boer War (1899-1902) was a clear attempt at establishing an imperial defence coalition to suppress the Boer insurgency and put Transvaal and the Orange Free State under the suzerainty of the Cape Colony, as well as maintaining this colony within the community of the British Empire.

During the 1890s the thriving British colonies of Rhodesia, the Cape Colony and Natal and the influx into Transvaal and the Free State of mainly Anglophone gold miners had made the Boers apprehensive with respect to their autonomy and way of life. As Boer legislation made life extremely unpleasant for the temporary immigrants, disaffection grew, and as social unrest developed in the wake of the so-called Jameson Raid in 1895, the scene was set for conflict on a larger scale.[48] In October-November 1899, Boer commandos totalling around 40,000 militiamen invested the British garrisons at Mafeking, Ladysmith and Kimberley and invaded Natal. War soon was a reality as civilians on both sides took up arms against each other, and as Australia, Canada and other British dependencies declared their intension of fighting for the Empire.

After the Black Week in mid-December 1899, when General Sir Redvers Buller's expeditionary corps was defeated at Stormberg, Magersfontein and Colenso, the garrison in South Africa was boosted to 180,000 under the command of Field-Marshal Lord Roberts of Kandahar, with Kitchener as his chief of staff. After almost a year of heavy fighting, Roberts defeated massive Boer forces and annexed Transvaal and the Orange Free State in September 1900. While he and Buller went back to England and Kitchener took over command, many Boers had no intention of giving up fighting. Kitchener's war, which is probably commonly remembered primarily for its use of blockhouse lines, barbed wire and

---

[47] Paul Kennedy, "Military Coalitions and Coalitions Warfare over the past Century" in Keith Neilson and Roy A. Prete, Eds., *Coalition Warfare: an Uneasy Accord* (Waterloo, Ontario, Canada: Wilfrid Laurier University Press, 1983) pp. 5-6.

[48] Norman Hillmer and J.L. Granatstein, *Empire to Umpire: Canada and the World into the Twenty-First Century* (Toronto: Nelson, 2008), p. 16.

concentration camps, ended with the peace accord of Vereeniging of 31 May 1902 by which the Boers became citizens of the federated British South Africa. 500,000 white soldiers from all over the empire had made up the fighting force of the coalition, and another 100,000 non-whites had assisted as labourers.

This, the imperial kind of coalition, conspicuously led by a great power, was not without hitches and frustrations, which we may see illustratively demonstrated by the Canadian example. Canada was among the British dependencies which decided to join the coalition bound for South Africa, but this had not been an easy decision for the Canadian political leadership.[49] As we have seen so often before as well as after, the true reason for war was the quest for benefit – for exploitation of natural resources. British Imperial politics was heavily influenced by the desire for access to the diamond and gold deposits in Boer territory, and since the Boers made this practically impossible, something had to be done. The reputation of the British Empire demanded action, and the English-speaking Canadians called for action too. However, the Francophone Canadians saw it differently. There was no love lost between the French Canadians and the British Empire and many of them even sympathised with the Boers, who they regarded as a maligned minority, like themselves, resisting the dominance of a larger group. With such contrary views across the French-English divide, the government of Prime Minister Sir Wilfrid Laurier found itself caught on the horns of a typical Canadian dilemma: commit an expeditionary force to southern Africa and risk alienating the densely populated province of Quebec with its Francophone majority, or do nothing and incur the hostility of the rest of Canada.

As it turned out, Laurier's decision was taken for him. On 3 October 1899, Canada received a telegram from the British Colonial Secretary, thanking Canada for its contribution of a volunteer contingent, even though no formal offer had yet been made. Laurier suspected that Canada's senior military commander – who was not a Canadian at all, but a British Army officer – Major-General E.T.H. Hutton, as well as Lord Minto, the British Governor General of Canada, had conspired to force him into a corner. Hutton was actually pleased with himself having forced: "the weak-kneed and vacillating Laurier Government with their ill-disguised French and Pro-Boer proclivities to take a part…in the great movement which has drawn the

---

[49] Parts of the following on the 2nd Boer War is largely due to a key note address given by Douglas Delaney at the Conference on Coalition Warfare in Copenhagen, 2011.

strings of our Anglo-Saxon British Empire so close."[50]

Not surprisingly, Quebec ministers opposed the dispatch of any military contingent to South Africa, while a good portion of his English-speaking ministers argued passionately for a 1000-man contingent. Eventually it was decided to recruit a battalion-size force and ship it to the Cape Colony, at which point it would become the responsibility of British military authorities. This offered both a demonstration of the jingoistic zeal of English Canada, and some assurance to French Canadians that they would not have to fund Imperial ventures.

Canada's contribution to the South African War was not without its problems. Recruiting was fairly easy as volunteers flocked to the first contingent – 1061 of them, including 62 officers.[51] The 2nd (Special Service) Battalion of the Royal Canadian Regiment (RCR), which was a miscellany of ex-soldiers, half-trained militiamen, and a good number of men who had never fired a rifle, drew overwhelmingly from urban centres and contained very few Francophones.[52] Two more contingents of volunteers, mostly mounted troops, were raised just as easily and sent to the South African *veldt* as well.[53]

In all, 8372 Canadians volunteered for service during the Second Boer War. Finding tough and brave soldiers was not a problem. The problems had more to do with army organisation and services – the medical and supply services first among them. More than half of the 244 Canadian fatalities in South Africa had succumbed to disease that adequate medication, a steady supply of drinkable water and functioning field hospitals might have prevented.[54] But the Canadian contingent in South Africa could provide none of these commodities on its own; it had to rely on the British for this kind of services, and they were sorely remiss. There was a growing sense among Canadian soldiers that they had to take greater responsibility for their military establishment and the expeditionary forces

---

[50] Doug Delaney in paper presented at the Coalition Warfare Conference in Copenhagen 2011.

[51] Ibid.

[52] Carman Miller, *Painting the Map Red: Canada and the South African War, 1899-1902* (Ottawa: Canadian War Museum, 1993) pp. 49-64.

[53] Doug Delaney in paper presented at the Coalition Warfare Conference in Copenhagen 2011.

[54] Ibid.

it generated. They also began to think about more systematic peacetime planning for future expeditionary force operations. Why did they not have a supply of reinforcements, for example? By June 1900, a combination of casualties and diseases had reduced the 1061 men of the first contingent to 438 all ranks.[55]

World War I

The First World War coalitions were branded by the unhappy relationship between politicians and the military professionals, which soon led to abysmal discrepancy between the political and ill-conceived aims *with* the war on the one side, and the military imperative of fulfilling these by setting realistic aims or objectives *in* the war on the other. Moreover, for the most part of the war there was an obvious dearth of unity of command, because of which unknown numbers of opportunities as well as countless human lives were probably unnecessarily squandered. Last but not least, there were cultural problems like language, distrust of allies, behavioural norms and lack of mutual reliance making collaboration amongst partners difficult.

Politics

Like the Napoleonic Wars from 1806 onwards, World War I was a conflict between two large coalitions opposing each other. However, this was a conflict between two peer-partners coalitions, making it different from the single great power led coalition type of the Napoleonic and Boer Wars. The opposing alliances of 1914 were both formed by great powers, whose individual national interests were believed to be best taken care of that way, and both sides believed that defeat of one or more of their coalition partners would be disastrous to their own well-being and ultimately their national survival. Moreover, at least on the side of the Anglo-French *Entente* there were coalitions within the coalition, in as much as the peripheral members of the French and the British Empires chose to join the struggle of their imperial centres, thus engaging themselves as enemies of the Central Powers, Germany, Austria-Hungary, Bulgaria and The Ottoman Empire.

While this practice was new to some of the warring states such as Germany and Austria-Hungary, others were well-prepared. As we have seen, in the early 1900s Britain and her dominions had already collected a

---

[55] Ibid.

lot of useful experiences painstakingly acquired in the South African War. For instance, it was on the basis of what had happened in 1899-1902 that Canada had developed plans on how to handle new emergencies requiring expeditionary forces to join coalitions overseas. Surprisingly, however, when war came in 1914 no one in Canada cared to use the plans.[56] Sam Hughes, the Minister of Militia, threw them out – and it is nearly impossible to explain why.[57] Hughes was a Boer War veteran, who believed fervently in the superiority of part-time citizen soldiers and for that reason he loathed the professional officer class, whose advice he regularly ignored. The result was a chaotic mobilisation in 1914, though reasonably soon its troops were moulded into an efficient fighting force.

Nonetheless, increasing military maturity coincided with a growing sense of national identity and the Canadian government's desire to have a greater say in the direction of the war. Prime Minister Sir Robert Borden, though a keen supporter of the British Empire, was determined that Canada would play an autonomous and influential rôle in the common war effort. Borden was also determined to retain ultimate control of how and when Canadian forces were employed, and the gruesome casualty lists from the battles of Second Ypres, St Eloi and the Somme only strengthened his resolve in this regard. Borden's policy and his constitutional position as a Dominion prime minister accorded General Currie, as Canadian Corps commander, what amounted to a *veto* on any matter concerning the employment and organisation of the Canadian Corps.

While within the British Empire coalition partners gradually extended their autonomy, the two major Central Powers went in the opposite direction. Well aware of the emotions engendered amongst the French over the German conquest of Alsace and Lorraine in the 1870-71 war, Germany feared French revenge, and as the Russo-French military affiliation was no secret the German Empire prepared for the contingency of a two front war with France and Russia. For this they needed their southern neighbour's support in the East and, as war came, this caused chaos and dangerous dispersal of Austro-Hungarian formations. Although in the late summer of 1914 Austria-Hungary had to fight both Serbs and Italians on their southern borders, Germany insisted that she also took responsibility for beating the Russians in Galicia on the eastern front, in order to take the pressure off the German Eighth Army operating from East Prussia. This caused

---

[56] The following on the Canadian World War I efforts is primarily due to a key note address given by Doug Delaney at the Conference on Coalition Warfare in Copenhagen, 2011.

[57] J.L. Granatstein, *Empire to Umpire: Canada and the World into the Twenty-First Century* (Toronto: Nelson, 2008), p. 19.

indescribable logistic chaos – the German request came so late that the newly mobilised Austrian troops were already on the move on the single track railroads towards the Balkans, where, in the end, this change of strategic concept would cost them unexpected casualties. However, the operations in the East soon produced a resounding German victory over the Russians at Tannenberg on 30 August 1914, and the German *Oberste Heeresleitung* [supreme army command] exploited the success to slowly but surely usurp supreme command on the entire eastern front including Austro-Hungarian forces.

Strategy

The *Entente* was originally a coalition of Britain, France and Russia, but it soon expanded to include countries like Japan in August 1914, Italy in May 1915, Portugal in Match 1916, Romania in August 1916 and Greece in June 1917. The United States entered in April 1917 followed by China and a number of South American countries. During World War I, the *Entente* collaboration clearly demonstrated that national pride and the incompatibility of doctrines and organisations might constrain unity to something happening at the strategic and operational levels only, if at all. Battlefield responsibilities were divided nationally based on the attitudes, capabilities and operational approaches that each nation brought to the coalition. On the Western Front, France and Britain retained individual supreme commanders who did not all the time agree on operational and strategic choices.

With Russia's withdrawal from the conflict in 1917 and Germany's subsequent ability to concentrate manpower on the Western front, the separate Allied armies found themselves outnumbered by a German army under the unified command of Field-Marshal Paul von Hindenburg and his Quartermaster-General Erich Ludendorff. The Western Front now assumed key importance the Central Powers enjoyed a seamless unity of command, while the *Entente* was still torn by conflicting aims. On 21 March 1918, the *Oberste Heeresleitung* launched Germany's spring offensive, the so-called Ludendorff, or Michael, Offensive, using intense bombardment and storm troops to gain breakthroughs through local numerical superiority. Focusing its attack on the critically vulnerable boundary between the British and French armies, within days German troops pushed the *Entente*'s lines to the brink of collapse. With the British supreme commander, Haig, wishing to ensure the security of the Channel Ports and his French opposite number, Pétain, wanting to cover Paris the Germans attempted to drive a wedge between the two, so as to deal with them individually.

Assurances and co-operation amongst *Entente* national commanders only went as far as providing a reasonably effective resistance. The coalition in a sense was still loose, and needed to be tightened under the imminent threat of disintegration. The Allied Supreme War Council, which had now been created in order to lift strategic decision-making out of the national supreme commanders' and top-level politicians' spheres, tried to detach it from the slow and cumbersome process and the egotistical interests of individual coalition partners. However, this Council made decision-making even slower, and on the insistence of the French Premier, George Clemenceau, after the poor performance during the Ludendorff offensive of March 1918, *Maréchal de France* Ferdinand Foch was appointed Allied generalissimo, with the task of unifying coalition efforts. Still, however, the actual running of operations was decentralised, hampering the co-ordination of effort. This was a patchwork approach, but it went some way to securing some degree of unity of command.

Command and Control

Since this type of alliance, where the defeat of partners would be detrimental to one's own interests, was primarily a coalition amongst great powers, the choice of a lead-nation was a delicate political matter unlike the cases of coalitions of one great power and a host of minor contributors. The command relationships within the *Entente* during World War I offer examples of difficulties arising from this challenge, because in this case the need for national survival acted as the fundamental bonding agent between states.[58]

Although Britain and France were already allied, as mentioned above, in 1918 the strategic situation required that they made further concessions as to sovereignty in order to combat the Central Powers effectively.[59] The appointment in the spring of 1918 of Marshal Foch as the Allied

---

[58] The following owes partly to a paper given by Patrick Cecil during a Conference on Coalition Warfare in Copenhagen 2011.

[59] For additional scholarship on coalitions and members seeking to survive, see John B. B. Trussell, Jr., *Birthplace of an Army: A Study of the Valley Forge Encampment* (Harrisburg: Pennsylvania Historical and Museum Commission, 1976). In extending coalition warfare from survival of a state to the survival of religion and identity, see the Holy Alliances of the sixteenth-century and such scholarship as: Frederic C. Lane, *Venice: A Maritime Republic* (Baltimore: The Johns Hopkins University Press, 1973); John Francis Guilmartin, Jr., *Gunpowder and Galleys: Changing Technology and Mediterranean Warfare at Sea in the Sixteenth Century* (Cambridge: Cambridge University Press, 1974).

Commander-in-Chief of the *Entente* armies on the Western Front was a necessary step in order to avoid the collapse of the Allied war effort against Germany. Despite almost four years of fighting against the Central Powers, Britain and France had lacked until then a unified command system, relying instead on assurances of support. Historians like Michael Neiberg, Elizabeth Greenhalgh and Richard Holmes note that army commanders guarded their armies' national sovereignty and control over their respective reserves, for they did not wish to handicap their own freedom of manœuvre by placing their troops under a foreign commander. The idea of an allied general staff had been floated in previous years, and in the summer of 1917 political leaders sought to create a Supreme War Council in order better to co-ordinate their military actions on the Western Front. However, Generals Douglas Haig and Philippe Pétain, as commanders of their respective national armies, resisted the Council's ability to control reserves. For these commanders, their coalition had not come under sufficient threat to give up their sovereignty and freedom of action.[60]

Then, on 26 March 1918, in order to ascertain unity of command and at the insistence of the British leadership, Ferdinand Foch, then acting as Chief of Staff to the French Army, was appointed to co-ordinate the allied military efforts – *de facto* generalissimo. Foch (1851-1929) is primarily remember for his doctrine of *offensive á l'outrance* (all-out offensive), which cost the French huge casualties in 1914, and for being the architect of the *Entente*'s final victory in 1918. He had taught theory of war at the *École supérieure de guerre* before the outbreak of World War I. In 1914 he commanded the *XX Corps d'armée*, later taking over command of the northern group of armies, before, in 1916, he became the French army chief-of-staff.

After the Ludendorff offensive in March 1918, Britain had realised that her armies' survival in the war effort on the Western Front required that her sovereignty in strategic and operational decision-making be subordinated to alliance obligations. Although Foch lacked the power to issue orders, his energy and ability to convince individual commanders to action, coupled by the German offensive losing steam, allowed the British and the French to

---

[60] On the Allied coalition in World War I and the appointment of Ferdinand Foch as a measure of survival for the British and French allies, see the following: General Sir James Marshall-Cornwall, *Foch as Military Commander* (New York: Crane, Russak & Company, Inc., 1972); William James Philpott, *Anglo-French Relations and Strategy on the Western Front, 1914-18* (New York: St. Martin's Press, Inc., 1996); Michael S. Neiberg, *Foch: Supreme Allied Commander in the Great War* (Washington, D.C.: Brassey's, Inc., 2003); Elizabeth Greenhalgh, *Victory through Coalition: Britain and France during the First World War* (Cambridge: Cambridge University Press, 2005).

reorganise and solidify their lines. Eventually, political and military leaders agreed to enhance Foch's powers to give strategic direction over all *Entente* military operations, including that of the arriving American army.[61] With their "backs to the wall," as described by Haig, the Allies finally sacrificed pride and sovereignty to achieve unity of command and contain the threat to the war effort and command structure.[62]

It was not only the *Entente* that encountered difficulties in determining their command and control arrangements. The Central Powers, too, were challenged and in particular so as far as the relationship between The Ottoman Empire and Germany was concerned. There was an abysmal gap between the two cultures, and the general German attitude towards the Ottoman Turks had been formed by earlier writings of key personalities like Field-Marshal Helmuth von Moltke the elder, who found Turkey 'a land consumed with laziness,' and Heinrich von Treitschke describing the Turks as being 'indescribably lazy' and their culture as a 'profound slumber of the soul.'[63] Moreover, much to the chagrin of Germany, Turkey had greatly procrastinated its entry into the coalition of the Central Powers, and when finally joining the Minister for War, General Ismail Enver Pasha insisted on launching three complicated campaigns against the advice of the German military advisor, General Otto Liman von Sanders.[64] One of these aimed at pushing forward through Afghanistan to India, totally ignoring geographical and meteorological conditions as well as enemy strength. This turned out a disaster costing 86 percent of the Turkish Third Army's manpower. Simultaneously, the Ottoman Fourth Army launched a futile attack against the Suez Canal, again against the advice by von Sanders, and since the equipment for crossing the canal could not be used properly because of lack of training, the operation had to be called off. The third was the defence of the Gallipoli peninsula in March 1915, in which the Turks succeeded not least because of German material assistance and von Sanders' able command over the Ottoman Fifth Army. The Turks, however, were not inclined to gratitude, and in meeting with Henry Morgenthau, the American

---

[61] National army commanders held the right to petition against Foch's orders if they felt it endangered the integrity of their armies.

[62] Greenhalgh, 206.

[63] Luft, Gal. Beef, Bacon and Bullets: Culture in Coalition Warfare from Gallipoli to Iraq. (no place: Gal Luft, 2009), pp. 37-38.

[64] Luft, Gal. Beef, Bacon and Bullets: Culture in Coalition Warfare from Gallipoli to Iraq. (no place: Gal Luft, 2009), p. 29.

ambassador to Constantinople, War Minister Enver noticed that there was no reason to be grateful, because the Turks had singlehandedly defeated the Royal Navy – which the Germans had not been able to do themselves – they had kept the Russian at bay in the Caucasus, and tied down huge British armies in Egypt and Mesopotamia. All in all, these three campaigns did little to boost reciprocal respect between The Ottoman and German Empires – actually, they generated more tension.[65]

Culture

In coalition warfare the linguistic ability to communicate is far from enough. Even if partners use the same language, more often than not, they use it differently. In 1887, Oscar Wilde wrote: 'We have really everything in common with America nowadays except, of course, the language,' and there are numerous examples where military terminology differs. In spoken American English one cannot hear the difference between 'route' (road, direction) and 'rout' (pursue, hunt, track down) and quite a few notions are designated by different words (such as "2 iC"/"executive officer," "aerial"/"antenna," "brigadier"/"brigadier-general," "commander"/"leader," "commando"/"ranger," "mobile"/"cell phone," "petrol"/"gas," "pw"/"pow," "section"/"squad"). The same trouble may arise when Austrians, Swiss and Germans try to communicate in the German language (such as "Brigadier"/"Brigadekommandeur"/"Brigadegeneral,"/"Divisionär"/"Generalmajor"/"Divisionskommandeur," "d.G."/"i.G.," etc). Moreover, one coalition partner might see the other as wanting in sophistication for the simple reason that his counterpart's cultural heritage is different from his own (different upbringing, manners, dress code, courteousness etc).

For example, prior to World War I the German military advisors to the Ottoman Empire sent home periodic reports with vivid descriptions of what they perceived as Turkish barbarism. These sentiments, conveyed by a handful of officers, pervaded the German military, and when German troops were massively deployed in Turkey most of them were already infected with a strong cultural bias. Prussian military culture, which had permeated most German armies since the formation of the 2nd Reich in 1871, had always emphasised virtues like precision, discipline, orderliness, detailed planning, efficient staff work, cleanliness and effective management. As these qualities were not strongly emphasised in the Ottoman army the

---

[65] Peter Tomsen, *The Wars of Afghanistan: Messianic Terrorism, Tribal Conflicts, and the Failures of Great Powers* (New York: PublicAffairs, 2011) p: 66.

impression of lacking enthusiasm, laziness and inefficiency conveyed by German authors and military advisors seemed to be vindicated. Consequently, animosity spread on both sides and contributed to poisoning the co-operation. In any event, the Germans had come to Turkey in order to promote German war aims. Similarly, the Turks had their own agenda and as long as the co-operation was seen to be beneficial to them they were willing to tolerate the superiority of German military theory and practice. However, sensibility to the partner's national pride was not amongst the dominant characteristics of German military tradition, and the Germans did little to accommodate their coalition partner's understandable self-esteem needs. Moreover, it was noteworthy that most German officers and men, who had been transferred to the Ottoman Empire, knew no Turkish at all and did not bother to learn. Being one of the very few, who did, General Liman von Sanders recognised that it was vital to communicate orders in good and acceptable Turkish, but most of his fellow German officers paid little attention. Moreover, the few Germans who actually spoke the language antagonised their Ottoman subordinates as well as their superiors by using badly chosen terminology.

This lack of cultural empathy led to scepticism towards the coalition partner's actions and, therefore, to failing support where this was most dearly needed: insight into the norms, languages and geography specific to the area of operations. As a result, an expedition commanded by Bavarian Lieutenant Oskar Niedermayer got off completely on the wrong foot. Niedermayer had been instructed to make his way from Iraq to Kabul and lure the Amir of Afghanistan into the war on the German side, but whenever local support was required, the Turks turned out to be either half-hearted in their support or simply refused to co-operate. As a result, most of his subversive activities were either delayed or never got off the ground. Lieutenant Niedermayer actually reached Kabul and eventually got back to Turkey, but in the process most of his associates were captured by the British.

Conversely, there was much more honest effort put into the Anglo-Japanese co-operation during the World War I siege of the small German colony Kiaochaw on the Shantung peninsula – not least by Japan. American researcher Gal Luft observed that:

> By and large, the Japanese exercised a great deal of tolerance and respect towards the British... The Japanese General Staff gave the Japanese commander Kamio specific instructions to place maximum effort in obliging British needs. He was advised to be attentive to issues like language, race, customs and religion. To make the British feel like equal partners [though they were actually a very small contingent], the Japanese commanders made sure

that British flags were placed alongside Japanese flags at coalition bases and positions.[66]

Nonetheless, communication between these two coalition partners, though all right at the senior level, caused troubles in the field, where an inability to communicate resulted in misunderstandings. Interpreters had difficulties in understanding dialects, especially those of the Sikhs and the Welsh, and occasionally British officers and men were taken to be Germans. This problem was solved, however, when the British began to wear distinguishing marks and take along interpreters of their own.

But the Anglo-Japanese co-operation entailed other cultural challenges. The Japanese were obsessed with security to the extent that they did not trust their coalition partner, and therefore rarely kept them informed on intentions for future operations to the point of forbidding communication with the British legation or other authorities outside the area of operations. Moreover, contrary to British military culture, which emphasised written orders, the Japanese relied on oral communication, using no written documents. Operation orders were supposed to be memorised, which was a totally unfamiliar practice to British officers used to receiving written orders. Moreover, while the Japanese perceived public showing of industry as undignified, the British behaved differently. They were businesslike and were not embarrassed to be seen as being occupied or even distressed.

Display of personal courage mattered a lot more to the Japanese that it did to the British. The Boer War had shown that face to face combat had given way to the use of machine guns, artillery and snipers, and unjustified personal risks were to be avoided. Not so for the Japanese, who found it gutless to shun personal danger. They had little appreciation for measures that British officers felt inclined to take as logical precautions to save lives, combat strength and equipment (such as not exposing their troops unnecessarily to enemy fire, going on the defensive when outnumbered or becoming a prisoner-of-war rather than facing annihilation). Nonetheless, although Japanese officers might have seen their British colleagues as spineless and weak products of a sheltered upbringing, and for this reason scorned them, this was not reciprocated. Unlike the Germans who spurned the Ottoman soldiery, most British officers and men held their Japanese coalition partner in the highest esteem, and it merits some interest that different views on religion did not seem to influence mutual understanding. Christian norms such as clemency in victory and care for one's soldiers'

---

[66] Luft, Gal. Beef, Bacon and Bullets: Culture in Coalition Warfare from Gallipoli to Iraq. (no place: Gal Luft, 2009), p. 85.

survival and well-being were not ideals shared by the Japanese 'State Shinto' religion, which sought to link religious belief with a militant nationalism hailing personal sacrifice above more earthly interests. However, these apparently incompatible beliefs did not lead to clashes among the allies.

Personnel

While in a coalition between great powers each of the partner states will normally provide its own complement of personnel, a coalition with a dominant lead-nation is different. This is especially the case when lesser coalition partners do not possess the full spectrum of skills or capabilities and have to be supplemented, mentored and supported to meet the common doctrine. An imperial defensive alliance will normally have a considerable need for lead-nation support to coalition partners – technologically and economically as well as with regard to instructors and staff officers.

Again, Canada's contribution to the defence of the British Empire's war aims provides illustrative examples. The mobilisation of 1914 demonstrated that Canada was still immature as an independent country with an equally unripe army. World War I, however, would contribute greatly to allowing the Canadian army to expand and mature. When the First Canadian Division arrived in Britain in late 1914, it was placed under the command of Major-General Edwin Alderson, a British Army officer of considerable experience. The General Officer Commanding was not a Canadian for the simple reason that Canada did not yet have officers with the requisite military experience and staff training for divisional command at that time. General Alderson arrived at First Canadian Division and brought along with him his own General Staff Officer First Grade (GSO 1) and one staff-trained major to act as the Brigade Major to the First Canadian Infantry Brigade.[67] But, by mid-1916, the number of battle experienced officers had increased, and when the Canadian Expeditionary Force (CEF) expanded into a corps of four-division, this resulted in a boost to the number and proportion of Canadian staff officers in the Force. The Canadians were keen to occupy as many command positions as possible within their own organisation. To make that happen successfully staff training was needed, and the British Army provided a sizeable complement of staff officers to mentor the new Canadian formations – a procedure which has found a

---

[67] Library and Archives Canada (LAC), RG 9, III, Vol. 871, Lists Imperial Officers in Canadian Corps (n.d).

modern equivalent in the mentoring process benefitting the Afghan National Security Forces of 2011. Over the years of war eighty-four British officers were attached to the various headquarters of the Canadian Corps. While British staff officers held most of the important staff appointments until late 1917, they were gradually replaced by the Canadians whom they had taught. However, right to the end of the war there were British staff officers employed with the Canadian forces, and Lieutenant-General Sir Arthur Currie, the first Canadian to command the Canadian Corps, retained British officers as his Brigadier General Staff and as his Assistant Adjutant & Quartermaster General.

This lead-nation seconding of mentors, staff officers and commanders was not an exclusively British phenomenon. The French officered their colonial troops and *la Légion Étrangère* – the Foreign Legion – and, as we have seen, a huge number of Germans served as military advisors to, and staff officers and even commanders of various units and formations of the Ottoman Empire – such as did for instance General Liman von Sanders at Gallipoli 1915. In most cases this worked satisfactorily, but not surprisingly to some degree national independence and pride of the lesser coalition partners were sacrificed.

World War II

Coalition problems in World War II were to be found partly in the same places as during the Great War, partly elsewhere. Within the British Empire, Dominions were not necessarily always in full agreement with London and domestic considerations within each Dominion State played their part in determining the degree of willingness to go along with British strategic choices. Amongst the western Allies there were disagreements not only among the states but also between the leading politicians and their top military practitioners. On the Axis side disparity of volume and capabilities of Italian, Hungarian, Romanian, Japanese and German forces, as well as the Italian quest for glory with as little effort as possible, made the Axis an uneasy coalition.

Strategically, there were conflicts of interests between Britain and the Soviet Union. While in 1941-2 the USSR wished the UK to declare war on Finland and accept revision of the Russo-Polish border, Britain's strategy aimed exclusively at defeating Nazi-Germany.

Politics

The opposing coalitions of World War II were generally alliances of great powers, albeit both supported by a number of smaller contributors. Although in the beginning, the British Empire fought alone against the major Axis Powers, the reality of the struggle was that an increasing number of states and 'militaries in exile', such as those of Poland and Norway, joined. In 1941, the Soviet Union and the United States, both coming under attack themselves, became parties to the wartime coalition, which from 1 January 1942 came to include twenty-six allies assuming the designation of 'the United Nations.' On the opposing side, during the 1930s a very threatening coalition of totalitarian states, the Axis Powers, had been aligning, and when war came they materialised as a wide-ranging menace to the states rallying around Britain.

While up to the start and in the early phases of the war the actual degree of intimacy and shared aims of the Axis states may be questioned, the fact remains that they appeared to pose a combined danger and to be working with each other.[68] Moreover, as soon as hostilities broke out it became obvious that this war could not be fought and won by a single state or empire, and during the six years of its duration it dawned upon politicians of both warring sides that coalition warfare was the single most adequate answer to their needs. From March 1941 onwards the world saw a long period of wartime coalitions and peace time alliances materialising.

In March 1941 the United States' Congress approved the Lend-Lease Bill, by which the British Empire got access to vital supplies provided by the United States' huge stocks of raw materials and its industrial capacity. Without this the British Empire would slowly but surely have ceased fighting, and for this reason the Lend-Lease Act may be seen as the budding recognition in the United States that, willingly or not, they were bound to defend the values of liberal democracy in a coalition with Britain. Moreover, with hindsight it is clear that even before the military alliance between Britain and the United States had become reality, there was already an economic coalition. Furthermore, at the first Roosevelt-Churchill summit at Placentia Bay in August 1941 – and even before then, at the ABC Conference – there was a preliminary understanding of a joint strategy.[69]

While the World War II alliance between the British Empire and the United States was a coalition formed with the aim of defending common values, ideals and a way of life, the world view of these two powers were disparate in many other respects. However, though there were significant

---

[68] Kennedy, "Military Coalitions and Coalitions Warfare over the past Century," p. 11.

[69] The ABC Conference took place in Washington, D.C. form 29 January to 27 March 1941, when American, British, and Canadian military staff initiated basic planning for possible US entry into the war.

political, and to a certain extent ideological, differences between the British Empire and the United States over India and the Asian colonies and over the policy towards Nationalist China, there was agreement on the supremacy of democracy over despotism.

At the strategic and operative levels there was even more incongruity. Churchill's peripheral preferences clashed with the Americans' desire to open a 'Second Front' in France as soon as possible, his wish for concentration of effort once Operation Overlord was launched as opposed to the American insistence to press ahead with the planned landings in southern France, Operation Dragoon, and various other key strategic choices often threatened to do lasting damage to the coalition. Moreover, not all the services were coalition minded, and in particular the American Admiral King was perceived by many European allied commanders and politicians as being obstructive.

While amongst the western allies differences were about minor – or at least nonessential – issues, there was a fundamental discrepancy in war aims between East and West. The sides agreed on the overarching aim of defeating Germany first. The Soviet Union fought to bring about the speediest possible eradication of Fascism just like the 'bourgeois governments in London and Washington' seeking the elimination of Nazi-Germany and Fascist Italy. Nevertheless, in the West many perceived the Soviet Union's primary aim to be weakening its 'imperialist' rivals and expanding its influence world-wide.[70] Very soon after Britain and Russia became co-belligerents, the question of a Second Front in France developed into a rock of Anglo-Russian discord. In spite of huge British transports of equipment and gold to Russia, along with the rapid consumption of supplies and the attrition of personnel in North Africa, the Russians could not understand why the British would not immediately throw all their small armies across the Channel and provide some relief.[71]

Each of the two opposing coalitions had problems of its own. While in the western, British-led World War II coalition Canada felt squeezed as a lesser partner both in terms of the size of her population and because of the French-English divide, Italy's rôle as an Axis power pretending to be great was of a somewhat different nature.

Italy had started her war completely on the wrong foot after being defeated by the French in the 'Alpine Line' and at the town of Menton near

---

[70] John Erickson "Koalitionnaya Voina: Coalition Warfare in Soviet Military History, and Performance" in Keith Neilson and Roy A. Prete, Eds., *Coalition Warfare: an Uneasy Accord* (Waterloo, Ontario, Canada: Wilfrid Laurier University Press, 1983) p. 110.

[71] Jenkins, *Churchill*, p. 661.

Nice.[72] While, until then, the French Army had fought alongside the British Expeditionary Corps in a vain attempt to stop the German onslaught in the North, and had a small force engaged in Norway, the Italian invasion was halted quickly and with moderate French losses. Moreover, Italy's attempts to conquer Egypt did not go too well either, eventually necessitating German reinforcement and leadership.

Italy had managed to defeat the Abyssinians in 1935-36, but Italian society and the Italian armed forces were in no way tuned to the requirements of modern armoured warfare. Italy's forces were relatively small and their equipment was not up to the standards of the British Empire. In 1940 Mussolini had hoped to be able to establish Italian hegemony in the Mediterranean basin and North Africa, but while British air and naval forces wore down the Italian navy and commercial shipping, Commonwealth troops based in Egypt drove Italian army forces back through Libya in February 1941 and forced the Italian garrisons of Abyssinia and Somaliland to surrender in May.

Amongst the primary troubles with the Italian military machine was the lack of a military culture and traditions – except perhaps in Piedmont-Savoy, which had been the leading power and military backbone of the *Risorgimento,* the 19th century movement to unite Italy. The Italian historian Mario Montanari claims that "Italy has never seemed well-inclined towards arms... In reality one must admit that Italy has never been well-inclined towards the state, whatever its leadership."[73] Recent experiences had been entirely negative: defeats during the *Risorgimento* by the Abyssinians in 1896 and at Caporetto in 1917. In spite of the notorious Fascist-futurist fascination by machines, speed and elitism of the 1920s and 30s, Italian society suffered from a universal lack of military expertise and enthusiasm in military affairs. The government had neither a military policy nor any inclination towards directing the development of the armed forces. The ultimate power remained in the hands of the king and the *Regio Esercito* [the royal army], the only power capable of throwing Fascism out. Mussolini, well aware of his own military shortcomings and anxious to uphold the popular impression of his dictatorial infallibility, habitually referred military matters to the professionals. He tended to intervene when their internecine quarrels became too heated, but avoided procurement and

---

[72] At the Côte d'Azure the Italians were held up by one French NCO and seven soldiers.

[73] Mario Montanari, *Politica e strategia in cento anni di guerre italiane*, Vol I, Rome 1966 in MacGregor Knox, *Hitler's Italian Allies: Royal Armed Forces, Fascist Regime, and the War of 1949-1943* (Cambridge: Cambridge University Press, 2000), pp 29-30.

organisational matters, however needed.[74]

Italy's issues notwithstanding, the search for the key troubles of the Axis coalition must start with a closer look at the general character of the Italo-German collaboration. The deficiencies of the Axis coalition emerged as early as 1939, three years after the promulgation of the Berlin-Rome Axis, as Mussolini called it. In May 1939, Nazi-Germany and Fascist Italy confirmed their special bond with a political and military treaty on mutual military assistance: the so-called *Stahlpakt* [Pact of Steel]. At the same time, high level military consultations began. No sooner had the *Stahlpakt* declaration been made than Hitler expressed his scepticism concerning the reliability and efficiency of the Italians. Issuing his *Weisung für die Kriegführung 1a, Fall Weiß* [War Directive No 1a concerning his plans for war against Poland], he explained to the inner circle of *Wehrmacht* leadership that for security reasons he intended not to inform the Italians about these plans. His reason for doing so seems to have been that, at that juncture, his confidence in his ally had been severely shaken by Mussolini's unsolicited and obstructive interference during the Czechoslovakia crisis in 1938. Moreover, the *Duce*'s comments on the Pact of Steel made Hitler realise that Italy would not be ready for war in the near future and that she lacked the necessary zeal for risking a major war at Germany' side. As if this was not enough, in early September 1939 Hitler received two unpleasant surprises. Not only did Britain and France, in March 1939, guarantee Poland's borders and integrity; they also honoured their pledge by first issuing an ultimatum that Germany withdraw her troops from Polish territory then, on 3 September, declared war on Germany. Moreover, on this occasion Mussolini resolutely refused Italian participation in the imminent conflict.

While Germany was understandably distrustful of its Italian ally, Italy, on the other hand, was no less suspicious of its coalition partner in the North. The Italians felt walked over by Hitler's reckless drive into war with two modern and well-equipped great powers, while their own designs were neglected: the aggrandizement of the Italian colonial Empire and the ultimate goal of a 'new Roman Empire.'[75] In particular, the Italian military had reservations about the danger of being instrumentalised by German interests, and abruptly suspended the talks with the *Wehrmacht* leadership.[76]

---

[74] MacGregor Knox, *Hitler's Italian Allies: Royal Armed Forces, Fascist Regime, and the War of 1949-1943* (Cambridge: Cambridge University Press, 2000), p. 31.

[75] Philip Morgan, *Fascism in Europe, 1919-1945* (London : Routledge, 2003) p. 144.

[76] König, *Kooperation als Machtkampf*, pp. 20f.

As a result, despite the fact that Mussolini had indicated his readiness to enter the war alongside Germany as early as in March 1940, as Hitler had sought an Italian military commitment on a combined offensive in the West no fruitful co-operation materialised.[77]

Germany's successful invasion of France in May 1940 reversed the previous situation – German interest in an Italian military commitment declined considerably, and it was now Mussolini who pushed for an involvement as he feared to fall behind in power politics and military glory.[78] When on 20 June, Italy finally entered the war, France had already been defeated, and Marshal Philippe Pétain had made peace with the Germans, who then occupied most of northern and western France, while the smaller part of the country close to the Alps and the Mediterranean remained under control of the new French government in the city of Vichy. Nevertheless, the French forces easily withstood the Italian invasion attempt and only due to last minute German assistance did the Italian campaign – which had been launched belatedly and precipitously across the Alps border and along the Mediterranean coast – not end in total disgrace. With the failure to achieve its political objectives, Italy's inelegant entry into the war caused the nation to lose face with its ally. Not only did Mussolini's insatiable political cravings make a bad impression, but the military dilettantism and inadequacy of his forces was a particularly humiliating revelation.

On top of this obvious flop, throughout the late 1940 further events completely unhinged the fragile inner balance of the Axis coalition. In the summer and autumn of 1940, and after the bad impression of his country's martial endeavours, Mussolini had eagerly desired to maintain Italy's reputation as a major power and to remain on a par with Germany. Aiming to fulfil his ambition of playing a key rôle in the European war, Mussolini forced Italian submarine and air forces upon his German ally in order to fight the British in the Atlantic and on the Channel coast. Nonetheless, Italian politicians had their primary focus on the idea of parallel warfare independent of the German activities in northern Europe, a notion developed by Mussolini in the spring of 1940. According to that concept, Italy should wage its own separate war, simultaneously with and independently of the unpredictable and un-beloved German ally's activities in northern Europe. Mussolini's cravings for expansion concentrated on the Mediterranean and North Africa, where Italian strategy-makers believed

---

[77] Ibid., pp. 21-26.

[78] Ibid., pp. 24f.

they had a particular obligation to promote Romano-centric prosperity, a notion which they had developed long before any agreement with Hitler had been conceived. However, the over-estimation of his own forces' capabilities soon resulted in the well-known military failures against Greece in the autumn of 1940 and against the British in North Africa in the winter of 1940-41. This reckless strategic adventurism harmed the relationship with the Germans. Although, due to the upcoming *Unternehmen Barbarossa* – the invasion of the Soviet Union – Hitler felt obliged to interfere in both these theatres of war, not only to avoid the imminent defeat of Italy, but also because Mussolini's Fascists required political stabilisation. Mussolini was forced to accept German assistance and German warfare expanded into the Italian sphere of influence, where Mussolini thought that his parallel war ought to have remained an exclusively Italian enterprise.

In February 1941, the ultimate humiliation materialised – the *Deutsches Afrikakorps* under Lieutenant-General Erwin Rommel began arriving in Italian Libya. The Italians had been chased out of Egypt, Abyssinia and Somaliland, they were under siege by the British, and the German intervention, therefore, marked a significant albeit temporary turning point in the record of the Axis.[79] But, through this involvement the Third Reich had initiated a creeping political guardianship of the Fascist regime in Rome, where the imperatives of wartime did not seem to have had any particular influence on the daily routines. Actually, as late as in 1943 Mussolini, despite entreaties from the Italian armed forces, maintained resistance to all attempts of imposing military discipline on the industrial workforce, and German Field-Marshal Albert Kesselring publicly expressed his worry about the peacetime working methods still prevalent in the Italian civilian dockyards.[80]

In the western desert things went from bad to worse for the Italians. Immediately upon his arrival, Rommel took over supreme command of all troops in the theatre and – arrogantly it might be said – completely ignored his Italian fellow generals. In many cases he did not even bother to inform them of his major operative initiatives. As a result, until the overthrow of Mussolini in the autumn of 1943, Italy was almost completely dependent on German power and a lesser partner whose only option was to follow the

---

[79] This turning point already can be seen with Hitler's letter to Mussolini on 20 November 1940. As a result of Italy's disastrous defeat against Greece, the letter shows a new level of communication between the two allies; cf. König, *Kooperation als Machtkampf*, pp. 38f.

[80] MacGregor Knox, *Hitler's Italian Allies: Royal Armed Forces, Fascist Regime, and the War of 1949-1943* (Cambridge: Cambridge University Press, 2000), p 33.

lead nation.

In early 1942, the Axis powers had had a rare opportunity to redirect their strategy.[81] The Red Army's unexpected resistance outside Moscow had forced the *Wehrmacht* on the defensive in its central and most crucial theatre of war. By contrast, the other main opponent of the Axis, Great Britain, had come under severe pressure from Rommel's campaign in North Africa and from the succession of Japanese victories in the Far East. For Germany and Italy, the prospect beckoned of expelling the troubled British Commonwealth forces from the Mediterranean and the Middle East – with potential far-reaching consequences for the global war.

The British island bastion of Malta posed the main threat to genuine Axis success in North Africa, however. Naval and air forces based on this isle hindered the transport of supplies across the Mediterranean to the Italo-German African Army. The Axis Powers therefore initiated a close siege of Malta running uninterruptedly from 1940 to 1942. During the siege, the fight for the control of the strategically important island pitted the air forces and navies of Fascist Italy and Nazi Germany against the Royal Air Force and the Royal Navy. The opening of a new front in North Africa in mid-1940 increased Malta's already considerable value. British air and naval forces based on the island could attack Axis ships transporting vital supplies and reinforcements from Europe. The Axis resolved to bomb, or starve Malta into submission by attacking its ports, towns, cities and Allied shipping supplying the island.

Success would have made possible a combined Germano-Italian amphibious landing. For this reason, the long harboured intentions of conquering Malta matured within the Italian military leadership, although they dared not carry out this operation on their own. Only in the spring of 1942 – when their German ally also seemed ready – the preparations for a combined and joint operation began. The Germans named the operation *Herkules* (Operation Hercules), whereas the Italians used the nickname *C3*.

Although the joint approach to Operation Hercules was quite promising, the Axis co-operation was cursed from the very beginning and therefore highly complicated. Eventually, the operation was not carried out – in fact, it was given up in the early planning stage because of political, strategic and operational difficulties. The preparations ended quietly after several months of hectic activity in the summer of 1942. The Axis powers failed to seize Malta at the crucial moment, a strategic blunder that caused German and

---

[81] The following paragraphs on Germano-Italian co-operation on Operation Hercules are largely due to a paper by Thomas Vogel presented at the conference on coalition warfare in Copenhagen, 2011.

Italian makers of strategy much regret, as the course of the war in the Mediterranean offered no further opportunities to take the island.

Although Operation Hercules never went beyond the planning stage, it deserves attention as an indicator of the shortcomings of Axis collaboration. It highlights certain noteworthy characteristics of the Axis coalition warfare during a crucial period of the Second World War. Although the eventual failure of the Italo-German common strategy did not occur until, in the autumn of 1943, Italy turned its back on Germany, there were much earlier precursors of the eventual outcome to be found in the structural and fundamental deficiencies of the Fascist-Nazi coalition warfare.

On the Allied side, during the early war years the troubles of coalition warfare were slightly different. The necessity of keeping the allies on board had to be juxtaposed with the problem of getting the political leadership to see not only the limitations of their military capabilities but also where the opportunities lay. This was the daily challenge to the military professionals – not least to General Alan Brooke, later to become Field-Marshal Lord Alanbrooke. As Chief of the Imperial General Staff (CIGS) Brooke had to argue continuously with Prime Minister Winston Churchill, who in his own imagination was as great a strategist as his great forbear, the first Duke of Marlborough, had been during the War of Spanish Succession.

In early December 1941, when Russia had entered the war and the United States still had not, Brooke had recently taken office as CIGS and saw it as his primary objective to "get the PM to see the advantages of a real North African policy."[82] Bearing in mind Britain's dependence on oil from the Middle East and communication with the far East, and realising that the extra tonnage required for shipping round the Cape of Good Hope would not be available, he was adamant that concentration of the limited capabilities of the British Empire was essential in order to keep the Mediterranean open as a supply line for the Allies. Churchill, however, was keen to please the Russians and wished to dispatch to the Soviets two divisions that could hardly be spared in the desert. Moreover, as if the struggle with the enemy and a truculent prime minister were not enough, Brooke felt inclined to confess to his diary that the Foreign Secretary Sir "Anthony Eden [behaved] like a peevish child grumbling because he was being sent off to Uncle Stalin without suitable gifts, while Granny Churchill was comforting him and explaining to him all the pretty speeches he might

---

[82] Danchev and Todman, eds., *War Diaries, 193-1945; Field Marshal Lord Alanbrooke*, p. 206 (3 Dec 1941).

make instead."[83]

Both on the political and military level there were considerations on how best to distribute responsibilities amongst the allies, but also worries of what might happen if the allies got too much influence. On 10 March 1942, at the Pacific Council, meeting Churchill announced that it had been decided that USA was to look after the Pacific, London after Burma, India and the Middle East and Mediterranean, while Combined Chiefs of Staff did Atlantic and land operations against Africa. And Brooke was worried: "Good in places but calculated to drive Australia, New Zealand and Canada into USA arms, and help to bust up Empire."[84]

It is self-evident that in alliances amongst great powers it is of paramount importance not to alienate alliance partners, not even the minor ones, a psychological subtlety which Churchill frequently chose to ignore not least when the French – and in particular de Gaulle – was involved. General Brooke was deeply apprehensive about the repercussions of the Prime Minister's orders to the Admiralty on 3 April 1942: "Discovered this morning's COS that First Sea Lord had agreed last night to the PM sending out instructions for the French battleship *Richelieu* to be attacked and destroyed if it attempted to enter the Mediterranean as it was expected to do."[85] Although, at the time, not all French forces could with any certainty be counted as allied, the potential damage done by sinking a major French warship might be far-reaching, and he was alarmed when "We found out in the morning that the PM had refused to withdraw the order to sink Richelieu! Luckily it had not come out."[86] However, notwithstanding the occasional condescending treatment of some allies, and in spite of a permanently bad relationship between Brooke and Churchill on the one side and General de Gaulle on the other, eventually the French were aligned with the Allied cause and de Gaulle was even persuaded to go and see the American partners, as Roy Jenkins wrote:

> De Gaulle saying that he quite understood that Britain would side with the United States. This was portrayed as being at least semi-conciliatory, although it was most certainly with more sarcasm than sympathy... De Gaulle slowly took Churchill's advice to go to America, which he did from 6 to 11 July

---

[83] Ibid., p. 207 (4 Dec 1941).

[84] Ibid., p. 238 (10 Mar 1942).

[85] Ibid., p. 244 (3 Apr 1942).

[86] Ibid., p. 245 (4 Apr 1942).

[1944]. He got on tolerably well with Roosevelt in Washington, better paradoxically than he had with Churchill in Hampshire.[87]

On the other hand, although allies should be pampered whenever possible, no rock solid promises should be made – not even to peer allies – which one could not fulfil with absolute certainty, and Brooke noticed with some satisfaction Churchill's oratorical skill when replying to Stalin on the issue of opening a second front in Europe:

> From the Russian point of view Turkey, Rhodes, Yugoslavia, and even the capture of Rome [all Churchill's pet objectives] were not important. Stalin asked him the direct and hostile question: did he really believe in Overlord? Churchill's reply was not bad, but not great either. It was a mixture of equivocation and rhetoric. Provided the conditions were met, he said, 'it will be our stern duty to hurl across the Channel against the Germans every sinew of our strength.'[88]

Similarly, in the discussion on what to do with the Germans after the war was over, "Stalin said that the German problem could largely be solved by rounding up and shooting the 50,000 foremost officers and technicians. Churchill recorded: 'On this I thought it right to say, "The British Parliament and public will never tolerate mass executions.".. I would rather be shot myself than sully my own and my country's honour by such infamy"'[89]

Strategy

In September 1941, in the only example of a direct Anglo-Soviet combined operation, Iran was occupied by British and Soviet forces, removing a pro-Axis government and securing a transportation base for supplies to Russia. Despite this token of the coalition spirit, collaborative frictions were not slow to appear: British reluctance to declare war on Finland and the issue of Soviet-Polish relations were cases in point. While the USSR wished to push back the border to where it had been prior to 22 June 1941 – i.e. the time when Germany and the USSR had shared Poland between them – at

---

[87] Roy Jenkins, *Churchill,* p. 743.

[88] Ibid., p. 723.

[89] Ibid., p. 723.

this juncture Britain was firmly committed to combat Nazi-Germany and to restore Poland to its former sovereign territory.[90] The disagreement between the two developed as from 20-28 May 1942 the Soviet Foreign Minister, Vyacheslav Mikhailovich Molotov (1890-1986), visited London. Molotov's instructions were to sign a treaty of friendship with Britain based on recognition of Russia's pre-invasion frontiers including into the USSR the parts of Poland and the three Baltic states which had been seized in 1939, and to get a promise of a 1942 Second Front in France.[91] Although Churchill was willing to give in on the border issue to avoid concessions in other more crucial fields, Roosevelt was adamant that there should be no such indulgence. Anthony Eden managed to negotiate an agreement, where the question of frontiers was left in abeyance in return for a twenty-year treaty. This, however, did not settle all disputes with Stalin.

Churchill's conference in Moscow in August 1942 displayed all the tension and division over strategic issues that were to persist: opening of a second front, Poland, convoy politics etc. Soviet anger showed itself quickly intensified because of the Mediterranean strategy and the 'peripheral approach' – attacking Europe through North Africa and Italy.[92] The Soviet position was that it was the USSR that gave the coalition shape and resilience and frustrated the enemy's hopes of defeating the Allies piecemeal. Moreover, Stalin seemed to be aware that the war aims of his allies were not identical: While Britain was worried about her colonial empire, in particular in the Far East, the USA cared about its markets, from which Axis victory would exclude it.

Negotiating with the United States, the British Strategic position remained the same: Germany first, and in this priority there was full harmony between these two. Not only did the political leaderships concur on the war aims, the chiefs of staff also managed to reach concurrence on the overall strategy. The American Chiefs – including the Anglophobe Admiral King – accepted a basic statement of position which had been drawn up by General Marshall and Admiral Stark: "notwithstanding the entry of Japan into the war, our view remains that Germany is still the primary enemy and her defeat is the key to victory. Once Germany is

---

[90] Erickson "Koalitionnaya Voina: Coalition Warfare in Soviet Military History, and Performance," p. 108.

[91] Roy Jenkins, *Churchill*, p. 688.

[92] Erickson "Koalitionnaya Voina: Coalition Warfare in Soviet Military History, and Performance," p. 109.

defeated, the collapse of Italy and the defeat of Japan must follow."[93]

The foundation for the strategic agreement across the Atlantic Ocean was laid at the Arcadia Conference held in Washington D.C. from 22 December 1941 to 14 January 1942. This was, in fact, the first meeting on military strategy between the heads of government of the United Kingdom and the United States following the latter's entry into World War II. The major delegations were headed by the British Prime Minister, Winston Churchill, and the American President, Franklin D. Roosevelt. The Arcadia Conference also had a wider international diplomatic and political aspect concerning the terms of the post-war world, which followed from the Atlantic Charter, agreed between Churchill and Roosevelt in August 1941. On 1 January 1942, the twenty-six governments attending the conference signed a declaration on the 'United Nations,' the framework for the inter-Allied co-operation for waging the war.[94]

Strategically, there were two approaches: a central and a peripheral one. While Stalin maintained the necessity of the central strategy aiming at opening a Second Front across the Channel at the earliest possible moment, Britain, and up to a point, in the early phases of their involvement and with some hesitation, the USA preferred the peripheral approach, attacking the Axis from North-Africa through Italy into the German heartland. Hopeful that the merits of the peripheral strategy might convince the Americans, General Alan Brooke wrote on 24 December 1941: 'Winston has arrived in Washington, far from the war, and is pushing for operations by the USA and ourselves against North Africa banking on further success of Middle East offensive towards Tripoli.'[95] Eventually, the US fully subscribed to the North African approach, but there were serious concerns about the sufficiency of the equipment due to both Soviet demands and the tragic losses in the Far East. Earlier in December 1941, Brooke had told his diary that, 'I said the best we could do [for the Russians] was 300 Churchill tanks by 30th June, and that although this offer might be acceptable to Russia, I did not recommend such a gift as we should be seriously denuding this country and prematurely disclosing a new pattern of tank.'[96] Moreover, he expressed his sincere apprehension about resources, because his hopes of

---

[93] Harry L. Hopkins, "White House Papers" in Jenkins, *Churchill*, p. 670.

[94] Jenkins, *Churchill*, p. 675.

[95] Danchev and Todman, eds., *War Diaries, 193-1945; Field Marshal Lord Alanbrooke*, pp. 213-14 (24 Dec 1941).

[96] Ibid., p. 207 (4 Dec 1941).

carrying on with the conquest and reclamation of North Africa are beginning to look more and more impossible every day. From now on the Far East will make ever increasing inroads into our resources. The loss of the American battleships and the *Prince of Wales* and *Repulse* will take a long time to recover and meanwhile we shall suffer many more losses in the Far East.[97]

It was not only the Allies who faced problems in the Mediterranean-North African theatre of war. At the turn of 1940-41, the Axis now embarked on a new strategic orientation which considerably changed the parameters for the solution of the increasingly virulent problem of Malta. Malta had been the main base of the British Mediterranean Fleet until October 1939, but due to its exposed position in case of war against Italy, the headquarters and major warships were moved to Alexandria. The remaining smaller units, submarines among others, and an increasing number of aircraft on the island made Malta a likely forward base of an offensive British strategy in the Mediterranean. These forces significantly contributed to British naval supremacy, sustaining its position in central Mediterranean and giving the island continued strategic importance. Until 1940-41, Malta had been considered a purely Italian responsibility. This constituted an obvious problem: the well-fortified British base at Italy's doorstep was a liability to Italy as well as to her African dependencies, and menaced the Axis' lines of communication. Moreover, the British bases in Malta, Gibraltar, and Egypt formed safe havens as well as bunkering facilities for the Royal Navy in the Mediterranean. Italy had long considered this a threat to its hegemonic aspirations and imperial notion of a *mare nostre* (our sea).

Initially, the clear distribution amongst the Axis powers of individual spheres of interest had made the Mediterranean theatre out of bounds for German strategic considerations. The head of the *Wehrmachtführungsamt* (Operations Staff), Major-General Alfred Jodl, was the first to cross this invisible line. Searching in 1941 for an alternative, and indirect, approach for the struggle with Britain, Jodl recognised that this theatre of war was a cornerstone in the British peripheral strategy. Soon afterwards, the German Supreme Naval Commander, Grand-Admiral Erich Raeder, decided that the strategic development in the Mediterranean Sea would be decisive for the eventual outcome of the war and he suggested a stronger German commitment in this area. But such considerations did not prevail.

At this juncture, when the fortune of war was still with the Axis Powers, Italy had not managed to find a workable military solution to her

---

[97] Ibid., p. 213 (20 Dec 1941).

Maltese problem. In 1938, Italy had actually considered an attack followed by an all-out invasion as a prelude to an official declaration of war. However, two years later Italy lacked not only the necessary military assets but the political will. No forces were available because the campaign against France received political priority. As a result, at that juncture the opportunity to conquer the island was missed. In 1940 Malta had been poorly prepared to repel a possible Italian attack, but soon thereafter the British reinforced their garrison to an extent offering little hope for the Italians, who were now unable – and indeed unwilling – to risk a joint operation alone.

Right from the beginning of 1941, the German dominance of the Axis coalition was extended to the Mediterranean region, putting the Malta challenge in a completely new perspective. To Germany's strategic decision-makers the Mediterranean, including Malta, was a side-show and of secondary importance. What really mattered was the planning for Operation Barbarossa, the attack against the Soviet Union, where strategically raw materials and a huge work force were to be found. The creation of *Lebensraum* [living space] and economic growth for the Germans was the overall political aim with the war, and Barbarossa was the top priority objective towards fulfilling it. Thus resources could be spared for action in the Mediterranean space only to the extent they were not needed for Barbarossa. At this point in time, the fate of the Malta project became a serious dilemma for the Axis and especially the Germans, because on the one hand the possession of the island would lessen the need for armed forces to protect the lines of communication between Africa and Europe, but on the other an amphibious tri-service operation of that kind would in itself require considerable allocations of troops, fuel, equipment and ammunition.

In February 1941, Germany sent what amounted to an armoured covering force to support the Italians in Libya: a *Panzer-Sperrverband* under Lieutenant-General Erwin Rommel. Initially, the task of this unit was purely defensive and gave no indication as to the future operational successes of the German *Afrikakorps* or the Italo-German *Panzerarmee Afrika*. On the eve of Operation Barbarossa, Rommel's formation was sent to Libya just to stabilise the Italian front, which was the weakest link of the Axis coalition's southern flank. Rommel's dispatch to Africa did not signify a change of strategy concerning Britain – it was a delaying action meant to allow Hitler temporarily to pursue his primary goals on the steppes of the Soviet Union.

However, the development in 1942 in theatres of war as disparate as the Far East, the USSR, and Africa opened up an unforeseen strategic opportunity. While the *Wehrmacht* had spread out its formations on an extremely wide front in Russia, and for that reason failed to break through both outside Moscow and in the Caucasus, Japan had had a resounding success at Pearl Harbour and Rommel's second offensive had come off to a promising start. The *Kriegsmarine*, the German navy, embodied the global dimension of German strategy. With the prize of the Suez Canal in mind, the dreams of the maritime strategists focused on pushing beyond this strategically important objective to link up with Imperial Japan in the Pacific. In February 1942, Grand-Admiral Raeder tried several times to persuade Hitler to approve a strategy focused – rather than on Moscow and the USSR – on the British Empire as being the true opponent or, at least, an opponent who would have to be defeated as a prerequisite to achieving the primary German war aim: *Lebensraum* in the East.

Had this development been prudently exploited, it might have allowed Germany to steer the war into leaving Britain only one choice: to accept a peace agreement. By April 1942, the Axis' situation in North Africa was developing in a favourable direction and the fate of Malta assumed renewed attention. Exploiting the success in North Africa required secure lines of communication, and within the German chain of command there were supporters of pushing ahead with the Italian plan for seizing Malta and neutralising its British garrison. Hitler had to make a decision on the Malta enterprise, and on 18 April he finally approved the operation. However, Hitler's resolve on the operation did not last long, as on 29/30 April 1942 at a Germano-Italian summit at Castle Klessheim near Salzburg he rescinded his earlier decision. Somewhat illogically, he now postponed Operation Hercules until such a moment when final victory had been achieved over the British in North Africa – or, in reality, he was effectively abandoning the plan.

On closer examination, this change appears far less inconsistent and contradictory than it seems. Despite the fairly rapid change of position, these choices were understandable given the specific strategic situation in the spring of 1942. The need for troops on the continent of Europe juxtaposed with the prospect of success in Africa put Germany in a strategic dilemma. On the one hand, a British defeat was beckoning in North Africa, which would be highly auspicious for the Axis. This, however, would demand the greatest possible support for Rommel, including securing the supply routes. On the other, Hitler doubted the viability of the plans for the conquest of Malta, particularly since he was justifiably suspicious of the efficacy of the Italian ally. Therefore he refused to allocate the extensive personnel and equipment resources needed for Operation Hercules, as such an allotment would happen very much at the expense of the upcoming summer offensive in the USSR or at the cost of the operational readiness for defence against possible allied landings on the Channel coast. On 18 April, the German Commander-in-Chief South, *Generalfeldmarschall* [Field-Marshal] Albert Kesselring, had managed to persuade Hitler because he had characterised the operation against Malta as merely a raid. Such an operation would have required relatively few German forces and might have been carried out rather quickly as Italian preparations appeared to be far advanced. Moreover, Kesselring's recommendation referred to consultations with Rommel, which made a deep impression on Hitler, who was made to believe that Malta was ripe for an attack because of the Germano-Italian air raids that had lasted for weeks. One further detail in favour of Operation Hercules was that the political situation in Italy demanded a boost to Mussolini's prestige.

Two weeks later, however, Hitler realised that, in fact, the situation was far from as favourable as predicted on the 18th. The Germano-Italian summit in late April offered Hitler a convenient occasion for revoking his earlier decision, overruling the Italian political and military leadership. This change of policy signified Hitler's response to the Italian supreme command's demands for extensive German personnel and equipment for the operation. At this juncture the Germans realised that the Italians possessed none of the means needed for a surprise combined and joint attack against Malta, for which the Italian *Comando Supremo* now realised that a large-scale landing operation against the strong British defences was essential. Understandably, Hitler chose to delay the operation indefinitely, although the planning was allowed to continue for some time. Thus, the German strategic choice remained a land-locked one as opposed to the *Kriegsmarine*'s concept of joining hands with the Japanese – the offensive in the Soviet Union retained top priority with the North African theatre coming second.

Unlike the rather clear-cut strategic decision-making in Germany, where the *OKW* (*Oberkommando der Wehrmacht* [the defence's supreme command]) as well as the service chiefs were directly subordinated to Hitler, the Italian strategy in 1940-43 was an unstable amalgam of Mussolini's sudden inspirations and more level-headed calculations by the general staff and service chiefs. It was characterised by myopia, dissipation of effort, passivity, logistical inefficiency and growing dependence on the Reich. Poor intelligence and lack of strategic imagination led to underestimation of British capabilities and the American potential and intentions in 1940-41.[98]

However, the Italian Army Intelligence made a major contribution to Axis war efforts when, in June 1942 they secured the US Army's Black

---

[98] MacGregor Knox, *Hitler's Italian Allies,* pp. 71-2.

Cipher Book through a clever embassy burglary. Then, decrypting a message from the US military attaché at the Cairo embassy analysing the British status of forces after the fall of Tobruk, the intelligence service greatly facilitated Rommel's final advance towards Egypt.[99]

Subsequent to the Japanese attack on Pearl Harbour, only a few in the Italian leadership saw the writing on the wall. The king, whose strategic horizon was defined by the traditional ambitions of the House of Savoy, mostly concerning Nice/Nizza and Corsica, was pleased with the early Japanese victories. Mussolini was delighted to be able, on 11 December 1941, to declare war on the United States without even discussing the consequences with the king and Marshal Ugo Cavallero, who had since 6 December 1940 been Italy's Chief of General Staff, or with the service chiefs.[100] By 1942, the king as well as the chief of the army's general staff, General Vittorio Ambrosio, believed that Russian collapse was near and would cause Britain and the US to sue for peace. Only Count Gian Galeazzo Ciano and his Foreign Office were convinced that, after US intervention, the Axis would ultimately be defeated, but their misgivings had little influence on political or military strategy.[101]

Perhaps the greatest strategic weakness of the Italians was their failure to estimate the intentions of their ally. Until the last moment Italy was totally unaware of the plans for Barbarossa. Then, as well as later, Italian assessment of Germany rested on the 'rational actor model' that states that a decision-maker will seek to optimise his war efforts in terms of the observers' own logic. In this context, however, the Nazi-racist reasoning totally evaded the Italians. That this ideological ferocity, which went beyond anything perpetrated by Fascism, in fact determined Germany's actions against the Soviet Union, as well as against the Jews, was a reality which only very gradually dawned upon the Italian leadership.

Dissipation of effort was perchance the epitome of Italian strategic folly. Mussolini's obsession with proving Italy a useful ally before she was ready and his continuous dispersal of forces precluded concentration of effort at any strategic centre of gravity. In the Mediterranean war such a centre of gravity could only be Egypt, the swift conquest of which would

---

[99] Ibid., p. 72.

[100] Marshal Ugo Cavallero, 1880-1943, was Chief of the Italian General Staff from 1940-43.

[101] MacGregor Knox, *Hitler's Italian Allies,* pp 74-75. General Vittorio Ambrosio, 1879-1958. Il conte Gian Galeazzo Ciano, 1903-44, was Foreign Secretary and son in law to Mussolini.

allow an advance towards the oil wells around the Persian Gulf.[102]

After Rommel's final failure to break the British positions at El Alamein in early September 1942, the *Commando Supremo* watched the disaster that was approaching as if they were mesmerised. In mid-November 1942, General Montgomery's victories in North Africa and the Anglo-American landings in West Africa did not manage to compel the *Duce* and the *Commando Supremo* to deviate from their hopeless belief in the 'principle of strategic dispersal.' They committed seven divisions to occupy southern France alongside the Germans, they never agreed on withdrawal of their divisions from Russia to defend the homeland, and they deferred the necessary decision of giving up North Africa completely. The increasing Anglo-American naval and air superiority in and around the Mediterranean was not allowed to influence Italian determination to fight a delaying action in Tunisia. The Fascist leadership refused to face the realities of the force ratio in North Africa, which clearly demonstrated the Italian need to disengage, and at a conference at Feltre near Venice on 19 July 1943, Mussolini therefore failed to make it clear to Hitler that Italy would have to leave the war. That failure at last moved the king to remove the dictator on 24/25 July, which in turn led to swift German intervention.

It is one of the ironies of the Mediterranean development that, in 1942, the British were so hard pressed that a raid against Malta might actually have succeeded. On 14 April, close to the date when Hitler first decided to press ahead with Operation Hercules, Alan Brooke wrote in his diary, "We were desperately short of shipping and could stage no large scale operations without additional shipping. ... To clear the Mediterranean, North Africa must be cleared first."[103] Moreover, in the summer 1942 the Allies were in no position to take the pressure off the Russians, as "Marshall admitted that he saw no opportunity of staging an offensive in Europe to aid the Russians in September. ...that season was such as to make cross-Channel operations practically impossible."[104]

While the Germans had trouble deciding whether to pursue a continental or a global strategy, the Allies were divided not only on the issue of a peripheral or a central approach, but also on the timing of their initiatives. During the American chiefs-of-staff's visit to London in July 1942, Brooke noted, "At 11 am met American Chiefs again. They handed

---

[102] Ibid, p 77.

[103] Danchev and Todman, eds., *War Diaries, 193-1945; Field Marshal Lord Alanbrooke*, p. 248 (14 Apr 1942).

[104] Ibid., pp. 283 (21 Jul 1942).

in written memorandum adhering to an attack on Cherbourg salient as the preliminary move for an attack in 1943. The memorandum drew attention to all the advantages, but failed to recognise the main disadvantage that there was no hope of our still being in Cherbourg by next spring."[105] Brooke put the disadvantages before the American chiefs, who then wished to put the matter before their president. Fortunately, the impasse was overcome as "Roosevelt wired back accepting that western front in 1942 was off. Also that he was in favour of attack in North Africa and was influencing Chiefs in that direction."[106]

In January 1943, the Casablanca Conference managed to bring together not only British and American decision-makers but also French top military leaders. This conference became famous for two things: first, it proclaimed the unwise doctrine of unconditional surrender of the enemy countries, quite likely stiffening the German resolve to fight to the bitter end, and, second, it procured a reluctant semi-reconciliation of Generals Giraud and de Gaulle, easing the passage into French North Africa.[107]

However, military top-level disagreement reappeared during the spring and summer of 1943, when the Americans pressed for larger allocation of forces and equipment to the Pacific theatre as well as for an early cross-Channel operation. The Trident Conference, involving heads of government as well as military staffs, was conducted in Washington on 12-27 May 1942. On 13 May, Brooke noticed that:

> Leahy began by stating the American conception of the global strategy, which differs considerably from ours in two respects: First in allowing too much latitude for the diversion of force to the Pacific. Secondly by imagining that the war could be more quickly finished by starting a Western front in France. It is quite clear from our discussion that they do not even begin to realise the requirements of European strategy and the part that Russia must play.[108]

And it was not only Leahy who caused Brooke serious concerns, as:

> It was quite evident that Marshall was quite incapable of grasping the objects

---

[105] Ibid., pp. 283 (22 Jul 1942).

[106] Ibid., pp. 284 (23 Jul 1942).

[107] The doctrine turned out to be unwise because it stiffened the German resolve to fight on to the bitter end. Roy Jenkins, *Churchill*, p. 705.

[108] Danchev and Todman, eds., *War Diaries, 193-1945; Field Marshal Lord Alanbrooke*, p. 403 (13 May 1943).

of our strategy nor the magnitude of operations connected with a cross-Channel strategy... To me the strategy which I had advocated from the very start, and which was at last shaping so successfully, stood out clearer and clearer every day. It was therefore not surprising that at this period my temporary inability to bring the Americans along with us filled me with depression, and at times almost with despair.[109]

Also the strategy for the Far East stumbled into heated arguments:

> we did not agree with their paper on the global strategy. I then had to make a statement on the Anakim operation [in Burma] and various alternatives. Stilwell disagreed with most of what Wavel had said and we left the problem more confused than when we started. He [Stilwell] is a small man with no conception of strategy. The whole problem seemed to hinge on the necessity of keeping Chiang Kai-shek in the war.[110]

While there was as much personal animosity as there were strategic differences in the trouble faced by Brooke at this meeting, it should be kept in mind that both of the personalities mentioned in his diary entry were decision-makers at the strategic level in important theatres of war. General Joseph Warren Stilwell (1883-1946) was a United States Army general known for service in the China-Burma-India Theatre and for his caustic personality, which was reflected in the nickname "Vinegar Joe." Lord Archibald Percival Wavell (1883-1950) was the British field-marshal who had been commander Middle East and Commander-in-Chief, India, until, in 1943, he was appointed Viceroy of India.

However, "during Trident the nearest approach to discord was caused by American (Marshall) suspicion that while Churchill was theoretically in favour of direct assault on France, his real hope was that German power might be worn down by peripheral attrition to the point of collapse, whereupon Anglo-American forces could perform a triumphal march from the Channel to Berlin..."[111] Despite all differences and recriminations Trident was successful and, unlike the Italo-German meeting at Castle Klessheim, this conference was one of truly equal partners. Eventually, Brooke was able to note with satisfaction that "the compromise which emerged was almost exactly what I wanted. We continued with the war in Italy with the aim of eliminating Italy. We forced dispersion of German

---

[109] Ibid., p. 411 (18 May 1943).

[110] Ibid., p. 403 (14 May 1943).

[111] Roy Jenkins, *Churchill*, p. 708.

forces under strategically bad conditions in Southern Europe."[112]

Although from June 1943 North Africa was securely in Allied hands, the Mediterranean space was still highly contested, which meant that at that moment equipment and troops could not be spared for the build-up for the invasion of France, then taking place in southern England. Nevertheless, much to Brooke's chagrin Marshall went on about France. "Marshall absolutely fails to realise what strategic treasures lay at our feet in the Mediterranean, and always hankers after cross-Channel operations. He admits that our object must be to eliminate Italy and yet is always afraid of facing the consequences of doing so."[113] It was no secret that ever since the hapless Gallipoli operation in 1915, and possibly before that, Churchill had been a keen believer in peripheral approaches as a means to avoiding the heavy losses that World War I-style frontal engagements were always likely to produce. The Americans saw it differently: "The Americans always suspected Winston of having concealed desires to spread into the Balkans."[114] Such development was anathema to the Americans, and Brooke's frustration over their disinclination to go along with his Mediterranean strategy was increasing by the day:

> The main trouble was the American desire to now swing priorities round to the Channel and in doing so to render it impossible to gather the full fruits of our present strategic position. The attitude of Ike's HQ was not encouraging. I knew that he never really appreciated the strategic advantages of Italy, and that the American blindfolded cross Channel policy must appeal to him as being easier to understand... The American outlook was unfortunately one of, 'We have already wasted enough time in the Mediterranean doing nothing, let us lose no more time in this secondary theatre...'[115]

The difficulties in Anglo-American collaboration were compounded by organisational issues, prominent among which were that of supreme command vis-à-vis component command. This surfaced as early as during the North African and Italian campaigns, and it would pop up again and again during the last years of the war. 28 Nov 1944 Brooke wrote: "I said that in my mind there were two main factors at fault, i.e.: a) American

---

[112] Alex Danchev and Daniel Todman, eds., *War Diaries, 193-1945; Field Marshal Lord Alanbrooke*, p. 403 (25 May 1943).

[113] Ibid., p. 433 (24 Jul 1943).

[114] Ibid., p. 459 (8 Oct 1943).

[115] Ibid., pp. 463-64 (27 Oct 1943).

strategy, b) American organisation. As regards strategy the American conception of always attacking all along the front, irrespective of strength available, was sheer madness. As regards organisation, I said that I did not consider that Eisenhower could command both as Supreme Commander and as Commander of the Land Forces at the same time."[116]

When at long last agreement on the cross-Channel operation was reached, new discord materialised over whether to concentrate effort in the North or make a diversionary landing elsewhere. The Americans were adamant that the Second Front should be supplemented by an Allied diversionary invasion of the south of France nicknamed 'Anvil,' although later re-baptised 'Dragoon.' They were as dedicated to this as the British were opposed. The Americans were absolutely unbudgeable on the issue and the British, now rapidly becoming the junior military partner, ultimately had no option but to accept their decision.[117] Additionally, during inter-allied discussions in 1943 and 44 the question was raised if the war was to be ended somewhere in Germany or as far east as possible. The Americans were essentially interested in smashing the *Wehrmacht* on the central front and, provided this objective were achieved, did not much mind where they joined hands with the Soviet army. The British, on the other hand, were increasingly concerned about keeping Communism as far east as possible. They wanted to stop Tito seizing Trieste, to get first to Vienna, to encourage the Americans to do the same with Prague and above all to see either Montgomery or Bradley into the capital of the Third Reich before the Red Flag was hoisted there.[118]

The two main points at issue were: first, Churchill's belief that he American strategy in the last stages of the war against Germany gravely underestimated the importance of the Western Allies getting to Berlin before, or at least as soon as the Russians; and second, how strong a western reaction was called for by Soviet breaches of the Yalta undertakings on Poland.[119]

---

[116] Ibid., pp. 629-30 (28 Nov 1944).

[117] Jenkins, *Churchill*, p. 749.

[118] Ibid., p. 755.

[119] Ibid., p. 785.

Operations

Planners have to plan for war, of course, but exactly that can be fraught with difficulties. Planning is an ongoing iterative process throughout a war, and at the military top-level there will often be fears that the political leaders might agree on something which does not make sense militarily. During the Second Washington Conference of 20-25 June 1942, General Brooke, whose planning priority was North Africa, noted that: "We made further progress towards defining our policy for 1942 and 1943.... We fully appreciated that we might be up against many difficulties when confronted with the plans that the PM and the President had been brewing up at Hyde Park [FDR's residence]."[120]

Certainly, planning with colleagues as disparate as the 'easterners' of the American Navy and the British sticking to their Mediterranean and European perspectives was not too easy:

> We received news today that Marshall, King and Harry Hopkins were on their way to discuss further operations. It will be a queer party as Harry Hopkins is for operating in Africa, Marshall wants to operate in Europe, and King is determined to strike in the Pacific... I found Marshall's rigid form of strategy very difficult to cope with. He never fully appreciated what operations in France would mean – the different standard of training of German divisions as opposed to raw American divisions.[121]

Despite their disparate strategic views, agreement was reached and during the summer of 1942, the American and British staffs were immersed in the planning for the two-front approach in Africa, preparing for Allied landings in French West Africa simultaneously with Montgomery's thrust from the East. On 29 August Brooke noted I his diary that:

> at 11 am PM held conference to discuss North African enterprise [Operation Torch] in the light of latest USA message. Finally agreed that we should examine possibility of doing Casablanca, Oran and Algiers instead of Oran, Algiers and Bône... The difficulty is that we shall require additional forces for it which can only be found by drawing on Pacific. This will not suit Admiral King.[122]

---

[120] Alex Danchev and Daniel Todman, eds., *War Diaries, 193-1945; Field Marshal Lord Alanbrooke*, pp. 267-8 (20 Jun 1942).

[121] Ibid., pp. 280-81 (15 & 17 Jul 1942).

[122] Ibid., p. 315 (29 Aug 1942).

One additional hindrance frequently encountered in coalition warfare planning is incompatibility of doctrines and organisations. This impediment often surfaced in the co-operation between the American and British staffs. The Anglo-Canadian collaboration during World War II, however, was an early example of successful avoidance of this problem. There was common training, almost identical organisation and the same equipment, and because of this Canadian and British formations were virtually interchangeable. In fact, Anglo-Canadian formations mixed and matched more readily in 1944 and 1945 than they had done in 1917 and 1918, even though the Canadians provided nearly all their own staff officers this time.[123] The First Canadian Army fought effectively under Sir Bernard Montgomery's 21st Army Group for the entire campaign, most of it with Sir John Crocker's 1st British Corps under command. That 1st British Corps had the 3rd Canadian Division from the D-Day invasion to 11 July, when it reverted to its place in 2nd Canadian Corps. The 2nd Canadian Corps, in turn, was part of the 2nd British Army from 11 to 23 July, when the First Canadian Army became operational in Normandy. For Operation Spring on 25 July, the commander of the 2nd Canadian Corps, Lieutenant-General Guy Simonds, had two British armoured divisions under his command – the 7th Armoured and the Guards Armoured. Two weeks later, the 51st Highland Division and the 33rd Armoured Brigade joined Simonds's 2nd Canadian Corps for Operation Totalise. In the Scheldt operations of October-November 1944, 1st Canadian Army had several British formations at its disposal. It had the 1st British Corps to clear north and east of Antwerp and it had the 52nd Lowland Division, the 4th Special Service Brigade and No. 4 Commando for the assault on Walcheren Island. For Operation Veritable and the battles of the Rhineland, Brian Horrocks's 30th British Corps joined General Harry Crerar's First Canadian Army for the first stage of the operation, which it fought with the 2nd Canadian and 3rd Canadian Divisions under command.

These inter-army arrangements were relatively simple because the British and Canadian armies were organised in nearly identical ways, and the commanders and staffs in both armies spoke the language of Camberley and Quetta – no small advantage when co-ordinating operations involving hundreds of thousands of troops.[124] There were still occasional problems.

---

[123] The paragraph on Anglo-Canadian collaboration is largely due to Doug Delaney's paper presented at the coalition warfare conference in Copenhagen in May 2011.

[124] Staff College, Camberley, Surrey, UK was a staff college for the British Army from 1802 to 1997, with periods of closure during major wars. In 1997 it was merged into the new Joint Services Command and Staff College. The Joint Services Command and Staff College (JSCSC) of today, located at Shrivenham, is a British military academic establishment

Montgomery and Crerar, for example, often did not get along.[125] But even during the rockiest periods of their relationship, the 'Q' [quartermaster] staff at 21st Army Group coordinated logistics with the 'Q' staff at First Canadian Army, who coordinated with the 'Q' staff at 1st British Corps, and so on. The same could be said for 'G' [general] staffs concerning operations and intelligence and 'A' [adjutant] staffs when it came to personnel matters. That level of interoperability was the result of decades of common staff training, officer exchanges, and Commonwealth coordination.[126]

Conversely, the co-operation within the Axis suffered from disparity in almost any field. First of all, there was a remarkable imbalance of power between Germany and Italy when seen against the similarities of the political ideologies of the two regimes. The Germans and Italians failed to create combined staffs which might effectively have reconciled their strategic interests. Inevitably, under these circumstances, there was no way to develop the top-level consultation and decision-making procedures that are essential to effective coalition warfare. Political summits gradually turned into Hitler's well-known one-man shows, where he overawed the *Duce* with his verbosity as well as with his often quite incisive knowledge of military detail. The political level also dominated the few military high-level conferences that did take place, and as the military participants did not have authority to adopt any major decisions, these meetings were merely exchanges of information and opinions with very little specific or prospective outcome.

Therefore, there was no effective co-operation at the military levels – either in planning, or command and control arrangements. The co-ordinating meetings of chiefs of staff like those conducted from 1941 onwards between the US and the UK, which were prerequisites for agreement on strategic and organisational matters, never really existed between German and Italian staffs. This kind of shortfall in consultation rendered the Axis' national supreme commands incapable of effective collaboration on planning and execution of combined operations.

---

providing training and education to experienced officers of the Royal Navy, Army, Royal Air Force, Ministry of Defence Civil Service, and serving officers of other states. Today, Quetta is The Command and Staff College of Pakistan. It was established in 1907 at Quetta, Balochistan, British Raj and is the oldest and the most prestigious institution of the Pakistani Army.

[125] See Douglas E. Delaney, "When Harry Met Monty: Canadian national Politics and the Crerar-Montgomery Relationship," in *The Canadian Way of War: Serving the National Interest,* ed. Bernd Horn (Toronto: Dundurn, 2006), 213-234.

[126] Delaney.

Convinced they were great captains, Hitler and Mussolini held on to their prerogatives of supreme command. The two dictators largely usurped military supreme command – Hitler even – on the dismissal in December 1941 of Generalfeldmarschall [Field-Marshal] Walther von Brauchitsch – took personal command of the Army. However, when the European war developed into a world war neither Hitler nor Mussolini mastered the art of war to an extent allowing them to manage two or three theatres of war at a time. The relative autonomy of army, navy, air force and regional commanders added a further dysfunctional element that compromised the effectiveness of the German and Italian armed forces. Moreover, Hitler as well as Mussolini exploited traditional inter-service rivalry in order to strengthen their personal power.

The byzantine chaos of competing power structures – a central feature of the German and Italian autocracies – hampered effective conduct of military operations. Especially in the Mediterranean theatre, where the Italian and German military bureaucracies were supposed to collaborate, the structural weaknesses exacerbated the inadequacy of high-level co-ordination. Here, of all places, integrated military structures were sorely needed, notably in staff organisation. The high commands in both countries, though well aware of these problems, proved incapable of creating a workable solution. Hesitant thinking and bureaucratic resistance joined with mutual distrust to harm combined joint planning at the operational level.

Lead-nation dominance is a logical consequence of coalition warfare by partners of unequal size and wealth. In some cases, minor coalition partners sense great power arrogance in the way they are treated by the leaders of their coalitions. This was indeed the case amongst the Axis partners in North Africa, where Italy almost immediately upon the arrival of German troops acquired the status of a lesser partner. Upset by Rommel's supercilious bearing, the Italians were justifiably reserved about German initiatives. Against this background, neither the frequent meetings of Kesselring and Rommel with the Italian senior leadership in Rome and Libya, nor the exchange of permanent military representatives as liaison between the High Commands could replace a combined planning entity. Such half-measures were poor substitutes for an integrated supreme command. The Mediterranean theatre of Axis operations demanded a synthesis of operational approaches in order to interact with the highest political levels. The absence of such integrated combined and joint bodies resulted in strategic problems that eluded any solution.

Compared with Axis collaborative difficulties, the co-ordination of Allied activities in the Mediterranean theatre ran comparatively smoothly. After a meeting 15 January 1943, i.e. during the Casablanca conference (14-24 January 1943), Alan Brooke wrote with satisfaction that:

After lunch Combined Chiefs of Staff meeting discussing...relative advantages between Western Front in France and Mediterranean amphibious operations. I made long statement in favour of the latter, which went down fairly well and remains to be argued further tomorrow... At 5.30 pm Combined Staffs, Eisenhower, Alexander and Tedder met President and PM, at which we did little except that President expressed view favouring operations in Mediterranean.[127]

However, not all inter-allied work was plain sailing. On 16 January Brooke "put forward all the advantages of our proposed Mediterranean [strategy] and counter arguments in favour of the French front plan. It is a slow and tiring business which requires a lot of patience. They can't be pushed or hurried and must be made gradually to assimilate our proposed policy."[128]

Having settled their business in Africa, the Allies' next major operations concerned Italy and not all the amphibious operations went according to plans – let alone wishes. On 14 September, Brooke, still mired in the conception of Britain being the lead-nation of the coalition, dotted down a wryly comment haranguing the inept Americans. "Salerno landings is going from bad to worse and I fear we are bound to be pushed back into the sea! It is maddening not to be able to get the Americans realise that they are going to burn their fingers *before* they are actually doing so.[129] However, on the 18th all seems to be well. 'Salerno landings now seem safe."[130]

Half a year later, the discord over Italy still existed. The political as well as the military top-levels were continuously in disagreement concerning the right strategy and amongst the generals this percolated through to operational and personal matters. Notwithstanding Eisenhower's political and diplomatic skills, Brooke, like many others, was despairing over his apparent unfamiliarity with the simplest of military theory. On 22 February 1944 he wrote, "It is quite clear to me...that he does not begin to understand the Italian campaign. He cannot realise that to maintain an offensive a proportion of reserve divisions are required. He considers that this reserve

---

[127] Danchev and Todman, eds., *War Diaries, 193-1945; Field Marshal Lord Alanbrooke*, p. 359 (15 Jan 1943).

[128] Ibid., p. 360 (16 Jan 1943).

[129] Ibid., p. 452 (14 September 1943).

[130] Ibid., p. 453 (18 September 1943).

can be withdrawn for a new offensive in the South of France..."[131] It seemed evident that Brooke's disagreement – and Churchill's for that matter – with the Americans concerning the cross-Channel operation vis-à-vis the campaign in Italy gradually became intertwined with his notion of Eisenhower as a military ignorant. The latter perception was aggravated during operations in Northern Europe during the summer and autumn of 1944, and there is lots of evidence to be found in his diaries. On 8 November 1944, Brooke wrote:

> This evening Cyril Falls, military correspondent of The Times, came to see me. He said that he was disturbed by the system of command in France with Eisenhower commanding in two planes, namely commanding the Army, Navy and Air Forces in his capacity as Supreme Commander and at the same time pretending to command the land forces divided into three groups of armies directly. He has hit the nail on the head and found the weakness of this set up... Unfortunately it becomes a political matter and the Americans, with preponderating strength of land and air forces, very naturally claim the privilege of deciding how the forces are organised and commanded.[132]

Brooke had now realised the inevitable: that the Americans had grown to become by far the strongest amongst the Allies and, thus, the natural lead-nation of the coalition. However, the issue of command did not go away. Montgomery, who shared Brooke's opinion on the matter, complained repeatedly about Eisenhower's inability to both decide at the level of supreme commander and at the same time command the land component. On 26 November Brooke committed to paper that:

> Monty flew over from Belgium and landed at Hartford Bridge Flats at 11.30 am... He came to discuss the situation in France and to try to arrive at the best way of putting it right. We decided that there were three fundamentals to be put right:
>
> > To counter the pernicious American strategy of attacking all along the line
> >
> > To obviate splitting an army group with the Ardennes in the middle of it, by forming two groups (northern and southern) instead of three as at present
> >
> > To appoint a commander for the land forces

---

[131] Ibid., p. 523 (23 Feb 1944).

[132] Ibid., p. 619 (8 Nov 1944).

The problem is how to get this carried out. What we want is Bradley as a Commander of Land Forces, Montgomery Northern Group of Armies with Patton's army in his group – by substituting 3rd Army for 9th Army – and Devers commanding Southern Group. Monty is to see Eisenhower on Monday and if he opens the subject Monty is to begin putting forward the above proposals.[133]

On 29 November, Eisenhower agreed, although only half-heartedly to parts of the British suggestions:

Received telegram this morning from Monty. He had had a talk with Eisenhower. The latter agreed that strategy was wrong, that results of offensive was strategic reverse, that front wanted reorganising, would not agree that commander of land forces was necessary. But prepared to put Bradley with a large group of forces north of the Ardennes under Monty's orders, leaving Devers south of the Ardennes.[134]

In the early days of December, Brooke saw Churchill to give vent to his frustration. Roy Jenkins notes that:

...he believed that the Eisenhower was neglecting the northern axis of advance through the Netherlands and Montgomery became increasingly critical of Eisenhower...Five days later Churchill complained to Roosevelt that...the time has come for me to place before you the serious and disappointing war situation which faces us at the close of the year. Roosevelt replied calmly: 'perhaps I am not close enough to the picture to feel as disappointed as you do... All this was before the German Ardennes offensive.[135]

Then, on 16 December, operations developed in an unforeseen direction as, by the last of their military ingenuity, the Germans launched Operation *Wacht am Rhein* [Guard on the Rhine – later to become known to English speakers as The Battle of the Bulge] in the Ardennes: "Rundstedt has proved how faulty Ike's dispositions and organisation were. Spread out over a large front with no adequate reserves and no land force commander to immediately take charge."[136] This sudden change for the worse apparently

---

[133] Ibid., p. 629 (26 Nov 1944).

[134] Ibid., p. 630 (29 November 1944).

[135] Jenkins, *Churchill*, pp. 765.

[136] Danchev and Todman, eds., *War Diaries, 193-1945; Field Marshal Lord Alanbrooke*, p. 636 (18 Dec 1944).

worked towards facilitating reconciliation of Brooke's and Eisenhower's operational preferences, and on 20 December the former noted that he:

received telegram from Monty which showed clearly that the situation in France was serious. American front penetrated, Germans advancing on Namur... First American Army in a state of flux and disorganisation, etc.... However, I got him to phone to Ike to put the proposal to him that Monty should take over the whole of the Northern Wing whilst Bradley ran the South. Ike agreed and had apparently already issued orders to that effect.[137]

Dominance

It remains one of the drawbacks of coalition warfare that, if one participant is a great power and the others are contributors of limited resources, because of the great power's 'investment' in the project as well as the sophistication of its equipment, organisation and training, this power will be the one telling the others what to do and – occasionally – how to do it. If the representatives of the great power are remiss in tact and understanding for the smaller partners this can lead to inter-personal discord and may, in the long run, occasion dissolution of the coalition.

This phenomenon was extant in both world wars, and in the second there was the added complication that the rôle of lead-nation – or at least the dominant nation – slowly but surely went from one power to another. As early as mid-1943, this was acutely felt by the British. On 18 November, Alan Brooke wrote:

First of all the new feeling of spitefulness which had been apparent lately with Winston since the strength of the American forces now building up fast and exceeding ours. He hated having to give up the position of dominant partner which we had held at the start. As a result he became inclined at times to put up strategic proposals which he knew were unsound purely to spite the Americans.... It was usually fairly easy to swing him back on the right line and to get rid of these whims. There lay, however, in the back of his mind the desire to form a purely British theatre when the laurels would be all ours... Austria or the Balkans seemed to attract him for such a front.[138]

Although it is both natural and fair that the main contributor decides how resources are employed, it sometimes happens that a lead-nation forces its

---

[137] Ibid., p. 637 (20 Dec 1944).

[138] Ibid., p. 473 (18 Nov 1943).

will upon lesser partners based on reasons completely irrelevant to the war at hand. Worried about the fate of the Italian campaign and the onward drive towards the Germans' southern flank, which was under-resourced because the Americans insisted on extracting forces from Italy to be able to carry out Operation Dragoon in southern France, Brooke saw an example of this trend when, in June 1944, he realised that:

> Owing to the coming [American] Presidential election it is impossible to contemplate any action with a Balkan flavour irrespective of its strategic merits. The situation is full of difficulties; the Americans now begin to own the major strength on land, in the air and on the sea. They therefore consider that they are entitled to decide how their forces are to be employed. We shall be forced into carrying out an invasion of southern France, but I am not certain that this needs cripple Alexander's power to finish crushing Kesselring.[139]

In some cases, though not indefinitely, the lesser ally does not accept its subordinate status and refuses to yield to lead-nation suggestions. In 1940, Mussolini had insisted on having Italian air forces on the Channel coast as part of the preparations for the planned invasion of England and was dismissive of any proposal that might leave the impression that Italy was not an equal partner. Even as Keitel and Hitler recommended that the Italian Air Force be withdrawn from the Channel area in order to exploit them better in Greece and Egypt, the *Duce* declined. This, however, was not vindicated by military arguments, but rather caused by simple national pride and petulance.

Italian arrogance was indeed unwarranted as successes in the war had so far been few and far between, and the reasons for the poor results of Italian warfare were political as well as military. Responsibility for the meagre performance on the battlefield lay with the military leadership that had not managed to complete what seemed to be a very modest task: the struggle in 1940 in the Western Alpine area against a France already defeated in the North by the Germans. On top of that shortcoming came political interference in the military conduct of the war by politicians who were, generally, incompetent. As the war in the Balkans was initially an Italian enterprise – which was doomed to failure – the Italians found themselves compelled to join the Nazi-German campaign because an exclusively German victory in Greece would expose Italy's helplessness.

Although Italy had been kept away from the planning of Operation Barbarossa against the Soviet Union, Mussolini insisted on participation by a *Corpo di Spedizione Italiano* in Russia [Italian Expeditionary Corps

---

[139] Ibid., p. 564 (30 Jun 1944).

(CSIR)], which was consequently deployed to Eastern Romania and placed under command of the 11th German Army. But, at the outset, Hitler and Keitel were reluctant to accept Italy's kind offer of operational support.

As the war wore on, the Italian pretension of being an equal coalition partner became increasingly difficult to sustain. In 1942 in the Mediterranean theatre of war, the Italo-German air power became subordinated to the combined leadership headed by Field-Marshal Albert Kesselring, and in consideration of the critical situation the Italian armed forces had to accept the humiliating offer of having their military sovereignty severely curtailed. On top of that, it must have been particularly annoying to the military leadership that Mussolini admired Rommel to the extent of letting the latter's views and appreciations prevail at the expense of his formal superior, the Italian Supreme Commander Africa, Ettore Bastico. Particularly after the Axis' successful conquest of Tobruk in June 1942, Mussolini's trust in Rommel was so sincere that he started to support his plans against the advice of the Italian *Commando Supremo*. Rommel was now foreseen as C-in-C in Egypt, after the prospective conquest of that country, with an Italian civil chief administrator, and as of 16 August 1942 Rommel's subordination to Bastico, was abrogated.

Personal Relationships

At top level during World War II, the UK-US special relationship was in many respects a personal matter, and although cultural differences surfaced every now and again there was more that united. From the very outset, Churchill made sure that representatives of President Roosevelt got the most exquisite treatment; Harry Hopkins, Roosevelt's close confidant, during his protracted visit from 9 January through 10 February 1941 was the first to receive what Roy Jenkins calls the PM's "daring embrace and calculated attention."[140] He was met by Brendan Bracken at Pool Harbour and taken to London for a conversation with the PM. Two days later, Churchill took the American by special train to Scapa Flow to acquaint him with the full splendour of the Royal Navy. From there, they went on to Glasgow and its surroundings to see the harbour facilities, which would become of such immense importance during the upcoming Battle of the Atlantic.[141]

Generally, on one hand, personal affinity mattered a lot, not only

---

[140] Jenkins, *Churchill*, pp. 647-50.

[141] Ibid., pp. 647-49.

between the two western Allies' heads of government and their trusted representatives, but also between the latter and the leaders' entourages of political and military advisors. Averell Harriman was one more of those Americans who came to play an instrumental rôle in the Anglo-American coalition. Moreover, it did the relationship no harm that he was almost infatuated with Pamela Digby (Mrs Randolph Churchill), who, since her husband went to Cairo in 1941, spent most of the war at Chequers.[142]

On the other hand, personalities were not always easily reconciled and some were constant sources of open animosity, which was most illustratively the case with General de Gaulle, the self-styled leader of the Free French. This trouble popped up as early as the Arcadia Conference, where the American Foreign Secretary Cordell Hull (1871-1955) took a very hostile attitude towards the general refusing to see the Free French as a member of the war-time coalition.[143] It is well-known that Churchill found de Gaulle a most tiresome person to deal with and that Brooke had great trouble bearing his presence. On 16 and 18 December 1941 Brooke noted in his diary that he had 'lunched with de Gaulle [who was] a most unattractive specimen. We made a horrid mistake when we decided to make use of him.'[144] In those days planning was going on concerning an operation aimed at taking the island of Madagascar, which Brooke found counterproductive, but "the General thought that this would be a splendid opportunity for the Free French to show their presence in Africa." Brooke feared that there might be "Further complications due to de Gaulle wishing to co-operate with the Madagascar operations. His support is more likely to be an encumbrance."[145] Additional animosity was generated as, at lunch on 3 February 1942, Brooke found de Gaulle a perfect spoiler of an otherwise happy party: "Lunch at Claridges, where I had all the Cs-in-C of Allied Forces. Sikorsky sat on my right and was in great form. The more I see of him the more I like him. On my left sat Chaney, the USA army commander, just back from seeing the USA troops arriving in Ireland. De Gaulle looking if anything more sour than ever, a most unattractive specimen."[146] Personal

---

[142] Ibid., p. 653.

[143] Ibid., pp. 675-6.

[144] Danchev and Todman, eds., *War Diaries, 193-1945; Field Marshal Lord Alanbrooke*, p. 212 (16 Dec 1941).

[145] Ibid., p. 212 (18 Dec 1941).

[146] Ibid., pp. 226-7 (3 Feb 1942).

relationships, however, have their ups and downs, and in March things ran more smoothly: "I dashed back for a lunch I was giving de Gaulle. As he had been getting deeper in disgrace during the last few days I thought situation would be strained. However, it went off all right."[147]

Personal relationships, evidently of key importance to coalition warfare, were not only a matter of the affinity of kindred spirits, but also one of compatibility of national characteristics. Although no Francophobe at all, Brooke seems to have had some general difficulties with the French: "General [Henri Honoré] Giraud (1879-1949) was reported this evening as having landed at Gibraltar in his submarine. He should be a great contribution to this venture, and may ultimately assist us in solving the de Gaulle impasse."[148] Giraud had been a prisoner-of-war in Germany but managed to escape. He was then secretly contacted by the Allies and agreed to support Operation Torch, the Allied landing in French North Africa, provided that only American troops were used, and that he or another French officer was the commander of such an operation. He considered this latter condition essential to maintaining French sovereignty and authority over North Africa.

On 5 November, he was picked up near Toulon by the British submarine Seraph, which took him to meet General Dwight Eisenhower in Gibraltar. He arrived on 7 November, only a few hours before the landings. Eisenhower asked him to assume command of French troops in North Africa during Operation Torch and direct them to join the Allies. But Giraud had expected to command the whole operation, and adamantly refused to participate on any other basis. Brooke noted that "This morning landing at Casablanca, Oran and Algiers... Giraud also turned sour and apparently refused to play unless he is given supreme command of all forces... I am afraid that his personal vanity may well upset some of our schemes."[149] Eventually, he was persuaded and in mid-November he took over command of all French forces in North Africa. Nonetheless, Brooke found the set up in North-West Africa complicated and less than satisfactory: "We had a strange set up in North Africa. A C-in-C deficient of experience and of limited ability in the shape of Eisenhower, and three possible French leaders, Darlan had ability but no integrity, Giraud who had charm but no ability, and de Gaulle who had the mentality of a dictator

---

[147] Ibid., p. 239 (13 Mar 1942).

[148] Ibid., p. 339 (7 Nov 1942).

[149] Ibid., p. 339 (8 Nov 1942).

combined with a most objectionable personality!"¹⁵⁰ In Brooke's case there was no Anglo-French 'national animosity' – it was simply a question of incompatible personalities and misplaced narcissism and, indeed, there seems to have been quite a few who had similar difficulties working with de Gaulle. "Unfortunately his [Churchill's] dislike of him [de Gaulle] has come rather late. He should have been cast over board a year ago and he gave plenty of opportunities for such action."¹⁵¹

Following the final clear-up of North Africa there was, in June 1943, a need to reconcile the various French 'prima donnas' with each other and Roy Jenkins tells us that:

> Then there was the business of putting some conjugality into the reluctant reconciliation, which Giraud and de Gaulle had accepted at Casablanca. Progress was made. There was a meeting which culminated in a convivial luncheon on 4 June even though de Gaulle was normally more than proof against such seductions. There was a feast of oratory. Churchill spoke in French (an excellent speech according to the critical pen of his nearly bi-lingual friend Alan Brooke).¹⁵²

The British relationship with the Chief of the US Army General Staff, General George Catlett Marshall (1880-1959), lay somewhere between the excellent rapport with Harriman and Hopkins and the frosty association with de Gaulle. Brooke found Marshall quite charming, "but not as sociable as the former gentlemen, and he was sent by Roosevelt, in April 1942, to convert Churchill to the idea of taking the pressure off the Russians by an early cross-Channel operation – a concept which did nothing to endear him to the PM."¹⁵³ Even worse was the impression made in Britain by the American Anglophobe Fleet Admiral Ernest Joseph King (1878-1956) as he visited London in July 1942. King tried even harder, in vain though, to persuade the British of the merits of Operation Sledgehammer (an early Channel crossing) at the expense of Operation Torch, the Allied planned landings in North-West Africa in support of the forces in Egypt.¹⁵⁴ Brooke was not convinced by King's arguments, and his semi-hostile attitude did

---

[150] Ibid., p. 363 (18 Jan 1943).

[151] Ibid., p. 427 (8 Jul 1943).

[152] Jenkins, *Churchill*, p. 712.

[153] Ibid., pp. 688-89.

[154] Ibid., p. 698.

little to help mollify the British. Brooke later commented that:

> With the situation prevailing at that time it was not possible to take Marshall's 'castles in the air' too seriously! It must be remembered that at that time we were literally hanging on by our eye-lids! Australia and India were threatened by the Japanese, we had temporarily lost control of the Indian Ocean, the Germans were threatening Persia and our oil, Auchinleck was in precarious straits in the desert, and the submarine sinkings were heavy. Under such circumstances we were temporarily on the defensive, and when we returned to the offensive certain definite steps were necessary.[155]

It was evident that the collaborative climate between the three was far from perfect. Marshall and King had political directives to carry out which did not always make sense strategically, and Brooke knew that he did in fact have a more comprehensive grasp of his trade as well as of the actual war than did his opposite numbers:

> I had Marshall for nearly two hours in my office explaining to him our dispositions. He is, I should think, a good general at raising armies and providing the necessary links between the military and political worlds. But his strategical ability does not impress me at all!! In fact in many respects he is a very dangerous man whilst being a very charming one. He has found that King, the American admiral, is proving more and more of a drain on his military resources, continually calling for land forces to capture and hold bases. On the other hand MacArthur [Commander South-West Pacific Area from April 1942] in Australia constitutes another threat by asking for forces to develop an offensive from Australia. To counter these moves Marshall has started the European offensive plan...but his plan does not go beyond just landing on the far coast... My conversation with Marshall that afternoon was an eye-opener! I discovered that he had not studied any of the strategic implications of a cross-Channel operation.[156]

Moreover, on 20 June 1942 Brooke realised that Marshall was not the only decision-maker whose strategic view did not match his own: "This dinner was my first opportunity of meeting Stimson, the Secretary for War. I found him an exceptionally charming man to meet, and certainly a fine administrative brain, but with a limited strategic outlook. He was one of the strong adherents of breaking our heads in too early operations across the

---

[155] Danchev and Todman, eds., *War Diaries, 193-1945; Field Marshal Lord Alanbrooke*, p. 248 (14 Apr 1942).

[156] Ibid., pp. 248-9 (15 Apr 1942).

Channel. Consequently a strong supporter of Marshall."[157]

Nonetheless, by and large Americans and British got on well together, and in particular President Roosevelt's helpfulness and understanding on hearing, during the Second Washington Conference 20-25 June 1942, about the fall of Tobruk helped oil the wheels of the relationship. Brooke remembered that "I always found that the Tobruk episode in the President's study did a great deal towards laying the foundations of friendship and understanding built up during the war between the President and Marshall on the one hand and Churchill and myself on the other."[158] Then, leaving Washington, Brooke had almost become fond of Marshall, though with some reservations: "It has been a very interesting trip and real good value. I feel now in much closer touch with Marshall and his staff and know what he is working for and what his difficulties are.... Then meeting the President was a matter of the greatest interest, a wonderful charm about him. But I do not think that his military sense is on a par with his political sense."[159]

General Eisenhower was the choice of President Roosevelt. Most diarists and historians speak well of his qualities as a political general and a reconciler of differences amongst his subordinate commanders, but his abilities as a strategist and as a tactician have been repeatedly doubted. He had no experience with commanding large formations and he knew very little of Europe and Africa, as Brooke wrote: "It must be remembered that Eisenhower had never even commanded a battalion in action when he found himself commanding a group of armies in North Africa! No wonder he was at a loss as to what to do, and allowed himself to be absorbed in the political situation at the expense of the tactical."[160] Conversely, his sense of decency and his eye for personal strengths and weaknesses of his collaborators are equally frequently praised:

> Clark has been creating trouble.... The news about Clark was a bit of an eye-opener and quite unexpected. However, from everything I gathered, there was no doubt that he was trying to discredit the British in the eyes of the French in order to obtain for himself the command of the Tunisian front. Eisenhower evidently became aware of this manœuvre and with his high quality of impartiality rid himself of Clark as his Deputy Commander and sent him back

---

[157] Ibid., pp. 267-8 (20 Jun 1942).

[158] Ibid., p. 269 (21 Jun 1942).

[159] Ibid., p. 272 (25 Jun 1942).

[160] Ibid., p. 343 (24 Nov 1942).

to command the reserve forces in Morocco. Through this action Ike greatly rose in my estimation.[161]

While in 1942-3 the British commanded in the Middle East and the advance from Egypt via El Alamein to Tunis, the Americans had led the landings during Operation Torch and the subsequent progress from the West. It was logical then that, for the onward mopping up of North Africa, one supreme commander of the entire theatre of war should be appointed:

> From many points of view it was desirable to hand this command [North Africa] over to the Americans, but unfortunately up to now Eisenhower certainly did not seem to possess the basic qualities required from such a commander. By bringing Alexander over from the Middle East and appointing him Deputy to Eisenhower, we were carrying out a move which could not help flattering and pleasing the Americans ... We were pushing Eisenhower up into the stratosphere and rarefied atmosphere of a Supreme Commander, where he would be free to devote his time to the political and inter-allied problems, whilst we inserted under him one of our own commanders to deal with the military situation and to restore the necessary drive and coordination which had been so seriously lacking of late![162]

It was less easy going with the Russian Ally. During Second Moscow Conference 12-17 August 1942, the British had a hard time:

> We finally arrived in Moscow at 8.30 pm [on 13 August], having been some 13 hours in the air and 15 hours travelling... At 11 pm we were due to go and meet Stalin in the Kremlin... I was much impressed by his astuteness and his crafty cleverness. He is a realist with little flattery about him, and not looking for much flattery either... The two leaders, Churchill and Stalin, are poles apart as human beings, and I cannot see a friendship between them such as exists between Roosevelt and Winston. Stalin...is ready to face facts even when unpleasant. Winston, on the other hand, never seems anxious to face an unpleasantness until forced to do so. He appealed to sentiments in Stalin, which do not I think exist there... Personally I feel our policy has been wrong from the very start. We have bowed and scraped to them... As a result they despise us and have no use for us except for what they can get out of us.[163]

A highly committed military decision-maker, Brooke at the same time

---

[161] Ibid., p. 356 (4 Jan 1943).

[162] Ibid., p. 365 (19 Jan 1943).

[163] Ibid., pp. 299-300 (12 Aug 1942).

respected politicians and loathed their ways. Their self-indulgence, their waste of precious time and their frequent deceitfulness were anathema to him and his diaries reveal several harsh judgements:

> He [Stalin] is an outstanding man, that there is no doubt about, but not an attractive one. He has got an unpleasantly cold, crafty, dead face, and whenever I look at him I can imagine him sending off people to their doom without even turning a hair. On the other hand there is no doubt he has a quick brain and a real grasp of the essentials of war.[164]

Although, generally, collaboration amongst the Western allies ran a lot more smoothly than with the Russians, there were clashes – and not only with Americans and French but within the British Commonwealth too. Canadian General Andrew George Latta McNaughton (1887-1966) was one who did not make life easy for the British coalition partner. Brooke found him quarrelsome and lacking an understanding of the way the coalition worked. During the difficult time of the early days of the Italian campaign he wrote:

> Came back to WO [War Office] to meet an infuriated McNaughton who had gone all the way to Malta to see Canadian troops in Sicily and had not been allowed by Alex (apparently owing to Monty's wishes)... I felt inclined to tell him that he and his government had already made more fuss than the whole of the rest of the Commonwealth concerning the employment of Dominion forces! The McNaughton incident was an excellent example of unnecessary clashes caused by failings in various personalities.... The troubles we had, I am sure, did not emanate from Canada, but were born in McNaughton's brain. He was devoid of any kind of strategic outlook...[165]

The problem was not quickly solved, although, "at 3 pm Ralston, Canadian Defence Minister, came to see me and remained nearly two hours discussing how we are to get rid of Andy McNaughton as Army Commander! No easy problem."[166] In due course these predicaments solved themselves, because Ralston was forced to resign and, due to pressure by critics and weakened by health problems, McNaughton relinquished his

---

[164] Ibid., p. 301 (14 Aug 1942).

[165] Ibid., pp. 431-2 (21 Jul 1943).

[166] Ibid., p. 435 (3 Aug 1943).

command in December 1943.[167]

As to Brooke's trouble with the Americans, some of it might be down to his frustration of being deprived of the job as Supreme Commander for Operation Overlord – the Second Front in Europe – which Churchill had promised him at an early stage but later rather highhandedly surrendered to the Americans. In late 1943, his main worry was how to make the Americans understand the importance of letting the Mediterranean and Europe remain separate theatres: "Suggestion by Leahy that Marshall should be made Supreme Commander of the European Theatre, to combine North Africa with cross-Channel! Luckily PM was entirely with us... The trouble is that meanwhile our proposal to combine command and control in the Mediterranean is being sidetracked."[168] Later on, Europe and the issue of the authority of the Supreme Commander vis-à-vis the Commanders of the army groups took up much of his attention, as:

> There is no doubt that Ike is all out to do all he can to maintain the best of relations between British and Americans, but it is equally clear the Ike knows nothing about strategy and is quite unsuited to the post of Supreme Commander... With that supreme command it is no wonder that Monty's real high ability is not always realised. Especially so when 'national' spectacles pervert the perspective of the strategic landscape.[169]

Irrespective of Brooke's praise of Montgomery and partial denigration of Eisenhower, there can be little doubt that the blame should not be laid at the latter's door alone. Mongomery was as difficult as de Gaulle and had a well-developed sense of self-promotion at others' expense as well as a faculty for annoying. Roy Jenkins describes how:

> Churchill on 3 January went to visit Eisenhower at his Versailles headquarters, and then on to see Montgomery at Ghent, a tedious night-train journey across the snow-bound plains of northern France... He had little good effect on Montgomery, who three days later gave a bombastic press conference, which

---

[167] Because of his support for a volunteer army, McNaughton remained friendly with Prime Minister William Lyon Mackenzie King, who wanted to make him the first Canadian-born Governor General of Canada. Instead, McNaughton became Minister of National Defence when Ralston was forced to resign after the Conscription Crisis of 1944, as Mackenzie King did all he could to avoid introducing conscription.

[168] Danchev and Todman, eds., *War Diaries, 193-1945; Field Marshal Lord Alanbrooke*, p. 467 (8 Nov 1943).

[169] Ibid., p. 575 (27 Jul 1944).

was immensely patronising of the Americans who had just borne the whole brunt of the last German offensive of the war. This caused so much offence that Eisenhower doubted whether he could order any of his generals to serve under Montgomery.[170]

Moreover, disagreement happened not only amongst coalition partner nations but between the services, too:

> There were rumours afloat that Monty was far too sticky, that he only thought of his own reputation, would never take risks, played for certainties et, etc. I discovered that these rumours emanated from two airmen, Coningham and Tedder, who were responsible for air support... I found several occasions in the war that airmen, entirely disconnected with the administrative problems of supply, which were mainly done for them, and with the vaguest conceptions as to the requirements of land tactics, were only too free in offering criticisms and accusing the army of moving too slowly.[171]

Korea

Upon the Japanese defeat at the end of World War II and the subsequent Communist take-over in China, East Asia had been turned upside down. While the USA had occupied Japan and stationed troops in various countries on the Pacific, Soviet Russia and Communist China dominated other parts of the hemisphere. Then, on 25 June 1950, the Communist People's Republic of Korea, a Soviet proxy, launched an attack on its southern neighbour, the Republic of Korea, which entertained a friendly relationship with the United States. Initially, the military numerical superiority of the North Korean Army was overwhelming. The North Korean Army had 130,000 men under arms – apparently well trained, equipped and motivated. Many of these soldiers were veterans of the World War II struggle against Japan and the Chinese Civil War. Between 1938 and 1949, thousands of Koreans fought with Mao's Army, in China or in the Red Army in Manchuria.

On 27 June 1950, however, the United Nations' Security Council launched an appeal to all the member states to come to the rescue of South

---

[170] Jenkins, *Churchill*, p. 773.

[171] Danchev and Todman, eds., *War Diaries, 193-1945; Field Marshal Lord Alanbrooke*, pp. 386 (2 Mar 1942).

Korea.[172]

Some member nations responded more enthusiastically than others. At the outset, France's reaction was symbolic as the French government decided to contribute merely the frigate *La Grandière*. At the time, this warship was in Saigon as part of French Indochina Naval Command. From the end of July to the end of November 1950, the *La Grandière* carried out troop transport duties, foremost amongst which was the Inchon Landing on 15 September.

However, dispatching a frigate was too modest a contribution even for a country having its own war to fight against Communism in Indochina. Not only was the war in Korea a welcome opportunity to train officers and soldiers of an army still in the process of reconstruction after its defeat by the Germans in 1940, it also offered a chance to reassert France's great power self-esteem. Thus, on 25 August 1950, the French government decided to dispatch a one thousand-man battalion. The French Battalion in the United Nations assistance force was officially created for this purpose, but because of various other commitments – such as occupation duties in Germany, North African defence, and the war in Indochina – as well as a certain downsizing in the wake of World War II, the Army general Staff tried to man this unit by reservists only. Most of these reservist-volunteers – totalling 1,050 all told – were veterans of World War II and the Indochina Campaigns. However, as the war went on, enthusiasm among the reservists declined and the General Staff had to infuse a number of professional soldiers. A total of about 3,400 French soldiers would serve in the French Korean Battalion between 1950 and 1953. The contingent left Marseilles on 25 October 1950 and arrived in Pusan on 29 November at a critical moment, a few days after the 300,000 so-called volunteers of the Communist Chinese Forces had launched their grand counteroffensive against the United Nations near the Manchurian border.

Like the other great powers, France had realised that since the end of World War II the technology and doctrines of martial activity had moved on, and the Korean War was the first modern conflict which constituted a genuine laboratory for western armies. The French army, whose reconstruction was in full progress, was trying to formulate modern military doctrines: The operations in Korea gave them the opportunity.

The French Battalion was attached to the American 2nd Infantry Division. The French battalion was then absorbed by the 23rd Regimental Combat Team as its 4th Battalion together with a Dutch (9th) and a Thai one

---

[172] The following owes a lot to a paper given by Capitaine Ivan Cadeau at the Conference on Coalition Warfare in Copenhagen in May 2011.

(38th).

In the beginning of 1951, and after a brief spell of training and familiarisation with US weapons, the French Battalion got an opportunity to prove itself. On 8-10 January 1951, in a determined defence of their positions at Wonju, two French Companies repelled a Chinese assault. The action was publicised by US War correspondents and won considerable acclaim from the lead-nation representatives. From then onwards, French troops had the unqualified trust of their US collaborators, a trust soon to be confirmed in the battles of the Twin Tunnels and Chipyong-Ni in early February 1951.

Alongside their coalition partners, the French took part in some of the toughest battles of the Korean War, such as Heartbreak Ridge or Hill 931 in September 1951 and Arrowhead in October 1952. The French suffered about 270 killed and more than 1,000 wounded.

Apart from re-establishing themselves as a great power with a global military reach, the French General Staff wanted to take advantage of the operations to learn lessons on modern combat with a view to adjust their tactical procedures, military doctrines and weapons use. In these respects the Korean War provided a test ground for western armies that might have to face a world war against the Soviet Union – or, at least, so it was assumed at the time. In order to exploit training and study opportunities to the full, France did what had been done in so many other conflicts over the centuries: they sent what amounted to military observers. As the French Joint Chiefs-of-Staff were eager to learn from the operations, France dispatched a staff with the battalion which was far beyond the need of a unit that size. This staff included thirty-four people, about ten of them with specialist expertise on infantry, artillery, corps of engineers, etc. The Chief-of-Staff was *général de corps d'armée* [lieutenant-general] Raoul Charles Magrin-Vernerey Monclar (1892-1964) a veteran of both World Wars and temporarily – for the sake of the mission in Korea – reduced to the rank of lieutenant-colonel. The staff officers' primary task was monitoring and analysing the operations and the enemy's methods of warfare, as well as the procedures of their coalition partners. The downside of the existence and activity of the 'staff of observers', serving more sophisticated goals than the simple needs of the commanding officer, was that it complicated the chain of command between the Americans and the French battalion.

Nonetheless, the battalion fought well and everyone derived some benefit from its being dispatched. The United Nations had wanted France to be involved, the Americans – the lead-nation – who, from a military point of view did not really need the French battalion, sustained a political need to see as many nations as possible to bandwagon in order to underpin the legitimacy of the enterprise, and the French waved the tri-colour and had their profile sharpened amongst the permanent members of the UN Security Council.

In the post-1945 period, it was harder for the Canadian Army to maintain similar levels of commonality with its closest senior alliance partner, in part because it wasn't always clear who the closest senior alliance partner would be. The Canadian drift into an American orbit had been happening for years. Economic ties between the two countries had been tightening for decades, and the Second World War nudged the North Americans closer still. In August 1940, when Britain seemed exposed to invasion, Canada and the United States signed the Ogdensburg Agreement, a bilateral arrangement that led to the establishment of the Permanent Joint Board on Defence to co-ordinate the protection of North America. The uncertainty of the early Cold War period pushed things along as well. The 1945 discovery of a Soviet spy network in Canada troubled Canadians deeply and led to even closer defence co-operation with the Americans, and the Soviet acquisition of an atomic bomb in 1949 only accelerated the trend. But the problem, as Canadians soon found out, was that dealing with an American superpower one-on-one usually amounted to accepting dictation. The Americans would do what they liked to ensure their own security, and were even less inclined to hear Canadian concerns than the British had ever been. In the past, Canadians had used their connection with the British Empire as leverage in their relationship with the United States, but in the late 1940s the British Commonwealth was not what the Empire used to be. Multi-lateral fora like the North Atlantic Treaty Organisation (NATO) or even the United Nations offered Canadians a much better chance of avoiding being forced into decisions they did not like. As Canadian diplomats used to say of NATO, "twelve in the bed means no rape."[173]

In spite of the commitment to the NATO military alliance, Canadian decision-makers had no real idea what they wanted their army, navy, and air forces to look like or do. In the era of the American nuclear monopoly, there was not much point in maintaining a large army establishment. Even after the Soviet atom bomb test in 1949, NATO's reliance on the American strategic arsenal still seemed to obviate the need for large standing forces. So Canadians shrunk their standing forces drastically. Demobilisation after the Second World War occurred at such a rate that, by July 1947, the Canadian Army was a mere 13,985, plus 33,704 reserves, down

---

[173] Cited in Desmond Morton, "Uncle Louis and Golden Age for Canada: A Time of Prosperity at Home and Influence Abroad," *Policy Options* (June-July 2003): 53.

significantly from its wartime peak of 495,073.[174] Most of the army was organised into three infantry battalions (with one company of paratroops each), two armoured regiments, and an artillery regiment, all of which was designated the Mobile Striking Force and nominally assigned to defend northern Canada against Soviet incursion. Beyond that, it was not at all clear to military planners how the Canadian Army would be employed in war, and under whose operational control it would fall.

In the absence of clear direction or commitments, the army hedged its bets. Understanding that Canada could not stay out of another European war, Canadian army leaders considered how they would return to Continental Europe in the wake of a Soviet invasion. Soviet conventional superiority in Europe led them to assume that another D-Day-type of invasion would be necessary, but no one was sure whether Canada would join such an enterprise under American or British command. Given that uncertainty, the Canadian Army, which was British in organisation, culture, equipment and staff procedures, had to get 'bilingual' – it had to learn the staff languages of both the British and American armies, especially as they pertained to amphibious operations, and it had to learn how to fight with US Army and US Marine Corps formations. The newly-established Canadian Army Staff College at Kingston was the crucible of its effort to 'bilingualise' the professional army between 1946 and 1956.[175] The staff college curriculum integrated training packages on American military organisations and equipment. American instructors were invited to deliver specific units on the conduct of amphibious operations, often for a week at a time. British directing staff from the School of Combined Operations did the same. American officers also joined, and soon outnumbered, British officers as exchange instructors in Kingston.[176] This was definitely a period of transition, but the Americanisation of the Canadian Army had only just started when the Korean War started, and the Canadian Army did well to have not jettisoned all things British. In 1950, the Canadian Army still had the weapons with which it had fought the Second World War, and it still operated very much like a British army in miniature. It made perfect sense

---

[174] Granatstein, *Canada's Army*, p. 316; and Stacey, *Six Years of War*, pp. 522-3.

[175] See Alexander W.G. Herd, "Preparing to Fight the Bear: American Influence on Canadian Army Staff Officer Education, 1946-1956" (PhD Dissertation: University of Calgary, 2011).

[176] In 1946-1947, there were three British and one American Directing Staff at the Canadian Army Staff College. By 1956, the numbers were reversed. See Howard Gerald Coombs, "In Search of Minerva: Canada's Army and Staff Education (1946-1995)" (PhD Dissertation: Queen's University, 2009), p. 141.

to place the 25th Canadian Brigade in the Commonwealth Division, where it fought with distinction between 1951 and 1953.

In deciding to send troops to Korea, the Liberal government of Louis St. Laurent followed a familiar pattern. Worried that support for the war would be shallow in Quebec, it moved cautiously, at first committing only three destroyers to the United Nations coalition. But pressure from the United States and English Canada induced the government to do more. Tearing a page out of Laurier's Boer War book, St Laurent authorised the recruitment of a Special Force brigade for service with the United Nations force, then commanded by the American General Douglas MacArthur. Seventy-five per cent of Canadians supported the UN action in defence of South Korea, but English Canadians were twice as likely as French Canadians to support the commitment of Canadian ground troops to the mission, even though the commitment was small by World War II standards.[177] Several Quebec newspapers echoed *Le Devoir*'s statement that Canada was now serving a new imperial master: "We are like little dogs that are eager to show their master that they adore him, that one gesture from him is enough to throw themselves into the water."[178] Still, there were no riots in Quebec streets and the debates were not nearly as heated as those of the past. South Korea in peril just didn't have the emotional pull of Britain in danger, and the Korean War affected far fewer Canadians than previous wars anyway. But the old divisions were still there and the deployment of troops picked a national scab that never seemed to heal. That was, and still is, the Canadian reality.

The first half of the twentieth century was a growing-up period for Canada. Canadians went from handing off their forces (and more or less washing their hands of them) in South Africa, to taking full responsibility for mounting and sustaining their own expeditionary forces and determining – for themselves – where, when, and how they would fight. It was not easy for the Canadian statesmen and soldiers who operated in coalitions. In fact, it was a series of balancing acts – of satisfying one part of the country's desire to do something, while accommodating the other part's unwillingness to get involved, just as it has been a matter of maintaining an ability to work with senior alliance partners while, at the same time, avoiding open

---

[177] Robert Teigrob, *Warming to the Cold War: Canada and the United States' Coalition of the Willing, from Hiroshima to Korea* (Toronto: University of Toronto Press, 2009), pp. 215-221. An August 1950 poll showed that, while only 21 per cent of Quebeckers offered unqualified support for the deployment of Canadian troops to Korea, 40 per cent of Ontarians supported the action.

[178] Cited in Ibid., p. 222.

commitments that could cause political problems at home.

Formally, The Korean War was a United Nations enterprise, but in reality it was driven by the United States. Deploying the overwhelming majority of the coalition's forces, the US was the unchallenged lead-nation, who decided the *modus operandi*, tactics and planning. The UK, Canada, France and other lesser coalition partners simply had to adjust to the best of their abilities to the standards and procedures set by the US. For all participating militaries of the future NATO countries this coalition deployment – the first combined operation since the end of World War II – formed the training ground for future Cold War co-operation and provided an eye-opener as to the needs in the fields of command and control as well as weapons technology and standardisation.

Chapter IV – The Crucible

HAVING UNDERTAKEN A *tour d'horizon* of key historical examples of coalition warfare – its benefits and its downsides – I will now endeavour to merge history with some theoretical ponderings on the matter. This way we may hope better to comprehend the mechanisms of the big melting pot that a fighting coalition always is, realising the challenges it poses and its influence on the men and women who struggle.

As far as actual fighting is concerned, in Europe the Cold War was a quiet period and coalitions did not tend to wage real war: the Soviet suppression of Hungary in 1956 and the Warsaw Pact occupation of Czechoslovakia in 1968 were some of the very few examples. Outside Europe, the War in Korea 1950-53 and the Suez intervention in 1956 were the only truly remarkable coalition enterprises seen. Since the end of the Cold War, coalition warfare engaging large parts of the international community as contributors has been much more frequent. It has been practised widely in order to share the burdens as well as to enhance the legitimacy of actions aiming at prohibiting lawlessness and atrocities in so-called failed states and states where ethnic or religious disagreement have ignited armed clashes. Allegedly, the long term goals have mostly been bringing tolerance and prosperity to the troubled areas, and it has become the norm rather than the exception that coalitions span a continuum from the technologically most advanced militaries to participants with rather limited capacities. Coalitions of the post-Cold War Era are generally large and varied as far as contributing nations are concerned, but as we have seen in the previous chapters, this kind of warfare is no invention of the 21st century – it can be traced back through the Modern Era and the Middle Ages all the way into Antiquity. And it is and always was a matter of prevailing morally and in equipment, and getting the upper hand as cheaply as possible. Clausewitz would agree that war is about forcing one's will upon the enemy, and powerful coalitions have often been the means chosen to bring this about.

Military Aspects

Doctrine and strategy

From coalition enterprises from *Fall Gelb* [Case Yellow], the German invasion of France in 1940, to Operation Corporate, the British re-conquest of the Falkland Islands in 1982, we know that joint operations represent significantly greater complexity than do single-service operations. During the German operations in the Netherlands, Belgium, and France in May 1940 the *Luftwaffe* supported the army's advance by close air support to the German panzer and infantry formations along with air interdiction and air offensives against the French airbases. Operation Corporate similarly was the British transfer and landing of troops in the Falklands following the Argentinean invasion in 1982 – a truly joint operation involving close co-operation amongst all three services employed in operations on the surface, beneath it and above.

While the expression 'coalition' signifies the outcome of a political agreement among two or more states to co-operate in war, in NATO parlance – at the armed forces level – the proper term for multi-national collaboration on the battlefield is 'combined operations.' In 'combined operations' the difficulties of 'joint operations' – martial activities where two or more services work together – are still extant, but with the added complexity due to the collaboration between the armed forces of two or more contributing nations. Because each coalition partner brings his separate national orientation and proclivities to the practice of warfare, the challenge grows.

In order to overcome the inherent difficulties and fight in a concerted way, a coalition must strive to agree on and operate using a common doctrine – a set of basic principles guiding the action on the battlefield. This doctrine must take advantage of the coalition partners' commonalties and reconcile their differences, creating a harmonious operational milieu. Such harmony is far from easy to bring about, but the effort will have to be made because doctrinal incompatibility amongst contributing militaries will always be a source of friction jeopardising the smooth running of operations.[1]

Like all military doctrines, a doctrine for coalition warfare is a set of basic understandings of the *modus operandi* in combined operations or – as described in the NATO Glossary – 'the doctrine is the sum of fundamental principles by which the military forces guide their actions in order to achieve their objectives. It should be authoritative but it will require sound judgement in its application.'[2] However, there are many more requirements that are just as important: the professional language in which coalition forces communicate, the battlefield missions, control measures, combined arms and joint procedures as well as command relationships, to name but a few. To achieve the full synergic effects of the combined and joint combat power, the war fighting doctrine should be common to all arms, because in the absence of a commonly understood doctrine planning and execution of military operations become extraordinarily difficult. Yet, the process of setting up a doctrine acceptable to all participants is frequently a cumbersome and time consuming task which has to be solved at the political level.

Apart from the doctrine's importance to the actions performed by the troops on the battlefield, waging war requires strategic concepts. This goes for coalition warfare as well as all other martial endeavours, and we can define such strategic concepts as notions laying down the aim and course of action which the coalition partners agree on after having come up with a sound estimate of the strategic situation. Frequently, it will be a short statement of what the aim is and what must be done to achieve it. It will be made in broad terms sufficiently flexible to allow its use in describing measures that should be taken in fields as disparate as military operations, diplomacy, economy and psychological warfare. Agreement among coalition partners on the exact strategy is fundamental to achieving well co-ordinated and successful military action. The strategy is derived from policy agreements between contributing nations and must be detailed enough to shape the direction of the upcoming military campaign, yet sufficiently broad to allow full exploitation of the capabilities of individual national forces. The disparity of the Franco-British strategic aims during the Crimean War is an example of how cumbersome the conciliation of two or

---

[1] Luft, *Beef, Bacon and Bullets*, p. xvi.

[2] *NATO AAP-6(S)*.

more partners' policy and strategic intentions can be. The development of an effective military strategy is a way of defining the aim with the war, and since it is difficult, even when military action is national, it is hardly surprising that it is even more trying in the crucible of hectic preparations for war undertaken by a number of nations with dissimilar cultures and traditions.

Strategy-making is an integral part of political life and it is therefore designed to accomplish political objectives – defining the aims *with* the war. Because it is found at the interface of political ambition and what is militarily realistic, it will be the point of reference for gaining consensus between military and political decision-makers – a consensus which, ideally, translates the aim *with* the war into aims *in* the war, or in other words: the military objectives. Nonetheless, history has presented us with a glut of examples that such consensus is difficult to achieve, and the Austro-Hungarian Emperor Karl's attempt in 1916 to negotiate peace with the Entente, which was duly sabotaged by his own Chief of the General Staff, Field-Marshal Franz Conrad Graf von Hötzendorf (1852-1925), and General Alan Brooke's frequent rows with Churchill during the Second World War over where to apply strategic concentration of effort are but two. Strategy is the single most important link between politicians and military leaders, but it is also the most likely centre of controversy in both political and military spheres.

Nations rarely enter a coalition with completely identical strategic views – let alone with the same intentions on all the ends to be achieved and the resources to be contributed by each partner. As a coalition increases in number of member nations, conflicting objectives and additional political constraints are added to the conundrum, but the common aim with the war remains central and must be the guiding notion throughout the conflict. Therefore the coalition's military commander-in-chief must strike a balance between compromise and military necessity because, on the one hand, he must achieve the continued backing of all coalition states and, on the other, he must preserve his ability to arrive at timely, realistic and adequate military decisions. At the same time, it is important to realise that in coalitions the will is strongest when the perception of threat is most urgent and this does not necessarily change simultaneously with all partners. As resources, threat level and conditions of the individual coalition partner change over time, so may his troops' fighting spirit, resolve and objectives. It may serve as a fitting example that after the Battle of Leipzig – the 'Battle of the Nations' 16-19 October 1813 – when Austria joined the anti-French coalition, Napoleon's German allies realised that it was no longer conducive to their individual aims of national survival and prosperity to stick with the French Emperor – and so in less than a year the whole Napoleonic alliance system fell apart.

While the strategic concept may well be a brief, general statement on what to do, the formulation of coalition strategy as a whole is difficult because of the sheer mass of specialties involved in the process. Coalition strategy requires merging and co-ordination of nearly every element of the multi-national pool of resources needed for accomplishing the military objectives. It calls for insights into various national industrial capabilities, mobilisation processes, transportation capabilities, and inter-agency contributions, in addition to military volume and capabilities. The coalition strategy must bind all these elements together with precision and care, and the various concessions to and the needs of individual coalition partners must be reconciled so that they do not hamper the achievement of the over-arching common aims. We must therefore expand our understanding of strategy to include not only the crossing-point of politics and defence, but also the borderland between international relations, diplomacy and military necessity amongst which it must seek to a establish reasonable harmony. In its entirety, the coalition strategy addresses issues as weighty as the end-state, or war aim, which partners have agreed to achieve, and as mundane as the rules of engagement to be applied at each stage of the conflict.

In order to put the coalition's strategy into practice it must be translated into achievable military objectives, and to obtain these the planning staffs must develop one or more campaign plans. The ability to plan an effective military campaign presupposes national and coalition concord on doctrine and strategy. Therefore, at the political level any possible disagreement amongst partners must be settled before the purely military campaign planning can be initiated.

At the operational level, the disagreements that might occur are mostly amongst the military professionals, but political ramifications, which might also crop up at this level, must be dealt with speedily and efficiently. We saw how, frequently, during many top-level conferences of the Second World War, American Admiral King suggested – contrary to the agreed strategy of 'Europe first' – that priority of force allocation and shipping should be given to the Pacific theatre instead of those of the Mediterranean and Western Europe, and how – almost immediately – this was put right either by Generals Marshall and Brooke or by the heads of state and government themselves.

As was so obvious during the build-up before Operation Overlord in 1944, a military campaign must be phased in concurrence with the availability of combat power generated from multiple national sources. In the case of Overlord, resources were accumulated in southern England over a long time as troops from North Africa became redundant there, as landing craft became available, and as the Allies consolidated their foothold in Italy.

Moreover, the campaign plan should provide a basis for defining and recommending national contributions: what kind, how much, how and when. Similarly, in the recent campaigns in the Helmand Province in Afghanistan, while the UK provided the lead-formation and an impressive array of versatile resources, the Danish contribution was limited but specialised. This was agreed and planned prior to deployment and it worked admirably well. Had there not been such consent, and had it not been put efficiently into practice, the combined commander, Commander Task Force Helmand, would have ended up with a force composition that was not rationalised towards the operational requirements.

The coalition's campaign plan should serve the purpose of integrating the perceived needs and the actual availability of forces and equipment. It serves as both a driver for force requirements and a schedule for generating such assets. It is the instrument which is applied in order to synchronise all elements of combat power, and we may use the German campaign plan for *Fall Gelb*, the 1940 attack on France, as an example. Initial preparations were commenced as early as in September 1939 when *Führerweisung Nr. 3* (Fuhrer directive No 3) ordered all forces that could be spared in Poland to assemble in the West. Then, on 19 January the draft for *Weisung Nr. 10* (Fuhrer directive No 10) laid down the campaign plan and the sequence of build-up in assembly areas close to Germany's western borders with a view to striking in the late spring of 1940 when sufficient army and air forces were expected to be available and ready to deploy.

The campaign plan provides combined commanders with the vital understanding of how operations, battles and engagements are to be linked to the coalition's strategic objectives. For example, during the Suez Crisis in 1956 the Anglo-French-Israeli campaign plan foresaw that Operation Musketeer/*Opération Mousquetaire* would be closely linked with the Israeli Operation Kadesh – an armoured thrust into the Sinai dessert. This operation was expected to be launched as a prelude to the Anglo-French amphibious landings. Moreover, the campaign plan is the tool of co-ordination of various activities with a view to achieving the end-state, the aim with the war, which has been laid down in the strategy. It must address a variety of choices concerning the approach to warfare – offensive or defensive, terrain- or force-oriented, direct or indirect approach – and in so doing, it becomes the instrument for the actual employment of coalition forces.

Planning

In the Thirty-Years War, the War of Spanish Succession, the Revolutionary and Napoleonic Wars, and to a certain extent during the two world wars as well, generally the lead-nation commanders-in-chief planned, and subsequently told the lesser coalition partners' individual commanders what to do. That was what usually happened if no partner was at a par with the lead-nation, but in the relatively rare cases of two or more almost equal powers working together no such one-way control could be wielded: in that case planning would be a matter of mutual agreement. Within modern coalitions, common planning processes are *de rigueur*. Therefore, the degree to which national commanders and staffs understand and are able to participate in combined planning, their diplomatic skills and tact, their grasp of inter-allied procedures, and their touch with global security issues all impact on their combined effectiveness. This efficacy matters with respect to the time required for planning, as well as the sharing of knowledge of every component of the operations that need to be considered.

The common planning process should facilitate unity of command, which is the basic precondition for integrating coalition forces in the combined multitude of arms. Unity of command is one of the most fundamental principles in warfare, but the single most difficult rule to implement in combined operations. It is dependent on many influences and considerations. Coalition partners are often very strict about preserving their operational independence. Because of the severity and consequences of decisions taken in war, relinquishing national command and control of forces is an act of reliance and confidence that is rare in relations between sovereign nations. It is a trust of human and material resources into another nation's care, and the blame will be substantial if they are squandered by some unforeseen imprudence on the part of the ally. In a coalition, extra-national subordination is achieved by constructing command arrangements and task organisations of forces to ensure that responsibilities match capabilities, contributions and efforts. For example, in the case of the Gulf War of 1991, command relationships between national commanders had to be carefully considered in order to ease the stress caused by disparate cultures and to ensure that authority vested in a commander matched the confidence he inspired among his coalition subordinate leaders. However, at the same time it is evident that arrangements made in command issues should first and foremost serve the war fighting requirements.

Training

The first priority in generating coalition combat power from a conglomeration of nationally disparate units is for those units to train together, emphasising their fundamental commonalties, common, doctrine, and the agreed compromises that might exist. The collaboration of British and Canadian forces during the world wars benefitted from almost identical organisations and staff procedures. When, during Cold War years, Canadians increasingly had to co-operate with American forces, new standards had to be acquired, and staff trainers both from the UK and the United States were invited to make sure that, militarily, everyone spoke the same professional language.

It is only through training that combined units can become capable of mastering and sustaining collective war fighting skills. This maxim was efficiently heeded by Montgomery in the western desert in 1942-3, as well as later on in north-western Europe. Similarly, the Danish ISAF Team 10 going to the Afghan province of Helmand in 2010 conducted extensive training with their British partners – in the UK as well as in Denmark – prior to deployment. This was particularly relevant because, as the coalition was brought together, staffs and commanders had to adapt rapidly to the common doctrine of the fighting organisation, which was in the process of being formed. It is through the training during the preparatory phases that the impediments and sources of friction become clear and the ways to solve the trouble are arrived at. Similarly, it is frequently during such pre-deployment familiarisation that personal bonds are tied and the solutions to future challenges are discovered.

There are fundamental considerations concerning the factors of mission, enemy, terrain, troops and time available on the battlefield that must be addressed during training. Some militaries rely heavily on armoured, cross-country transport, while others tend to prefer small, agile vehicles. Some, like the Danish ISAF Team 10, deploy battle groups with the full panoply of weapons including tanks, while others choose lightly armed and lightly armoured multi-purpose means of transportation. Therefore, factors like mission and terrain will dictate the alignment and tasks of differently equipped and talented forces on the battlefield. While lightly armed forces can perform military operations in urbanised, densely foliaged, or mountainous terrain, heavy and more mobile forces are suitable in open environments, and airmobile or motorised forces in virtually any terrain. In any event, pre-deployment training must take place for each of the categories of troops, but it should also facilitate co-operation amongst them.

Command, Control, Communications, Computers and Intelligence

As soon as a combined doctrine is agreed upon, it must be put into practice with an aim towards training and familiarising the troops with the tasks foreseen and the prospective conditions on the ground in the deployment area. The application of this dogma – the doctrine of the coalition – relies on a command, control, communications, computers and intelligence architecture, C^4I in modern military lingo, which is supposed to fuse the endeavours of the disparate units and formations of all prospective coalition contributors. However, the C⁴I assets are in short supply with many of the countries which are keen contributors of contingents to today's coalitions-of-the-willing actions around the world. These shortcomings in the electronic 'brave new world' risk disrupting the highly coveted 'seamless co-operation of forces' on the battlefield. Troops not in possession of the state-of-the-art electronic systems are left in the dark as far as situational awareness is concerned. When it comes to continuing to improve capabilities for collecting, analysing and disseminating intelligence, managing the vast amounts of information upon which decisions are made, and incorporating more and more computer aids to the battlefield decision and execution processes, a coalition requires all partners to be in the electronic loop. Unless a coalition maintains the ability to share with, and in turn receive from, all coalition partners, its battlefield will not be as transparent and seamless as it would wish it to be, and significant additional risks and challenges will materialise.

Nonetheless, achieving integrated C⁴I within a coalition comes with a number of caveats. First, there is the language barrier. Communication problems are among the primary sources for tension, confusion, misunderstandings and, consequently, inefficiency in multinational coalitions. While spoken and written language is a vehicle to convey ideas, desires and feelings, it is also a source of frequent miscommunications. Bygone times' huge cells of military interpreters are not an adequate response to this challenge. In modern coalition warfare speaking through interpreters is an inadequate and dated method, and adhering blindly to this notion will seriously hamper the smooth conduct of coalition operations. Direct communication by means of an agreed common language has been practised for centuries, and today, when the demands for speedy decision-making are as acute as never before, this is of crucial importance. In the 17th and 18th centuries, French was the *lingua franca* spoken by all officers and diplomats. Similarly in the 21st, English and, in some cases, French are the preferred languages used by military staffs when engaged in coalition warfare, with translation into national languages happening only at the lowest tactical or combat levels. In this context it is worthwhile noticing that during the Danish ISAF Team 10's stint in the Helmand Province in 2010-11 under British lead, even at platoon and section level communication was conducted in English as British, Danish and Afghans patrolled together, although individuals speaking with a thick local accent caused occasional trouble.

C⁴I resources are essential parts of the so-called network-centric-warfare concept, which aims at enabling military units to engage targets quickly and accurately. This can be done because all sensors and means of communication are linked electronically and will always be capable of pinpointing targets irrespective of fog, darkness, snow or other encumbrances. However, troops with no or limited access to such means are left out of the loop. This is a trouble which stands out in many research papers and quite a few theses on coalitions and networked collaboration submitted by students at professional educational institutions, notably military staff courses, 'higher command course' equivalents etc. While there are a great variety of themes among papers submitted by military students, many writing on issues concerned with their personal experiences of coalition warfare participation, a recurring theme seems to be a general concern that Network Centric Warfare – driven by the US military's keenness to integrate as much information technology as possible into its operational concepts – threatens the cohesion of coalitions and the efficiency of their operations. This kind of challenge to coalition cohesion posed by the overwhelming use of information technology appears to exist at all levels of warfare.

As the USA's principal military partner, the UK has a strong interest in keeping up with America's military developments. Even with the third largest military budget in the world, at least until implementation of the recently announced cuts in military expenditure, the British concern about not being able to keep pace with the United States clearly puts the problem of coalition interoperability into stark contrast. And if the British cannot manage, who can?[3]

Logistics

Supplying armies is and has always been crucial to warfare. Lazare Carnot as well as his keen student, Napoleon Bonaparte, realised that by dropping the time-honoured system of fixed depôts their armies could live off the country they passed through. This made army columns much shorter and enhanced the troops' operational flexibility. Upon introduction of the internal combustion engine into army transport, the need for vast quantities of fodder for riding and draught animals slowly vanished and made logistics even less complicated, and with the advent of helicopters and satellite based information systems allowing focused supply of the forces, the burden of supplying the front line was once again eased.

Coalition forces' management of logistics is a matter depending on a

---

[3] Paul T. Mitchell, *Network Centric Warfare and Coalition Operations: The New Military Operating System* (Oxon: Routledge Global Security Studies, 2009) p. 111.

wide variety of variables: lines of operation, access to strategic raw materials such as fuel and minerals, force protection, means of transport, management systems and procedures, etc. Moreover, practical matters like national arrangements made by individual coalition forces, host nation support agreements with the receiving country and equipment compatibility amongst allies are parameters that are important to coalition planning.

Some coalition forces will enter the coalition with the intention of, and means to, provision themselves. In these cases, coalition control may be no more than a need to co-ordinate or provide ports of entry, offload capabilities, storage sites, and routes and means for pushing sustainment forward. Others will arrive with the need for more extensive support. This may be solvable through bi-national agreements between two member nations providing support to one another, or may require active coalition management. As a rule, actual execution of tactical logistics support to alliance members should be decentralised. At the coalition headquarters level, the focus should be on measuring the requirements of executing the campaign plan, providing advance estimates of these requirements to national units, and ensuring that proper controls are in place to de-conflict and permit movement and processing of combat power to units.

Constraints: Culture

Coalitions are brought together by partner states, each with its individual cultural background, operational traditions, language, history, norms, and sensitivities of political or religious natures. Even as smooth a co-operation as that within the British Empire during the world wars had its drawbacks, as in the first year of World War II, for instance, when the British were left in the midst of fighting with no options for getting out. The British Dominions, however, were – for some time – still discussing whether or not they should take part in the Imperial defence. Similar examples could be found in the 2000s' War on Terror, where many nations, but certainly not all, joined the tussle. During the war in Iraq from 2003-7, Germany and France categorically refused to come along. Similarly, in Afghanistan from 2001 onwards countries like the UK, the US and Denmark accepted the serious costs of fighting in Helmand, while others restricted their forces to the more peaceful North.

Decisions to join coalitions are always political matters, and national decision-makers often differ as to size of their countries' contributions, their rôle and the rules of engagement. However, as soon as a decision to join has been taken, the forces committed must merge into one single fighting entity and work as seamless together as at all possible. The ability to do so is rarely as easily achieved as it happened in the case of Britain and Canada in the world wars, and in most cases it grows slowly. The more familiar the coalesced military organisations are with each other, the quicker is the learning process. Therefore, the level of exposure of military organisations to other cultures in the pre-coalition stage determines their ability to minimise cross-cultural tensions and misunderstanding with fellow coalition partners. Nevertheless, even close partners may be disparate in one or more respects, prone to occasional clashes of interests as well as to differing views on and approaches to strategy and operations. Their strategic interests are never completely identical, their ways of fighting and commanding may differ considerably as may their views on international treaties and agreements, and each of them will strive to preserve their operational independence and the national identity of their troops.

Understanding the behaviour, norms and sensitivities of partners of dissimilar cultural backgrounds is a precondition for smooth co-operation and flawless interoperability within a coalition, and it goes without saying that the more used a military is to co-operating with foreign partners the easier will it adapt to fighting alongside new ones. The frequent NATO exercises and the annual turns of the defence planning cycle during the Cold War years served exactly this purpose, and they have contributed enormously to making the combined fighting against terrorism in the new millennium a lot easier than it might otherwise have been. It is also of no small importance to realise that the more educated coalition officers are regarding the world beyond their national borders, the more painlessly they mingle with foreign colleagues. The challenges in this field are to gain awareness with respect to the differences and to analyse them concerning their potential for causing insult in order to make sure that due respect is paid to everyone and that misunderstandings and embarrassment are avoided. Different cultures mean different doctrines and disparate approaches to planning and execution of operations; and understanding why will require a few observations concerning the mental layout of human beings.

Every person has a vast number of ideas in their head. A network of such ideas within the individual's brain may be termed a mental model. For all of us, mental models provide explanations of how things around us work, and they will therefore influence our judgements, reasoning and decisions. Mental models tend to spread when people of any given society come into contact with each other, and when they are widely distributed and long lasting they become parts of the cultural fabric of the community – *cultural* models. One of the characteristics of cultural models is that they are shared by many people across space and time. As contact is of essence in this process, we may assume that the more geographically separated coalition partners live, the less contact they can sustain and, consequently, the fewer are the shared cultural models.

We may claim that mental models are domain specific because they are explanations on how instruments, processes and groups work. Decision-making, negotiation and team-working are some of the domains which are of profound importance to coalitions, and it seems obvious that partners with cultural models as different as those of Danes and Afghans require considerable adjustment before smooth and well-organised co-operation becomes the norm.

This is as true today as it was in the nineteenth century, when the British envoy to Afghanistan, Sir Pierre Louis Napoleon Cavagnari, was forced to realise this reality. On 3 September 1879, he and his entire entourage were murdered by citizens of Kabul regardless of the peace agreement between the Afghan Amir Yakub Khan and the British-Indian government signed in Gandamak earlier the same year. On a somewhat different note, the Danish forces in Afghanistan in 2010 training the Afghan National Security Forces experienced behaviour and attitudes with respect to precision, task-orientation, sexual preferences and hygiene which was – to put it politely – rather foreign to their own conception of modern, military efficiency and bourgeois decorum.

Culture may be defined in various ways. We often see it as a combination of traits such as language, religion, customs, laws and philosophy. But we may also define it in much broader terms as an ethical pattern distinguishing one group from another, comprising elements like shared values, experiences, myths, perceptions and goals, which may have been passed on through generations. The culture of a society manifests itself in a shared worldview and common attitudes, codes of conduct and conceptions of self and strangers.[4] As far as coalition co-operation is concerned, some of the cultural models which matter the most are methodical planning, analytical approach to decision-making, language, social behaviour and perceptions of dignity.

Casualties, 'friendly fire,' collateral damage, and respect for human life are other prominent domains where cultural models frequently differ to the detriment of coalition co-operation. This became especially obvious as a parliamentary debate was raised over Danish troops' rendition of prisoners to the Americans in Afghanistan in 2002. A former prisoner-of-war later complained that he had been badly treated by the US troops, and questions were raised concerning the legality of rendering prisoners to a country whose norms for treatment of captured enemies was known or believed to

---

[4] Francis Fukyama in Luft, *Beef, Bacon and Bullets*, pp. 1-2.

be more harsh than the one practised by one's own military personnel.

Military organisations are sometimes dominated by members of a distinct confessional or sub-cultural group and do not necessarily carry the cultural traits of the community they are bound to defend. To a large extent, society's religiosity determines its ability to stomach casualties. Militaries from West European societies with a high sensitivity to casualties are likely to reflect their sensitivity by relying on firepower and air power, reluctance to engage in high risk operations, conservative training programmes, and an adherence to often multifarious safety standards in the daily life of the troops. This can cause tension regarding certain ethical and moral issues applicable to military life such as rules of engagement, treatment of civilian populations in enemy territory, environmental issues, and the use of certain types of weapons, torture and abuse. Unlike countries like Britain and Denmark, the US appears willing to tolerate casualty figures considerably higher than most coalition partners, which is part of the reason why the Americans operate in a more daring manner than do most of their European partners.

As we briefly touched upon in a previous chapter, the American criticism of Montgomery's cautious and thoroughly rehearsed operations in Europe in 1944-45 got a parallel in the recent Anglo-Danish counter insurgency operations in Afghanistan's Helmand Province. In December 2010, WikiLeaks disclosed that high-level American and Afghan politicians had expressed severe criticism of the British – and by implication the Danish – strategy, resource allocation, risk avoidance and allegedly poor ability to create security in Helmand. Be that as it may, the criticism shows a remarkably different attitude to losses – a disparate cultural norm.[5]

The fear of casualties and the possible repercussions for politics at home are some of the causes for the significant operational restrictions on what European coalition partners will permit their ground forces to do during deployment.[6] Moreover, this is true, not only as far as own troops are concerned, but equally as much as to the opponent. To most European powers casualties, also among the opposition, should be minimised and civilian fatalities must be avoided at almost any cost. Vis-à-vis the American coalition partner this leads to differing rules of engagement, and may be one of the causes for disagreements over the handling of hostile combatants, human shields and prisoners-of-war.

---

[5] *The Guardian*, 3 December 2010, "WikiLeaks cables expose Afghan contempt for British Military."

[6] Mitchell, *Network Centric Warfare and Coalition Operations,* p. 115.

It is not only doctrinal prescriptions and ethical standards that are rarely identical – partners normally emphasise the display of independence and sovereignty of their own forces. Since they have to collaborate, this is a challenge which must be addressed. In 1815 at Waterloo, Blücher and Wellington attacked along converging, though clearly separate, axes making it possible for the Prussians to claim victory in their own right. In the Gulf War of 1991, the coalition-of-the-willing set up a system of parallel commands. The Saudi commander, Prince Khaled, and the American commander, General Norman Schwarzkopf, each commanded his own separate headquarters and his own designated units. The American command included all American units as well as British and Canadian ones. The Joint Forces Command (JFC) headed by Prince Khaled was responsible for all Arab parties – the Egyptian and Syrian contingents being the largest – as well as forces from France, Senegal and Czechoslovakia.[7] This was indeed a sensible arrangement since, as in most coalitions, the war aims of coalition partners are seldom exactly the same. Khaled had to defend Saudi territory and ring military forces around towns on his country's border. Unrealistically, he expected the US to do the same, but soon had to realise that safeguarding Saudi territory was not the only purpose the European and North American partners had come to achieve. The Americans were hesitant to spend military efforts guarding what they saw as 'the vast empty.' They appeared to be interested mainly in offensive operations and defending targets of strategic importance such as oilfields, airfields, and main transportation routes.[8]

One of the most decisive cultural challenges to coalitions is that of a working language, and since France left NATO's integrated military structure in 1966, in most coalitions English has been almost without competition in this respect. However, using the same official language among coalition partners does not necessarily mean avoidance of all misunderstandings, as languages acquired late in life rarely gives the user the perfect grasp of the vernacular and especially not of dialects mired in the individuals' different social backgrounds. Although clashes of interests and misunderstandings of intention did materialise between Generals Prince Khaled and Norman Schwarzkopf, the Gulf War turned out to be the first modern war in which no major incidents of miscommunication due to

---

[7] Luft, *Beef, Bacon and Bullets*, p. 155.

[8] Ibid., p. 158.

language gaps were recorded.⁹

Across history, cultural disparity and lack of delicacy at the encounter of two different religious systems, as well as different views as to how to handle this conundrum, has led to strange results and the Gulf War of 1991 was no exception. Shortly before Christmas a popular French singer, Eddy Mitchell, arrived in Saudi Arabia with his band to entertain the French contingent. This, however, conflicted with the Saudi *wahabi* piety instituted by the 18th century Muslim theologian Muhammad ibn Abd-al-Wahhab, who had preached against what he saw as the moral decline and political weakness in the Arabian Peninsula. For this reason the Saudi king forbade the performance, but the French Defence Minister, Jean-Pierre Chevènement, who strongly objected to this unsolicited interference, instructed his commander in the Gulf, Lieutenant-General Roquejeoffre, that Eddie Mitchell must sing. Since the French had refused to serve under American command and opted for that of Prince Khaled, they now faced veto by the latter. Caught between Khaled's and Chevènement's conflicting orders – or disunity of command – General Roquejeoffre was experiencing one of the most typical dilemmas of a coalition commander: whose orders to follow.¹⁰ In the end President François Mitterrand ruled in favour of Arab sensitiveness.

The need to reconcile diverse cultural traditions or religious practises is exemplified by the precursors of the Sepoy Rebellion of 1857. The sepoys were the Indian infantrymen of the British Honourable East India Company's armies. In 1772, the Company's armies saw rapid expansion and sepoys were recruited primarily from among the high-caste rural Indians. In order to forestall social unrest, the Company sought to adapt its military practices to the requirements of their religious rituals. Consequently, these soldiers dined in separate facilities and were exempted from overseas service, considered polluting to their caste. Moreover, the army soon came officially to recognise Hindu festivals. These seemingly wise concessions to local tradition also bore the seed of trouble. Some Indian soldiers misread the presence of missionaries as a sign of official intent of mass conversions of Hindus and Muslims to Christianity. Also, changes in the terms of their professional service may have created resentment. A financial grievance stemming from the general service act, which denied retired sepoys a pension, caused considerable trouble, and while this only applied to new recruits, it was widely suspected that it

---

⁹ Ibid, p. 168.

¹⁰ Ibid., p. 181.

would also affect those already in service. However, while these minor disagreements and misunderstandings all added to a general and increasing tension, the major cause of resentment arising ten months prior to the outbreak of the rebellion was the General Service Enlistment Act of 25 July 1856.

Until then, men of the Bengal Army had been exempted from service overseas. Specifically, they were enlisted only for service in territories to which they could march. Although the Act required only new recruits to the Bengal Army to accept overseas deployment, serving high caste sepoys were fearful that it might eventually be extended to them. The final, and – in terms of cultural differences – most decisive controversy that sparked the explosion was over the ammunition for new Pattern 1853 Enfield Rifle. To load the new rifle, the sepoys had to bite the cartridge open. Many Sepoys believed that the paper cartridges were greased with lard (pork fat), which was regarded as unclean by Muslims, or tallow (beef fat), an abomination to Hindus. On 27 January 1857, the Company ordered that all cartridges issued from depots were to be grease free, and that sepoys could grease them themselves using whatever lubricant they might prefer. This, however, caused many sepoys to be convinced that the rumours were true and that their fears were justified. The Sepoy Rebellion started on 10 May 1857 and lasted until the signature of a peace treaty on 8 July 1858. During that period the defence of the British Raj was in great jeopardy. The trouble caused by these cultural and religious misunderstandings could not be allowed to repeat themselves, and for this reason the Crown took over from the East India Company sole responsibility for British interests in India, including all military matters.

Conciliation of cultural and religious differences must be overcome to achieve optimum collaboration among coalition partners, and a lead nation will go out of its way to accomplish precisely that. However vague the common cause, however few the shared values, sensible coalition partners will hardly let irrelevant factors like ethnic or devout sensibilities stand in the way of completion of the military enterprise upon which they have embarked.

As in other cases of cross-cultural coalition relations, the effect of mutual racial antipathy or arrogance cannot be discounted. The famous naval theoretician, A.T. Mahan – one of the intellectual leaders of the yellow peril movement in America – referred to the Chinese as pitifully inert and dangerously barbaric. Similarly, the Sino-American co-operation in World War II was hampered by a strong reciprocal racial prejudice. Most Chinese and Americans viewed each other as inferior races and had a low regard for their partner's culture. Moreover, the strategic and technological incompatibility between the American troops and their Chinese counterparts was severe and produced a coalition that suffered right from the beginning from almost every possible disadvantage: divergence of aims, poor logistic co-ordination, lack of trust, and operational disagreements. However, despite many cultural disparities, there was one otherwise important cultural element that played a very small rôle in the flawed Sino-American coalition: religion. Although the Chinese adhered primarily to Buddhism and Taoism and the Americans were Christians, the relationship between them was, like the Japanese and the British during the First World War, one of mutual respect and religious tolerance. On the part of the Chinese, this tolerance may have been furthered also by their traditional inclination towards Confucian philosophy, which does not assign importance to the different religious persuasions of fellow human beings.[11]

Coalition Partners and Dominance

In any coalition the lead-nation will hold considerable sway over strategy, campaign plans and the rôles of lesser contributors. During the Napoleonic Wars there could be no doubt that France was such a dominant coalition leader. On the opposite side – although various other great powers did have a say in coalition war policy – Britain, for the simple reason that she was paying the most, eventually came out as the *primus inter pares* – the first among equals – in the final battle at Waterloo. She also managed to place herself in a decisive position at the Conference of Vienna in 1814-15, where the rearrangement of the European political layout in the immediate aftermath was negotiated under the aegis of Prince Metternich, the Austrian host.[12]

Similarly, during the war between France and the North German states in the late nineteenth century, Prussia took the obvious lead being not only the victor of two previous wars – one against Denmark in 1864 and another against Austria in 1866 – but also demographically larger than the rest and economically at the forefront of German industrial and political development. Although this did not happen without protests from various South-German Catholic principalities and kingdoms, notably Bavaria, the Prussian dominance was generally accepted as unavoidable and as a precursor of German unification, which came about at the coronation of

---

[11] Ibid., pp. 143-44.

[12] Klemens Wenzel Nepomuk Lothar, Fürst von Metternich-Winneburg zu Beilstein, 1773-1859.

Prussian King Wilhelm [William] as German emperor in 1871.

Not only the British and the new German Empires realised the need for coalitions. In 1887, General G.A. Leer, the father of Russian modern strategic thinking, insisted that 'we are now entering on a period of great coalition wars' making the political preparation for war even more important.[13] At the time, opinion over the issue of principles governing this kind of warfare was divided, in Russia as much as in other European military circles, and while Leer was preoccupied with establishing general and unchanging principles, his many opponents seemed more interested in predicting the possible duration of such wars. Taking a keen interest in the relationship between war and economic performance – a theme that was becoming increasingly prominent in the late nineteenth century – a professor at the Russian General Staff College, Colonel Gulevich, claimed that any future war would engulf the whole of Europe, drawing in some nineteen million men and drawing on the whole strength of the states.[14] This theme was thoroughly analysed by Warsaw banker Jan Gotlib (Bogumil) Bloch (1836-1902/1901), a keen student of modern industrial warfare. War, he predicted, would become a duel of industrial might, a matter of total economic attrition. Severe economic and social dislocations would result in imminent risks of famine, disease, break-up of societies' cohesion and, consequently, revolutions from below. His contemporary, Russian General Mikhnevich, in his *Strategiya* (Vol 1), opined that coalition military forces are always less than the sum of the component armies, and allies will strive to shift the heaviest burdens onto other shoulders. Moreover, diversity of aim was an inherent problem with all coalitions. Coalition strategy would have to be flexible, abstaining from operations risky enough to alarm allies, but decisive enough to hold the alliance together. Countering enemy coalitions, he believed, meant seeking out weak spots, both political and military.[15]

Since, in World War I, the Entente's war aims were dictated by diverging national interests, the coalition's strategy materialised rather as a succession of compromises than as a genuinely harmonised concept – let alone a preconceived campaign plan. Later Soviet military historians have claimed that the bourgeois military thought had 'failed to absorb the lessons of the first wars of the "imperialist epoch" and failed equally to understand

---

[13] Erickson "Koalitionnaya Voina: Coalition Warfare in Soviet Military History, and Performance" in Keith Neilson and Roy A. Prete, Eds., *Coalition Warfare*, p. 85.

[14] Ibid., p. 87.

[15] Ibid., p. 88.

the new socio-economic conditions which had come to prevail: more specifically, too little attention was paid to economic and moral factors, which led in turn to disinterest in command and control problems of coalition warfare and the military integration of the allied armies. Strategic plans were worked out, but on purely national lines without reference to the requirements of existing coalition agreements.'[16]

The Central Powers – the World War I coalition opposing the Entente – included Austria-Hungary, Germany, the Ottoman Empire and Bulgaria. It was clear from the outset that the Ottoman Empire was the weak link in need of much assistance and advice, which was, however, frequently ignored. Bulgaria was a minor partner, although she was of key importance to the coalition because her geographical position made her the link between the Habsburg and the Ottoman Empires, allowing the Central Powers to operate on 'inner lines.' As we have briefly touched upon in Chapter II on World War I politics, the Austro-Hungarians experienced trouble from the first day of mobilisation, because the concentration of effort had to be shifted at short notice from the Serbian to the Russian front. Prior to the outbreak of war, it was obvious to the two chiefs of the Austro-Hungarian and German general staffs, Franz Conrad von Hötzendorf and Helmuth von Moltke the Younger respectively, that Germany would be seriously engaged on the western front. Austria-Hungary agreed to launch an offensive through Galicia to take on the major Russian forces themselves. To facilitate this, Germany offered substantial support to operations on the Eastern Front by means of an attack by the entire German Eighth Army from East-Prussia in a southward direction. However, both coalition partners tried to evade their obligations.

While Austria-Hungary preferred to deal with Serbia – constituting a graver risk to her security – before turning on Russia, Germany, believing in a quick victory in France, moved eight-ninth of her forces west and provided only minimal contributions to the war in the East. Austria-Hungary had prepared for war with a group of thirty divisions called the *A-Staffel* in Galicia, one of ten division prepared for defence in the Balkans called *Minimalgruppe Balkan* and a flexible force of twelve division called the *B-Staffel*, which could be dispatched to reinforce either of the former groups. However, primarily due to poor rail communications, when first set in motion towards one of the theatres *B-Staffel* would face severe difficulties if ordered to shift to the other one of these. Therefore, as Germany ignored her promise to launch substantial attacks from East-Prussia, Austria-Hungary had to shift the considerable forces of *B-Staffel*,

---

[16] Ibid., p. 92.

already mobilising and partially on the move, from the Serbian front to fight a lonely battle against the Russians. At the same time, the German high command realised that no easy victory beckoned on the Western Front. This prohibited reinforcements to be sent east until the spring of 1915, and in the meantime the Austrians suffered massive losses. As a result, in spite of the successful, albeit defensive, battle of Tannenberg where the Russians suffered a resounding defeat by the Germans commanded by Field-Marshal Paul von Hindenburg, a highly mobile attrition campaign wore on in the East until 1915.[17] Also, Conrad von Hötzendorf's first offensives against Russia were remarkable for their lack of effect combined with massive human cost. Moreover, successive strategic mistakes led to a disastrous first year of war that crippled Austro-Hungarian military capabilities.

The most devastating defeat came in 1916, in Russia's Brusilov Offensive, where the Austro-Hungarian forces under Conrad's command lost nearly 1.5 million men and were never again capable of mounting an offensive without German help. Thus, most of Austria's later victories were possible only in conjunction with German armies, on which the Austro-Hungarian army became increasingly dependent. From the Brusilov fiasco onwards, Germany was the obvious lead-nation of the Central Powers coalition, and the entire Eastern Front – including Austro-Hungarian forces – came under German operational control by Field-Marshal Paul von Hindenburg. For the rest of the war all major policy and strategic decisions were taken by Berlin and the German *Oberste Heeresleitung*.

From the end of World War II onwards, the USA has been the dominant power of most formal as well as many less regulated coalitions. This was the case in the Korean War (1950-53), in the post-French Vietnam War (1955-75), and during the Cold War (1947-91). Even in conflicts fought by coalitions not including America, the US has been able to wield considerable influence on their duration, termination and outcome. This happened during the Suez crisis in 1956 and to some extent in the various Arab-Israeli Wars.

After the end of Cold War, the re-unification of Germany, and the re-definition of NATO's rôle, many things have changed and – as we have seen with the bombing campaign in Libya in 2011 – US lead is no longer a given. While NATO partners were generally supportive of the US after the atrocities of 11 September 2001, agreeing to the UN sanctioned action in Afghanistan to eliminate the Al Qaeda menace, there was no unison in

---

[17] Norman Stone, "Moltke and Conrad: Relations Between the Austro-Hungarian and German General Staffs, 1909-14" in Paul M. Kennedy ed., *The War Plans of the Great Powers, 1880-1914*, pp. 222-26.

2003 when the George W. Bush administration tried to persuade old allies as well as potential new ones to back their war in Iraq. Severe divergence emerged between the United States and some of its major partners, notably Canada, France and Germany, indicating a decline of the American hegemony. Unlike in 1956, when the US was able to force France and Britain to back down over Suez, in the post-Cold War environment of 2003 Washington was unable to mollify partners and make them adapt their policies to American needs. Indeed, as the dispute went on, each side became more intransigent.

The Downsides of Coalition Warfare

Coalitions have both advantages and downsides. While the benefits include burden sharing, increased legitimacy, and frequently strategic and operational gains – such as access to resources, encirclement of the opponent and large reserves – the drawbacks also loom large on the horizon. As mentioned above, among them are differing war aims, disagreement on contributions, cumbersome decision-making processes, different sensitivity to casualties, and troubles arising from uneven access to the information technology on which today's Network Centric Warfare is based.

Unlike permanent or semi-permanent alliances – such as NATO, SEATO from 1954 to 77 and METO/CENTO from 1955-79 – whose *raison d'être* are generic, intangible threats, the war aims of coalitions set up for specific occasions are usually limited to achieving what the partners have originally set out to do – nothing more. At the end of the seventeenth century, the Grand Alliance of Protestant countries threatened by French King Louis XIV's expansionist policy had the sole purpose of reining in aggressive France. Similarly, the slightly smaller coalition-of-the-willing, comprising Britain, Denmark, Brandenburg and a large contingent of exiled French Huguenots, fighting for King William III of England and Scotland in 1689-91 was formed exclusively in order to rid Britain of the Jacobite menace in Ireland, along with the longer term prospect of invasion of Scotland. King James II was King Louis XIV's proxy and the conflict in Ireland aimed at tying up British forces, preventing their transfer to Flanders and their interference with his wars in Europe. Combating this scheme was what the coalition had agreed upon. Should lesser partners need forces for protection of their homeland, this was not part of the deal, although the treaty text was amended by a clause allowing partners to withdraw own troops should an emergency occur at home.

In the 1689 coalition-of-the-willing there was disagreement on troop contributions. Like in any other war, the forces were worn down, casualties and fatalities from sickness as well as desertions constantly tore up the troop contingents, and while the lead-nation frequently reminded Copenhagen of the force goal set up in the treaty, the Danish King Christian V never filled up his depleted units.

The decision-making process of a single power with its own well-rehearsed command system runs a lot more smoothly than that of a coalition, where commanders of different backgrounds and traditions have to go out of their way to avoid misunderstandings and secure agreement amongst themselves and with their individual capitals. In Kosovo in 1999, when NATO fought the army of the Former Republic of Yugoslavia to stop atrocities being committed by both of the opposing sides, Yugoslav President Slobodan Milosevic was able to engage NATO's military forces greatly superior to his own because he enjoyed the advantage of being a unitary actor confronted by a complex coalition of powers, only loosely held together by a broadly defined common objective. Moreover, NATO fought under considerable constraints, which the Yugoslav forces did not share. Intense political pressure was applied to minimise casualties – friendly, enemy and civilian – curtail attacks on local infrastructure, and put an end to the ongoing ethnic cleansing. The tensions between NATO's wartime objectives were a product of the tangled negotiations that ultimately had brought the alliance into the conflict. Because of vetoes by Russia and China there was no Security Council mandate in place, nor was NATO able to agree internally on a single legal basis for the war. This left each force contributing nation applying its own legal and political justifications.[18]

While Entente commanders on the Western Front during World War I were generally insensitive to casualties, later wars have shown a far more varied picture. The trend is towards greater unwillingness to take losses, but opinions are not identical amongst nations. Smaller countries tend to be more careful than great powers, and the relative political importance of casualties, especially those caused by fratricide, increases within a coalition. While the US seems to be willing to tolerate casualty figures in terms of hundreds of dead and wounded soldiers per month, similar figures, even in reduced terms of the scale of military commitment, would prove politically ruinous for the governments of most coalition partners. The repercussions of casualties and losing major weapons platforms, such as battleships and aircraft carriers, are severe, and they tend to make politicians place

---

[18] Paul T. Mitchell, *Network Centric Warfare and Coalition Operations: The New Military Operating System* (Oxon: Routledge Global Security Studies, 2009), p. 27.

significant operational restrictions on what they will permit their forces to do once they have been deployed alongside coalition partners.

Today, coalition warfare is largely conducted by lead-nations possessing an overwhelming capability as far as information technology-based command and control systems are concerned: the US, Britain, France and a few more perhaps. If smaller coalition partners wish to co-operate at a par with their lead-nations, they must adapt to the highest degree possible. They must be able to find their rôle within the triangular relationship among Network Centric Warfare, information release and coalition strategy – coalition partners who do not possess optimum Network Centric Warfare capabilities are permanently at risk of being sidelined.[19]

Coalition Disunity and Dissolution, Defection of Coalition Partners

The greatest dangers, always intertwined with the advantages of having allies, lie in disagreement, disunity and defection. These are perennial, albeit latent, risks to coalition strategy and success. Subsequent to the retreat from Moscow in 1812, the Prussians – who had never been anything like enthusiastic allies of Napoleon's cause – defected from the French-led coalition, and in 1813 the seed of dissolution spread throughout the alliance system which the emperor had so carefully nurtured since its inception in 1806. And once the rot has set in it is difficult to prevent it from spreading. Nonetheless, in the spring of 1813 Napoleon managed to re-establish the *Grande Armée* at slightly less than half a million men, including allied contributions of around 88,000 German, Polish and Italian troops. In the spring of that year he conducted two modestly successful battles at Lützen and Bautzen. However, his newly raised troops were not up to the usual standard, and the cavalry was particularly inexperienced. This caused a number of opportunities to be lost due to the cavalry's inadequate ability to rout and pursue beaten enemy forces.[20] After the French re-conquest of Dresden a truce was agreed by both sides, during which Napoleon hoped to be able to reorganise and replenish his armies and persuade the Austrians to stand by him. Eventually, however, Austria chose to join the anti-French coalition; Marshal Joachim Murat, King of Naples, conspired with the Austrians, and Marshal Jean Baptiste Jules Bernadotte, former Prince of

---

[19] Ibid., p. 119.

[20] Owen Conelly, *The Wars of the French Revolution and Napoleon, 1792-1815* (London and New York: Routledge, 2006), pp. 189-97.

Pontecorvo and now the crown prince of Sweden, encouraged by British subsidies and promises of receiving Norway at the expense of Denmark, agreed to command allied formations against his former sovereign.

Then, in October, Napoleon concentrated his host at Leipzig where, during the days of 16-19 October, the armies clashed in the battle later known as the Leipzig Battle of the Nations. On the 18th, the Saxon Army deserted the French cause, and on the 19th the anti-French allies stormed the city. Four French *corps d'armées* were trapped. With the remnants of his once magnificent armies, Napoleon now withdrew post-haste towards France, shedding his remaining German allies en route as the Confederation of the Rhine quickly disintegrated. His European empire and alliance had collapsed due to diverse interests, newly generated nationalism and, last but not least, British gold. A modicum of martial exhaustion on the part of the lesser coalition partners might have played a rôle, too.

Exhaustion and fiscal strain as well as rising opposition at home against the whole project may have been among the reasons for allies to depart from the American-led coalition-of-the-willing in Iraq in 2007. As the British government announced that UK forces would withdraw from the Basra Region, their Danish coalition partners had no option but to follow suit. Similarly, in the summer of 2010, the worldwide financial crisis, which had been going on since 2008, and long term perspectives of severe drains on their fiscal and human resources, made Prime Ministers Cameron and Rasmussen announce the end of the presence of British and Danish combat troops as part of NATO forces in Afghanistan by the end of 2014. In both these cases the Americans – now in situations similar to that of Napoleon – faced desertions by coalition partners who no longer believed in the viability of the cause. The forging of these coalitions, which had at the outset aimed at increasing the legitimacy of the enterprises as well as saving American taxpayers' money, threatened to become a stone around their neck.

Coalition Interoperability

Although the 19th century Russian General Mikhnevich claimed that coalition military forces are always less than the sum of the component armies, and that allies will strive to shift the heaviest burdens onto other shoulders than their own, coalitions are normally entered in the hope of sharing the costs of war with allies and achieving synergic gains from marrying-up with partners of different qualities. However, at the same time many factors combine to hamper coalition interoperability. Cultural factors, as we have already seen, can make serious hindrances if partners are not willing to – or capable of – adopting a common working language and operational doctrine, or if they do not share views on operative imperatives such as planning, timeliness and training standards. Similarly, national pride, tradition and ethical differences may stand in the way of effective co-operation. However, among the more tangible matters there is always a risk that incompatible equipment may cause a coalition to malfunction.

For many years now, the US driven development of Network Centric Warfare has forced those who might wish to co-operate at a par with technologically advanced militaries to develop or purchase such equipment themselves. Frequently, this happens at great costs to their treasuries and with an immanent risk of buying the wrong items – potential coalition partners may have moved on long before the new equipment has been delivered. The network-centric concept builds on the fact that information is vital to decision-making, and the party possessing predominance in this field holds enormous advantages over its less fortunate opponent. Network Centric Warfare as an operational concept has been devised to take advantage of technological developments from the tactical right up to the strategic levels, and this is now determining strategic issues such as military co-operation between major coalition partners as well as their relationships with the lesser ones. Today's computer technology has reinforced the marked discrepancy between technologically advanced countries and those which can afford only modest modernisation of their forces. Because the recent evolution of military technology and doctrine is based on the assumption of information supremacy, it presents issues that will be difficult for minor coalition partners to resolve.[21]

It is well recognised that Network Centric Warfare is changing how militaries operate, both in battle and in 'operations other than war', and that the sharing of information can only grow in importance as armed forces continue their never-ending quest for competitive advantages. It is also axiomatic that the potential for failure in coalition operations exists should partners diverge too greatly in terms of their ability to operate together. This is nowhere more apparent than in Afghanistan in the attempted co-operation between NATO and its Afghan coalition partners, viz. the Afghan National Army and the Afghan National Police. There remains hope, however, that in the long run technical means may obviate this problem – at least as far as the relationships between the greater and minor western militaries are concerned.

Major events – not least sudden changes on the global political stage or the unexpected manifestations of looming menaces – may bring nations

---

[21] Mitchell, *Network Centric Warfare and Coalition Operations*, p. 46.

together despite a possible basic inclination to avoid conflict. The extreme danger represented by Hitler, the Soviet Union, or international Islamist terrorism forced a level of co-operation among the western states that was in many ways unprecedented. The slow collapse of the Soviet threat to the West and the ultimate disappearance of the Soviet Union set up reverberations within the Western alliance that have yet to be resolved. Still, it is readily apparent that the problems of incompatibility, when committing to new political objectives, have become more and more blatant, as Paul Mitchell writes:

> Technological incompatibilities can be solved through 'fixes', generally established from sharing technology of the development of long-term policies that 'bridge' the gaps between two nations' technical systems with non-technical means. More typically, however, work-arounds are developed amongst partners. These are solutions that seek to reduce the impact of incompatibilities rather than eliminating them entirely.[22]

The use of liaison teams, inserted as far down organisationally into a partner's land forces as resources allow, can lessen the impact of incompatibility and this is, in a way, what happens in Afghanistan as the Operational Mentoring and Liaison Teams from NATO countries are attached to Afghan army units. The time to train is essential, however, and there is reason to believe that the process will be a long lasting one.

While the disparity between the United States and the rest of the world is partly a product of the Americans' own design in order to reinforce their unipolar military status, in its turn it creates growing problems for the US' own operations.[23] American military units are so lavishly equipped that few other countries – except perhaps Britain and France – are willing to do likewise. This leaves the Americans in the situation they probably wanted to avoid, namely having to bear the overwhelming majority of operational costs themselves. Some smaller partners have, nonetheless, chosen to follow suit technologically, but at the expense of numbers. While Denmark today has armed forces which are at a technologically level almost as advanced as that of Britain, over the last decades the force structure has shrunk to a level allowing hardly more than one battalion at the time to be deployed on overseas missions.

---

[22] Ibid., p. 116.

[23] Ibid., p. 117.

Chapter V – The Message of History

FROM THE WARS in Ancient Greece to the ISAF collaboration in Afghanistan in the 2000s and beyond there is a whole host of similarities of purposes as well as of experiences which gives us a fairly precise notion of coalitions – their main characteristics and their *raisons d'être*. But there are differences, too. While we are now familiar with the coalitions of yesteryear as well as with those of the immediate past, it is time so try to fathom what has actually changed, what remains unaltered as well as the message this leaves us with.

Seen in a historiography perspective, authors have characterised coalitions primarily by the personalities and achievements of their leaders – Lysander, Philip of Macedon, Alexander the Great, Hannibal, Wellington, Napoleon or Foch – but the key to combined success was never a commander's extraordinary abilities and personality alone. Many more parameters must be considered such as purpose, resources, inter-allied agreement or disunity, conciliation of doctrines as well as personal relations amongst decision-makers at various levels.

Interests

There is much merit in Lord Palmerston's observation that in politics there are no permanent friends or enemies, but 'interests [which] are eternal, and those interests it is our duty to follow.' States joining coalitions do not do so because of friendships – they do it because they share some interests: getting rid of a potential or real threat, acquiring access to resources, or stabilising the immediate neighbourhood or more distant hotbeds of trouble. They do not necessarily do so to the same depth or for the same length of time, and they may agree in some vital fields but remain competitors in other areas. We saw this clearly demonstrated in the Anglo-American alliance during the Second World War, when Britain strove to re-establish world order by reining in Nazi-German aspirations and needed a strong democratic partner to help achieve this goal. But while Britain aimed at consolidating her empire, the Americans were less than enthusiastic about this particular part of the war aim, and though Britain remained committed to combating dictatorships, whether Nazi or Communist, the US became slightly more reluctant to go on to Moscow when the end of a long and costly struggle beckoned as defeat of Germany drew closer.

During the Korean War, the French contribution to the UN operations was operationally insignificant but UN and American as well as French interests were served by France's joining the coalition. To the United Nations and the lead-nation, the United States, the French contribution mattered in terms of legitimacy of the enterprise undertaken by the international coalition, and to France there was an important aspect of familiarising her forces with modern combat under Cold War conditions, as well as a boost of French great power self-esteem. In Afghanistan in 2010-11, the UK and Denmark were NATO allies as well as coalition partners sharing the same political aims with the war – combating terror and stabilising the Helmand Province of Afghanistan – but being, at the same time, very disparate associates as far as the sizes of contributions were concerned. While the leading partner needed to assert her position of great – or at least medium – power, to the lesser contributor it appeared important to be internationally and domestically discernible, to be seen to be willing to fight as well as sustain significant losses for the sake of common ideals, and thus gain recognition and goodwill with major allies. For this reason it was essential not only to participate with an independent battle group, however small, but also to occupy key posts within the coalition's decision-making machinery.

There seems to be a characteristic to coalition warfare which has not changed over time – namely that the overarching war aims in terms of political or ideological notions apply more or less uniformly to all partners. Conversely, there are various more particular wishes and goals specific to individual coalition partners. These, however, vary a great deal and might be as diverse as territorial claims, economic aspirations, training opportunities for military forces, favourable terms with the lead-nation and major coalition partners as well as commerce or protection. Coalitions are prudent, egotistical security arrangements, but they are not necessarily friendships. Twentieth century coalitions like the Austro-German alliance in the First World War and the Anglo-American one in the Second were based on assumptions of common values, which were genuinely believed to be of decisive importance at the time – and therefore must carry historic weight – but in general coalitions are formed to protect practical interests rather than merely ideals and friendly feelings.

Although coalitions are entered into for motives of national self-interest, usually for economical or self-protection reasons, they normally combine like-minded states agreeing on a variety of key goals. This was as true in the Middle Ages and the Early Modern Era as it is today. The Italian League was true to the notion of keeping strangers out of Italy, the Christians of the centuries right up to the siege of Vienna in 1683 were determined to put the kibosh on the Ottomans, the Grand Alliance fought French hegemonial aspirations in the 17th and 18th centuries, and the Western Allies plus Russia struggled against the expansionist Germanic powers in two world wars.

To be slightly more specific: The 15th century Italian League was the manifestation of a shared wish amongst the league member states to maintain territorial *status quo* in Italy thus protecting against non-Italian aggressive powers – and particularly against France. Similar reasons for convening a coalition in order to combat an aggressor with aspirations to dominate were in essence what was seen again, but on a grander scale, during the final battle against the Ottoman Army at Kahlenberg outside Vienna on 11 and 12 September 1683 and during the War of Spanish Succession in the early years of the 18th century.

Conversely, disparate war aims, lack of military manpower, differing global security interests, and an underlying hostility toward the coalition partner all hampered the development of a manifest coalition strategy for the Crimean War. The Anglo-French coalition did not have much in common and they did not agree on the aim with the war. While primarily Napoleon III wanted his army to look successful and glorious, the British nurtured some unclear conceptions of bottling up Russia in St Petersburg and the Black Sea. This led to strategic compromises which must have sounded familiar to the older statesmen of the day, who might have recalled the muddled British war planning of 1803–6 while the Third Coalition existed. A coalition war plan where two or more equal partners are concerned is inevitably a compromise and it is not always the optimal solution.

In many respects the war aims of the First World War – though it was primarily a conflict between two great coalitions opposing each other – were not very different. It was merely yet another example of a form of coalition warfare fought because individual national interests were believed to be best taken care of that way, and because defeat of one or more coalition partners was seen as potentially disastrous to one's own security and commercial interests.

However, it was even more complicated than most of the previous conflicts involving coalitions due to its magnitude and because, at least on the side of the Anglo-French *Entente,* there were coalitions within the coalition, in as much as the peripheral members of the French and the British Empires chose – or were coerced – to join the struggle of their imperial hubs, engaging themselves as enemies of the Central Powers. In many respects this happened again during the Second World War, although the interests of Britain's peripheral partners had by then become increasingly independent of the centre, especially so in the cases of South Africa and Canada. Although Churchill was an imperialist at heart and, up to a point, took the Dominions' willingness to join for granted, he had to accept the standpoints of the Dominion prime ministers. Eventually, at the political level co-operation functioned reasonably well and Churchill came to appreciate the advice and collaboration of the Dominion prime ministers, and especially so that of South African Field-Marshal Jan Christiaan Smuts (1870-1950).

Political Purpose

Like all other military activity, coalition war is waged in order to fulfil the political purposes which have laid the foundation for the formulation of the aim *with* the war – in this case the common aim agreed amongst coalition partners at the outset. It is the duty of the military commanders of the coalition contingents to strive towards fulfilling the aims of their individual governments while co-operating constructively with their foreign partners to achieve the overarching war aim of the entire coalition. However, in war it often happens that pressure is imposed on a commander by political or military superiors in order to achieve aims which are politically opportune, although completely irrelevant or outright detrimental to his objectives on the ground. The Swiss-French general and military philosopher Antoine Henri Baron de Jomini called such aims 'political decisive points,' and he observed that the imposition of such aims were among the reasons for the Third Coalition's lack of success fighting Napoleon in 1805. Similarly, the World War I coalitions were bogged down by the unhappy relationship between politicians and the military professionals, which soon led to abysmal discrepancy between the political, and badly conceived, aims *with* the war on the one side, and the military imperative of fulfilling these by setting realistic aims or objectives *in* the war on the other.

Across history we find examples that politico-military opportunism has forced commanders to undertake tasks which have not fitted with the reality on the battlefield as presented by their intelligence branches. This happened, for instance, during the Allies' Operation Market Garden on 17–25 September 1944, when General Browning and Field-Marshal Montgomery refused to accept last minute intelligence updates, and it might as well have happened in Afghanistan in 2010 when politicians began looking for obvious successes and justifications for reducing the level of their military presence. Similarly, although the swollen staff of the French Battalion in Korea was not only a matter of political opportunism – but rather a vehicle for extracting lessons from fighting during that war – it was an example of political orders which hampered the commanding officer's operational agility. The downside of the existence and activity of this 'group of observers' on the staff of the French battalion, led by an officer much senior to the commanding officer and serving more long-term goals than those of the struggle at hand, was that it complicated the chain of command between the Americans and the French battalion.

Prestige is no rare companion of political reasoning about war, and the brief Danish leadership of the Protestant forces during the Thirty Years War as well as participation in the British led coalition fighting against James II in Ireland 1689-91 are but two well-known examples. Geo-strategic ambitions are among the primary drivers of political decisions to join coalitions, as we saw when Austria-Hungary as well as Denmark seriously considered joining Napoleon III during France's struggle against the Prussian led German onslaught in 1870-71. While Austria-Hungary wanted revenge for the defeat of 1866, France hoped to secure its eastern borders by advancing at least as far as the Rhine, and Denmark wished to regain the lost Duchies of Schleswig and Holstein. Eventually France's defence collapsed before any such coalition was convened. Coalition war has always implied considerations and caveats concerning political purpose, risks, mutual trust, national wealth and pride, compatibility of military forces, and a glut of intangible forces and effects characterising human interaction in combat. The Danish contingent with the Williamite army in Ireland had achieved considerable concessions because of caveats concerning the return of the forces should the situation at home necessitate it, Danish generals' participation in William III's war councils, the place of Danish forces in the coalition array, etc. A modern example might be the German caveat that the country's forces should not be deployed outside the north-western parts of Afghanistan – and particularly not to Helmand.

Since the signature of the Peace of Westphalia in 1648, legitimacy has been an important issue of warfare. In today's globalised world this remains so, and the more participants from amongst the UN member nations the more solid the legitimacy. In launching coalition actions there are issues that will invariably be raised: is it necessary, is it in accordance with international law and are the human rights of the opponent considered when an army ventures abroad for purposes not directly relevant to the state's territorial defence, is it backed by a reasonable number of democratic states, etc. Most wars of the late twentieth and early twenty-first centuries have been mandated by the United Nations and conducted as coalition enterprises led by an alliance or a great power. This has helped boost the legitimacy and ease the burden on the individual nation's treasury, and we may take the French participation in the coalition fighting in the Korean War as a fitting example. The United Nations had wanted France to be involved; the Americans – the lead-nation – who, from a military point of view, did not really need the French battalion, had a political need to see as many nations as possible join the bandwagon in order to underpin the legitimacy of the enterprise and save some of the American tax payers' money; and the French were able to boost their prestige by waving the tri-colour as well as having their profile sharpened amongst the permanent members of the UN Security Council. Moreover, the French battalion fought well and everyone in the coalition derived some benefit from its being dispatched.

Economy is a matter which always carries enormous weight in political considerations on whether to join a coalition, how to do it, and with how much. In a Medieval Italian context, the balance of power was not a general aspiration, nor was the war aims of coalitions based on universal acceptance of the need for peace and harmony or any kind of political enlightenment. The agreement on setting up the Italian League was rather a consequence of economic exhaustion and a realisation that the days of cheap conquest were over, even though the inclination to promote national interests remained alive. In 1689, the Danish Treasury was determined to make the war with the Williamite colours a low budget one, so the troops were left to themselves as far as the daily needs were concerned and no extra personnel were dispatched to complement the establishment during the years abroad. Similarly, there are economic reasons – among others, of course – for NATO's decision to withdraw combat troops from Afghanistan by the end of 2014.

Basically, coalitions come in two types: one composed of a great-power lead-nation and a number of lesser states contributing contingents of varying sizes and qualities, and another combining a number of more or less equal partners. The Second Boer War (1899-1902) was an attempt at establishing coalition of the former type in order to suppress the Boer insurgency and keep Transvaal and the Orange Free State under the suzerainty of the Cape Colony, as well as maintaining this colony within the community of the British Empire. Canada was among those British dependencies which decided to join the coalition bound for South Africa, but this had not been an easy decision. There was opposition amongst the French Canadians and groups holding the opinion that colonial warfare was none of Canada's business. Moreover, it could be argued that Canada's militia was not a force at a par with the British army, having only limited forces and no general staff of its own. As to the latter type, we find examples in the Franco-Russian treaty agreed in 1807 on the raft off Tilsit by the French and Russian emperors, Napoleon I and Alexander I. The Tilsit Treaty included a secret clause on a joint enterprise of invading British India from the west, but although France started negotiations with Persia on this issue it never materialised, and a few years after the coalition fell apart as France invaded Russia. While in the beginning of World War II the British Empire fought alone against the Axis Powers, the reality of the struggle was that an increasing number of states and 'militaries in exile' joined, such as Poland's and Norway's, forming a coalition of the former type. However, as the Soviet Union was invaded by Germany on 22 June 1941 and the United States was attacked by Japan on 7 December of the same year, it mutated into a multi-lateral alliance of the latter type which, since the Declaration by 26 states on 1 January 1942, assumed the designation of the 'United Nations.'

War Aims

Seen against the backdrop of the coalitions of more than two millennia, it will be reasonable to downplay any fancy of altruistic or idealistic aims behind the creation of such defensive arrangements – war aims always were egotistical aspirations and they remain so. To the Danes three strategic aims made joining the British led coalition-of-the-willing of 1689 a tempting notion. First, there was the common European aspiration of eliminating the French menace. In this respect Danish war aims were identical with those of Britain, the Netherlands and Brandenburg. Secondly, King Christian V of Denmark needed funding for the upkeep of his armed forces, which were relatively larger than those of other European countries measured by the soldiers-to-population ratio, and he required training opportunities for officers and men in order to maintain and develop his army for later action. Finally, Denmark would have to retain a credible force at home to deter Sweden from further attempts of aggression. Similarly, some have argued that the War in Iraq 2003-7 was not waged for altruistic reasons but, simply and straight forwardly, for oil.

The aims with a war are not always as clear as those of the World War II Allies: to fight on until the unconditional surrender of the Axis powers and to secure all free nations – but not the British and French colonies – the right to decide for themselves. The World War II alliance between the British Empire and the United States was a coalition formed with the aim of defending common values, ideals and ways of life which they both cherished; but in many other respects these two powers held disparate world view. However, although there were significant political, and almost ideological, differences between the British Empire and the US over India, the Asian colonies, and policy towards Nationalist China, there was agreement on the supremacy of democracy over despotism. In Afghanistan today, a very similar agreement can be observed: Al Qaeda must be prohibited from operating from a sanctum in Afghanistan and the Afghans must be supported in their endeavour of creating a society based on law and order with respect for human rights. Nonetheless, all contingent commanders come with more or less specific tasks imposed by their governments, which they then need to make tally with those of ISAF as well as the Afghan government. The task force commander of the UK Herrick 13 team in Helmand saw his war aim as identical with the central NATO intention, which must, however, coincide with that of the Afghan government: denying Afghanistan to international terrorism and promoting a stable community and a reasonably well functioning state, giving the Afghans confidence in their government, and bringing about a sense of security and opportunity.

The aims *in* the war, too, have grown in complexity over recent years, but they still lay within the realm of military operations. In 2010-11, Task Force Helmand had two such aims – namely contributing to the enterprise of supporting societal development, and propping up the improvement of the Afghan National Security Forces.

There are also examples of coalitions which have not agreed on common aims *with* the war they are about to fight: in the Second World War, Italy fought primarily for Mussolini's vague ideas of a new Roman Empire around the Mediterranean, while at the same time Germany strove for hegemony in Europe and a completely new world order. Italy failed miserably from the beginning being defeated by the French, and her attempts to conquer Yugoslavia did not go too well either, eventually necessitating Germany to divert considerable forces which were dearly needed elsewhere prior to the launch of Operation Barbarossa against the Soviet Union.

Strategy

Coalitions have come a long way to arrive at the relatively smooth collaboration that we see in Afghanistan today. The realisation of the need for strategic agreement has seen considerable development since World War I, when distrust of allies, absence of common doctrines, and lack of a common supreme leadership made co-ordinated action extremely difficult. In the Franco-British-American coalition during The Great War, unity of command and decision-making happened, if at all, at the strategic level. Battlefield responsibilities were divided nationally based on the attitudes, capabilities, and operational approaches that each nation brought to the coalition. Although a French generalissimo was appointed to unify coalition efforts, the actual running of operations was decentralised, hampering the co-ordination of effort. While Haig preferred to ensure the security of the Channel Ports and Pétain wanted to cover Paris, the Germans – employing the well-tried Napoleonic stratagem of a *force de rupture* – appeared to be driving a wedge in between the two, where they could be dealt with individually. The Entente coalition was too loose and needed tightening under the threat of disintegration.

During World War II not everything was plain sailing, but at least the aims with the war and the main elements of a coalition strategy were in place: Europe first, then Japan. Negotiating with the US, the British maintained their strategic position: Germany was the primary aggressor and the enemy without whose defeat the war could not be successfully concluded – therefore she would have to be beaten first. The political leaderships concurred on the primary war aims, and the chiefs of staff also managed to iron out their discord and reach agreement on the overall strategy. However, they were not always in agreement over details. The American Chiefs – including the Anglophobe Admiral King – frequently opposed British views, although they accepted the basic statement of position which had been drawn up by General Marshall and Admiral Stark: 'notwithstanding the entry of Japan into the war, our view remains that Germany is still the primary enemy and her defeat is the key to victory.' Nevertheless, at the strategic and operative levels disagreement did occur. Churchill's well-known peripheral preference clashed with the Americans' desire to open a 'second front' in France as early as possible, his wish for concentration of effort once Operation Overlord had been launched ran counter to the American insistence on pressing ahead with Operation Dragoon – the planned landings in southern France – and various other key strategic choices similarly threatened to do lasting damage to the coalition. Moreover, with respect to the Russo-British collaboration, frictions appeared right from the outset: the British were unwilling to declare war on

Finland and the issue of Soviet-Polish relations caused deep concern. While the USSR wished to push back the border to where it had been prior to 22 June 1941, at this juncture Churchill was firmly committed to combat Nazi-Germany and restore Poland to its former sovereign territory.

While disagreement happened occasionally on the Allied side and was invariably overcome through reasoning discussions, the problems were more constant among the Axis partners. At the turn of 1940-41, the Axis had embarked on a new strategic orientation which changed the parameters for the solution of the increasingly irritating problem of supply lines across the Mediterranean Sea. Until this moment, the Mediterranean Theatre of War had been exclusively in the Italian sphere of interest. The well-fortified British base in Malta was a liability to Italy as well as to her African dependencies, and constituted a severe menace to the Axis' lines of communication. Moreover, the British bases in Malta, Gibraltar, and Egypt formed safe havens as well as bunkering facilities for the Royal Navy in the Mediterranean Sea. On 11 December 1941, and subsequent to the Japanese attack on Pearl Harbour, Mussolini was delighted to make a clear gesture of merging his strategic endeavours with those of Germany by declaring war on the United States, but he did so without even discussing the consequences with his king, Marshal Cavallero, or the service chiefs. The resulting dissipation of effort was perhaps the epitome of Italian strategic folly.

Strategic disagreement on that scale has not been seen in the coalitions-of-the-willing which have sprung up since the end of The Cold War. Although in the Gulf War of 1991 there were two independent commands, one led by Saudi Arabia and one by the US, the strategic aims and approaches were co-ordinated. Similarly, the Danish battle groups deployed to Afghanistan's Helmand Province were meant to advance the Danish official strategy, but this was well within the bounds of what had been agreed to in NATO as well as with the UK having the lead-responsibility in Helmand.

Doctrine

Many, but certainly far from all, coalitions are merged hurriedly together in crisis or conflict. As a result, they may not provide an unfailing model upon which a doctrine may be founded. Although the *Entente Cordiale* had existed since 1904, Anglo-French staff talks had been conducted for some years and the UK had set up a British Expeditionary Force consisting of six infantry divisions and five cavalry brigades, no true development of either Anglo-French strategy or doctrine for coalition warfare had taken place. Co-ordination with the prospective coalition partner, France, had happened only in the form of a limited number of staff talks. Conversely, within the British Empire doctrinal matters did not cause any trouble. In both the First and Second World War Britain and the Dominions agreed on war aims and strategy, and their forces were almost identically equipped and fought to roughly the same doctrine.

During the Cold War, before the advent of international terror, and as a consequence of many years of combat in their shrinking colonial empire, the UK had a well-tried counter insurgency doctrine based on a long history of combating riots and civil unrest. This doctrine, however, was rendered at least partially irrelevant by the September 2001 atrocities, and a new doctrine was conjured up in order to cope with the challenges of emerging asymmetric threats. The Danish planning for operations in Afghanistan took this as its point of departure, and was largely inspired by the new UK doctrine on counter insurgency. As a result, while employed in the Task Force Helmand, contingents of either country experienced little difficulty working with the other.

Planning

While politicians normally restrict themselves to laying down war aims and strategy, the planning that must happen to translate these loosely stated goals into action is done by the professionals of various government departments, notably that of defence. Being a complex procedure involving a wide variety of experts from the lead-nation as well as from lesser partners, planning takes time, requires compromises, and it is an ongoing iterative process throughout the war. Once set in motion on the basis of the political decisions, at the military top-level there is constant apprehension that the political leaders might all of a sudden agree on something which does not make sense militarily and upsets the planning process.

During the Second Washington Conference 20-25 June 1942, General Alan Brooke, whose planning priority was North Africa, noted that: 'We made further progress towards defining our policy for 1942 and 1943. ...[and] we fully appreciated that we might be up against many difficulties when confronted with the plans that the PM and the President had been brewing up.' This military policy planning was particularly important because it would constitute the starting point for the planning of campaigns, which should – if successful – combine to achieve the goals set out politically. Sudden changes infused by the politicians would not only delay the odd campaign, but also upset the mutual harmony prerequisite to a successful outcome.

While the eventual result of major conflicts like the world wars rested on mutually supporting campaigns, it is much harder to see an undertaking like the international assistance to Afghanistan in terms of traditional campaign planning. The mixture of counter insurgency, training of local security forces, influence operations and humanitarian aid does, in many respects, call for *ad hoc* initiatives, which can hardly be fused into a conventional notion of a 'military campaign.'

A well-led coalition force does not initiate operations just for the sake of being seen as pro-active – it does so because it has conceived a well-thought out and nicely concerted campaign plan. So indeed did the Danish element of Task Force Helmand, the ISAF Team 10, and with reasonable success.

However, a campaign plan does not always survive the most daring of enemy initiatives unchanged. Concurrent adaptation to the tactical development is always necessary to minimise losses and retain (or regain) the initiative. When on 16 December 1944, the operations in northern Europe developed in an unforeseen direction as the Germans launched Operation *Wacht am Rhein* in the Ardennes, Eisenhower's nicely planned campaign had to undergo drastic revision. This sudden change for the worse apparently worked towards facilitating reconciliation of Brooke's and Eisenhower's operational preferences: the organisations of the army groups were adjusted, Eisenhower put Montgomery in overall command of the armies and air support employed in the joint operation relieving the pressure, and within weeks the last German offensive of the Second World War ground to a halt.

Over the years, planning and conducting joint operations within a coalition have been made easier by the advent of increasingly smart electronic equipment. In Afghanistan in 2010, a company commander with the assistance of a single forward air controller was able to conduct a joint strike against a team of Taliban insurgents. At one point in time, all the company's bases came under assault, and after a while the officer commanding requested close air support. As was described in Chapter II, an F/A-18 Hornet fighter aircraft on patrol overhead, was ordered to drop a GPS-guided bomb, despite the target being as close as 185 metres from own troops in contact. However, as the GPS was unaffected by the dust the pilot was instructed to go ahead. The target was destroyed without any collateral damage and the company was relieved of the immediate threat to its continued existence.

Air support is quick and mostly precise, but even today it can be difficult to direct flawlessly. Friendly fire – *blue on blue* in military parlance – still happens as it has always done. It does so because the friction of war is as abiding a companion of coalition warfare as it is in any kind of war. Throughout history we hear about troops being shot at by their own. It happened in Ireland in the 1690s, in the world wars, in Iraq, and in Afghanistan. Today, we have electronic gadgets which are supposed to distinguish between friend and foe, but nonetheless troops and aeroplanes continue shooting at the wrong targets. On the ground visual recognition is difficult because uniforms and equipment are much the same the whole world over, and this predicament was indeed extant at the era of horse and musket as well.

Logistics

To readers of military history it will come as no surprise that operations are intimately linked with logistics. In the Middle Ages, a castle under siege could be starved into submission, and even today a military force can still be cordoned off if one can only cut the supply lines to and close the airspace over the opponent.

In 1942, this almost happened to the Axis forces in North Africa because the British naval forces in the Mediterranean had safe havens in Malta and Gibraltar. But it was one of the ironies of the development in that theatre that the British were so hard pressed for naval as well as for commercial vessels that a raid against Malta – had the Axis had the necessary means available – might have succeeded. Phrased in another way, logistics need close attention so as to make sure that all necessary transport, supply and protection are available where the operational priorities lay.

Disparities

While military necessity and scarceness of resources are the prime movers in forging coalitions, there have always been various inhibitors as well, and among these cultural dissimilarities figure prominently.

The First World War's Central Powers were challenged in the cultural field as soon as their troops were deployed outside western Europe, and in particular where the relationship between the Ottoman Empire and Germany was concerned. There was an abysmal gap between the two cultures, and the general German attitude towards the Ottoman Turks was condescending, having been formed by the writings of key personalities like Field-Marshal Helmuth von Moltke the elder, who found the Turks excessively lazy, and the author Heinrich von Treitschke, describing the Turkish culture as a soul in constant hibernation. Moreover, since sensibility to a partner's national pride and customs was not among the dominant characteristics of German military behaviour, the Germans did little to accommodate their coalition partner's natural self-esteem needs.

Not only the Germans had their trouble with trans-cultural collaboration – the British World War I relationship with their Japanese coalition partner's troops ran into problems, although of a slightly different nature. The British and Japanese military cultures were worlds apart, and while written orders were a matter of routine with the British, the Japanese preferred oral instructions which were supposed to be memorised by the recipients. Moreover, the Japanese perceived public showing of industry as undignified, which was no longer a worry to the British officer class. Compared with the Japanese, and unlike their notion of military courage and dignity of Crimean War vintage, the British were now business-like and not too embarrassed being seen to be busy.

Probably, no two wars see the same cultural encounters among alliance partners. Around a century has elapsed since the First World War, and today's coalitions-of-the-willing are different from those who collaborated then – their cultural clashes are different too. Still, each nation has peculiarities of its own, and while Danes, nowadays, tend to believe in the merit of extensive informality, others might prefer a slightly more ordered leadership style. Over recent years, Danes have developed a rather liberal attitude towards commenting and criticising the wisdom, decisions and prerogatives of superiors, and it is fairly common for them to raise doubt about the ingenuity of a commander's plans – although only until these have been transformed into orders. From a non-Danish point of view this might, and did actually in Afghanistan in 2010-11, cause occasional apprehension. Moreover, today Danish soldiers and NCOs tend to have a rather casual relationship with many of their officers, which frequently caused considerable surprise to coalition partners.

It was obvious from the outset that the culture the ISAF forces found in Afghanistan in the beginning of the 21st century was very different from what they believed was required of a modern civilised society. However, while everyone agreed that the troops were there to support development, they were not expected to do so by trying to change the cultural norms to fit those of their own. The policy of the coalition was that the Afghans would have to make changes in their own way and at their own pace because, eventually, Afghan issues would have to be solved independently of the coalition. Today's cultural pitfalls are of a different nature from those of the two world wars, but ISAF coalition forces seriously endeavour to avoid them, doing what they can not to offend their Afghan collaborators: they abstain from commenting on hygienic, sexual and other controversial issues, they respect religious festivals whenever this is operationally possible, and they do what they can to include Afghan commanders in decision-making prior to planning combined operations.

Apart from the cultural hitches already mentioned, a number of practical disparities normally exist amongst the coalition partners. On D-Day, the British made extensive use of various fancy machines to facilitate movement across the Normandy *bocage* landscape, but the Americans had decided to do without most of them. In the Helmand Province in 2010-11 there were inhibiting disparities, many of which had to do with dissimilar weapons and equipment, while others pertained to tradition, outlook and educational aspects. As to equipment, cross-country mobility was where Danes and British differed. The Danes drove armoured vehicles and tanks, the British did not. Disparate standards of dedication to mission was seen off and on in collaboration between Danes and Afghans. The feeling of being alone in shouldering an important task led to questions about the meaning or futility of the whole enterprise. The apparent Afghan indifference to training schemes set up to improve the Afghans' professional skills, as well as their unreliability when on parade, was de-motivating to instructors and collaborators alike.

The technological limitations of various partners may be aggravated by the political nature of coalitions and their management. In the age of American military primacy, influence will be tightly restricted to the very few partners who are capable, willing, and trusted to make meaningful contributions to US-led operations. The Danish Battle Group in Afghanistan suffered from various shortcomings in communication because with no encryption equipment of their own they had to rely on external support. Communication – both voice and data transmission – was of immense importance to operations, but many coalition partners did not enjoy the same advantages in equipment.

Disagreement

Commanders do not agree on every aspect of warfare all the time, and the relationship between Montgomery and Eisenhower is the obvious example. Spreading the forces thinly along the entire front was not Montgomery's prescription for a quick break-through, nor was it half a century later the way the commanding officer of the Danish Battle Group intended to manage his area of responsibility in Helmand. Having liaised with his Afghan opposite number, the commanding officer of 3rd *Kandak*, he realised that in the long run more troops would be needed as a mobile reserve, and could be employed more effectively around the town of Gereshk rather than being spread out in a large number of patrol bases. In his view, bases which the Afghan commander had declared to be useless would have to be abandoned. Moreover, giving up such bases nicely fitted with the Afghan plans concerning which bases should be taken over and which should not when ISAF eventually left the country. However, as this needed the Commander Task Force Helmand's approval, as well as acceptance by Afghan authorities and by the Regional Command South West, the process was time consuming, but over the months of their collaboration these matters worked out to the satisfaction of all parties concerned.

Lead-Nation

Whether a coalition is one of equal partners or composed of one great power and some lesser contributors, there will generally be one which is the lead-nation. This is necessary for co-ordination purposes as well as to ensure the unity of command, without which the war aims are unlikely to be achieved in a cost-effective manner, or be achieved at all. The lead-nation phenomenon is no invention of the 21st century – it has roots as long back as written history allows us to gaze.

The Revolutionary and Napoleonic Wars seem to be an aberration as, during this protracted conflict, France fought consecutive coalitions in which it is hard to pin-point clear leaders. For long periods, England was the primary contributor as far as funds were concerned, but it was Russia and Austria that seemed the most dominant powers on the Continent of Europe right up to 1814 – except in the Iberian Peninsula. It was at Waterloo in 1815 that Britain emerged as the coalition leader with the Duke of Wellington as the supreme commander, and even then it can be argued that the various armies fought more or less under individual national commanders.

We may find one more anomaly if looking at the Crimean War in the mid-19th century, when the Anglo-French coalition fought together but apparently with none of them in the rôle as the lead-nation. While France was the dominant land power, Britain 'ruled the waves,' and there was a strange lack of agreement on war aims and strategy. In the three wars of German unification from 1864-71, Prussia gradually emerged as the lead-nation of the North-German coalition.

Being a 'lead-nation' of a multilateral coalition presents a challenge that is compounded by the need for doctrine to conduct joint operations in a combined environment. Nonetheless, coalition warfare is generally handled *ad hoc,* as no commonly accepted doctrine for coalition warfare has survived until today. Any multi-national operation will require planning by all the participants, interoperability, shared risks and burdens, emphasis on commonalties, and diffused credit for success, but in the cases where the lead-nation is contributing disproportionately more than anyone else it will also wield considerable dominance – a logical consequence of coalition warfare by partners of unequal size and wealth. Nonetheless, it happens that minor coalition partners sense this as great power arrogance and, frequently, there will be some partner nations which are upset by the way they are treated by the leader of their coalition.

This was indeed the case among the Axis partners in North Africa. Upset by Rommel's arrogant bearing and the lack of agreed combined planning, the Italians were justifiably reserved about German initiatives. Against this background, neither the frequent meetings of Kesselring and Rommel with the Italian senior leadership in Rome and Libya, nor the exchange of permanent military representatives as liaison between the High Commands, could replace a combined planning entity or create the mutual trust on which the smooth working of any coalition depend. Similar feelings developed amongst the western Allies, and by 1943 General Brooke had realised the inevitable: that the Americans had grown to become by far the strongest amongst the Allies and taken over the rôle as the natural lead-nation. This embittered the relationship amongst the leading generals, and Montgomery, who shared Brooke's opinion on the matter, complained repeatedly about Eisenhower's inability to decide at the level of supreme commander while at the same time command the land component.

We may conclude that it remains one of the drawbacks of coalition warfare that if one participant is a great power and the others are contributors of limited resources, because of the great power's 'investment' in the project as well as the sophistication of its equipment, organisation and training, this power will be the one telling the others what to do – and occasionally even how to do it. If the representatives of the great power are remiss in tact and understanding for the smaller partners – as was the case in the Germano-Italian coalition – this can lead to inter-personal discord and may, in the long run, occasion dissolution of the coalition.

Leadership

Historically, leadership of coalition forces has implied issues like who should decide what and at which level direction should be given when putting officers and soldiers into harm's way. During the War of Spanish Succession Marlborough and Prince Eugene managed to get along without disagreement, in the First World War it was France that had the claim on supreme command when eventually this was established, but in the Second it upset General Brook, who had been promised the post, that Churchill – without informing him – gave away the position of supreme commander of the Allied Expeditionary Forces in Europe to the Americans. Since then, supreme command of coalitions has always been in the hands of the biggest contributor. In Afghanistan the command structure allows for an American supreme commander in Kabul, and regional commanders of various coalition partners and lead-formations in the provinces, such as the UK in Helmand.

Since the days of the elder Moltke, *Auftragsbefehl* – or mission command – has become a basic principle in military leadership. Today most NATO nations subscribe to this principle, allowing subordinates to solve their missions as they see fit. Therefore, commanders should try to avoid micro-management. However, the present day's excellent means of communication might tempt impatient commanders to direct their troops in a rather detailed fashion. The commander of Task Force Helmand maintained that he managed to avoid this trap, but he realised that, at all levels of command, invariably from time to time there would be wishes for specific actions which would make interference in subordinate commanders' dispositions unavoidable. Although this was not desirable, it was a logical consequence of tactical imperatives, which should be ignored by no one.

The appointment of *Maréchal de France* Ferdinand Foch as the Commander-in-Chief of the Entente armies on the Western Front in the spring of 1918 was a necessary step in order to avoid the collapse of the Allied war effort against Germany. Though already allied against the Central Powers for four years, Britain and France lacked a unified command system, relying instead on assurances of support. Historians note that army commanders guarded their army's national sovereignty and control over their respective reserves, for they did not wish to handicap their own freedom of movement or to place troops under a foreign commander.

Choosing to place a number of coalition partners' forces under the command of one commander-in-chief is not only a matter of unity of command, but also one of co-ordination. However, circumspect composition of this commander's staff is prerequisite to coalition success. The leadership tier must be acceptable to all concerned and their decisions must appear relevant and just. The war councils of British King William III seemed to be inclusive and acceptable to the nations contributing to his coalition in Ireland, and – if for a minute we leave the rivalry between Monty and Ike alone – the staff of SHAEF (Supreme Headquarters Allied Expeditionary Forces) also appeared well balanced and sensible. Compared with Axis collaborative difficulties, the co-ordination of Allied activities in the European Theatre of War ran smoothly. And, recent experiences give optimistic proof of the possibility to co-operate, in as much as Task Force Helmand operated in close co-operation with the Afghan 3rd Brigade.

Personal Understandings

Since coalition warfare means working together, inter-personal understanding amongst the commanders is of key importance. Personalities, and especially those of ambitious and successful leaders, do not always match, and over the years many coalitions have faced trouble because of the colossal egos of *prima donna* commanders. Frequently, relations assume the form of personal envy, because amongst commanders there will always be some whose ambitions and self-importance stand in the way of smooth collaboration and amicable communication with their peers.

In 1690, Lieutenant-General the Earl of Marlborough, having just arrived in Cork, believed himself destined to take overall command of a combined operation in which his colleague, Lieutenant-General the Duke of Württemberg, had already been engaged for some time and claimed the same right. Eventually the found an agreeable *modus vivendi*, but had they not, this would have jeopardised the whole enterprise.

At top level during the Second World War, the Anglo-American special relationship was in many respects a personal matter, and although there were cultural differences surfacing every now and again as well as personal animosity, at the end of the day there was more that united. From the very first, Churchill made sure that representatives of President Roosevelt, such as for instance Averell Harriman, Harry Hopkins and Admiral King, got the most exquisite treatment. However, there were key co-operation partners with whom not everything went as smoothly as that, and the British relationship with the Chief of the US Army General Staff, General George Catlett Marshall, lay somewhere between the excellent *rapport* which Churchill had with Harriman and Hopkins and the frosty association most British leaders had with French General de Gaulle. General Alan Brooke found Marshall quite charming, but not as sociable as Harriman and Hopkins, and deplorably remiss in strategic understanding. However, clashes of personalities occurred aplenty. Churchill found de Gaulle a most tiresome person to deal with, and Brooke had great trouble bearing his presence. Montgomery was as difficult as de Gaulle and had a well-developed sense of self-promotion at the expense of others, as well as a faculty for annoying his closest collaborators. As a result, Brooke took much trouble to reconcile him with Eisenhower, as well as with Churchill, and quite a few others whom he happened to upset. Nonetheless, by and large Americans and British got on well together, and in particular President Roosevelt's helpfulness and understanding on hearing, during the Second Washington Conference of 20-25 June 1942, about the fall of Tobruk, helped grease the wheels of the relationship. Notwithstanding the attempts to assuage allies, the relationship with the Russians was often difficult, and in particular during the Second Moscow Conference of 12-17 August 1942, the British had a hard time.

Such clashes are still a concomitant risk in coalition warfare, and we saw a striking example in Afghanistan in 2010. This time the quarrel was not between commanders of different nationalities, but caused by the American General Stanley McChrystal scoffing his boss, American President Barrack Obama. Not unnaturally, the president soon found a new incumbent for the job as commander-in-chief in Afghanistan, General Petraeus, but although this worked out well, all high-level disagreement is potentially damaging to the coalition.

Though we have little evidence of inter-personal vindictiveness in Ancient and Medieval times, it seems reasonable to assume that this is a shortcoming which is characteristic of the dealings of mankind. The plethora of examples from modern coalition warfare gives us reason to believe that in this respect little has changed over the years.

Languages

Although over the years many coalitions have included partners using the same language, many others have had a broader recruitment. Moreover, in past centuries it was not uncommon that an officers' career included service for many sovereigns, requiring him to command a professional working language, which would normally be French.

The degree to which national commanders and staffs understand each other and are able to participate in combined planning impacts the time required to plan and the sharing of knowledge of every component of operations. Language remains a prime concern, and working through interpreters is a cumbersome affair. For centuries French was the language spoken by all people of a certain social standing, and military literature was mostly written in either French or German. Today French and English are the official languages of NATO, but English is by far the most commonly used, by officers and politicians of NATO member countries as well as by almost all those who join multinational coalitions. In his excellent account of the atrocities committed by opposing militias during his tenure of the post as UN commander in Rwanda, Canadian Lieutenant-General Romeo Dallaire deplores the trend towards excluding French as the operational language used when performing tasks in Francophone countries, but at the same time he realises that this is a lost cause.[1] In Helmand in 2010 English was the *lingua franca,* not only between Brits and Danes but also with the Afghan officers, of whom some had a reasonable command of oral English and a somewhat more modest ability to write.

Asymmetry

Until recently, coalition warfare has not been characterised by asymmetry in the way we have seen it during the war on terror. Most coalitions have fought armies or other coalitions equipped with roughly the same weaponry. However, in their early phases, many wars have had asymmetric traits because of novel and surprising doctrines and organisation. Seen in this perspective, the Revolutionary and Napoleonic Wars were asymmetric until the opponents learned the lesson that conscription was the foundation of modern mass armies, and that armies would have to live off the country to be flexible and retain surprise and initiative. The German armies of 1866 and 1870 fought asymmetrically because of their superior exploitation of modern means of communication, notably railroads and telegraph, as did the German army of 1940, effectively employing close air support and massed armoured formations against the Anglo-French coalition, whose tank forces were spread thinly across their armies. Similarly, we may suggest that the Cold War ended in part because of increasing asymmetric advantages for the western alliance due to the Network Centric Warfare concept, which the Warsaw Pact had neither the means nor the know-how to match.

In the post-Cold War Era, asymmetry has assumed a new meaning. While organised coalition forces are now invariably equipped with huge quantities of information technology, the perpetrators of terrorism work with a fundamentally different, and low-tech, *modus operandi*. Satellites and sensors frequently miss their targets because these do not emit any electronically perceivable signal, they blend in with local citizens, and their command structure is of a highly volatile and intangible nature. In 2010, in the Danish sector of the Helmand Province the Taliban held both a physical and psychological sway that extended from Gereshk in the south to Qaleh ye Gaz in the North. Their command and control was based on a number of key nodes in various locations that they used for projection of both soft

---

[1] Roméo A. Dallaire, *J'ai serré la main du diable* (Montréal : Libre Expression Canada, 2003), pp. 435 and 452. (Also published in English as *Shake Hands with the Devil: The Failure of Humanity in Rwanda* (Toronto : Random House Canada, 2003)).

influence and violent activity. Moreover, the insurgents had apparently put much effort into perfecting a multi-layered system of improvised explosive devices based on caches as well as large production and storage facilities, which were all difficult to spot electronically. The coalition's combat against this activity could not possibly happen symmetrically and had to be based on human intelligence rather than modern technology – it depended heavily on both contacts to local civilians and the Afghan National Security Forces.

Terrorism

Although toppling dictatorships and combating terrorism worldwide are coalition war aims typical of the 21st century, we might ask ourselves if this is really a novelty. As far as dictators are concerned, two world wars were fought to rid Europe of autocratic and non-democratic leaderships, Napoleon was indeed a practical dictator, although in this respect he did not differ too much from main-stream continental heads of state at the time; King Louis XIV of France was indeed an autocrat and a menace to all neighbouring countries, and we might continue the list even further back without too much difficulty. As for terrorism, there does not seem to be as many of the small scale 'private terror' actions in the past as there are at the present. However, examples of terror on a grand scale perpetrated by states, organised hordes, or powerful leaders exist aplenty. Attila, Genghis Khan, Robespierre and the Revolutionary Tribunal, as well as the Nazi-German SA and SS figure notably among them.[2] Coalitions are entered into for reasons of state security, and that does indeed include indirect security by combating terrorism. Today we speak of terrorism mostly as something connected with the September 2001 atrocities or later acts of unwarranted violence against innocent civilians, and it was in response to this kind of modern vindictiveness motivated by religious fantasies that in 2001 the multi-national military coalition, the International Security Assistance Force or ISAF, moved into Kabul on a mandate by the United Nations' Security Council. In 2006, this coalition expanded its area of responsibility to cover most of the country and, from then onwards, the establishment and training of a reasonably efficient Afghan security apparatus gathered momentum. That means that, possibly for the first time ever, a coalition has deployed to a war torn country with development of local military forces and hand-over to indigenous authorities as some of their primary 'war aims' (though with

---

[2] The Reign of Terror (*la Terreur* 5 September 1793, to 28 July 1794).

some approximation we might claim that the intervention in Iraq in 2003-7 shows some distinct similarities).

Since 2007 – under the auspices of the British Task Force Helmand as the lead formation – a Danish-led battle group has been part of that effort, and we have seen a hitherto rare combination of very stubborn counter insurgency combat alongside influence operations and humanitarian aid. Politically, military engagements in coalitions in far-flung countries serve security indirectly – the present Afghanistan mission is no exception – but they also send messages of solidarity with partners wishing to keep international law and order and combat terrorism. Although the United States is by far the most significant contributor to the endeavours towards a peaceful Afghanistan, and a single battle group seems of little consequence, it has been clearly demonstrated that in terms of legitimacy this contribution is of no small importance. Moreover, the willingness to share the burdens, including casualties, does indeed matter to solidarity within the coalition as well as to the wider world. Similarly, the development effort and the partnership with the Afghan Army have been improving by the day and, in particular as far as operations were concerned, disagreements were solved and issues of importance were co-ordinated concurrently: the novel 'war aim' of development and hand-over has moved closer. Operating in this fashion, the coalition forces have contributed decisively to establishing law and order while planting the seed of democracy in large parts of Afghanistan. Over the years of the coalition's toil in that country, remarkable improvements of both security and community organisation have been seen. As early as in 2010, the Danish Battle Group 'ISAF Team 10' was able to launch the process of training the Afghan National Security Forces, and slowly but surely start handing over responsibility to the indigenous security authorities.

The End of Coalitions

While many coalitions have survived with almost the same combination of partners throughout and right to the end of conflict, there have also been many instances of dissolution due to defection from the common enterprise. The Roman Empire split up and its military domination of the continent – and Britain for that matter – came to an end as the army dwindled and new threats arose in the East. Napoleon's *grande armmée* collapsed after the defection of allies from The Confederation of the Rhine, and Austria-Hungary gave up in 1918 when the Polish, the Czech, the Ukrainian, and numerous other nations of the multi-national empire went their individual ways.

Similarly, on 25 February 1991 at a meeting in Hungary of Warsaw Pact defence and foreign ministers, it was decided to discontinue collaboration. On the first of July 1991 in Prague, the Czechoslovak President Václav Havel formally ended the Warsaw Treaty Organisation of Friendship, Co-operation, and Mutual Assistance and so disestablished the Warsaw Treaty after 36 years of military alliance under Soviet leadership. Five months later the USSR disintegrated.

While in the summer of 2010, the British and Danish prime ministers agreed on 2014 as the year in which to terminate the presence of combat troops in Afghanistan, the UK Ministry of Defence bureaucracy believed that the mentoring of Afghan forces and partnering between these and the western military units might lead to immediate redundancy of troops in Afghanistan. However, since these activities would require more rather than fewer western mentors and partner units, the British Army was adamant that all 9,500 troops would be needed – at the very least until 2012. Seeing the writing on the wall, the United States, anxious not to be left alone with the responsibility for security, law and order in Afghanistan, began, in concert with President Karzai, negotiation with parties within the Taliban to bring about a compromise allowing for a coming inclusion of them in the future governmental structure of the country. The Coalition-of-the-Willing at last began to come to an end.

Epilogue

FROM ANTIQUITY TO the present, coalitions have arisen because of the need to justify military enterprises, economise resources, and achieve goals of common interest which might have been hard to achieve by any warring state fighting on its own. One ever-present reason for joining such coalitions has been the need to never let one's armed forces stand idle – the necessity of always keeping the military instrument sharp and prepared for action. The advantages gained by joining coalitions have been the pooling of resources, underpinning of legitimacy, and the prospect of making wars as short as possible. Conversely, some features have turned out to threaten the survival of coalitions or diminish their achievements, and these should not be forgotten. Among them are cultural disparity, personal envy, great-power arrogance, decreasing zeal, dwindling resources and disparate technological levels, all of which contribute to hampering operations and, ultimately, may lead to the dissolution of any given coalition. Friction in war has been the abiding companion of military coalitions fighting in the past, and this remains the case. Bad relationships between politicians and military professionals will always lead to unhealthy discrepancies between the political aims *with* the war on the one side, and the military imperative of fulfilling these by setting realistic aims or objectives *in* the war on the other.

Coalitions may be constituted by a number of great powers of almost equal military muscle, or by a number of small or medium powers rallying around the colours of a single warring state. However, coalitions of peer partners are not necessarily more effective than those gathered around one lead-nation with abundant resources for the enterprise and a unified and well-rehearsed command and control system, although the long term consumption of fuel, ammunition, manpower and funds may be better served by the former.

While weaponry, organisation and doctrine have changed over time, the fundamental features of coalition warfare have remained unaltered. The need for determined leadership, agreed war aims, strategies and doctrines, as well as reconciliation of cultural disparities including language are of paramount importance and have always been so.

It is arguable if the relative influence of the lead-nation and its followers has mutated as large parts of the world have become more democratic. The dictatorial dominance by the Spartans or the Romans is gone for good. As early as during the Second Boer War, Britain could no longer just order participation and effort from her Dominions, and during the world wars this development was accelerated. NATO-led coalitions-of-the-willing feature similar constraints, as the world has seen in the case of Afghanistan and, even more pronounced, in the action to protect civilians in Libya in 2011.

The motives for joining coalitions may be the most obvious province of change. While the Lakedaimonians fought because Sparta told them to, the two parties of the Thirty Years War combated each other over whether it should be the local princes or the Emperor – albeit on behalf of the Catholic Church – who should be the primary decision-maker in Europe. Denmark joined Britain to fight James II in Ireland and Louis XIV on the Continent because the English monarchs paid, and the Danish troops got valuable training for free. The Entente came together because a number of western democracies wished to curb the authoritarian empires' menacing cravings for expansion and commercial dominance, and the World War II anti-Fascist coalition fought because this was the only option short of total extinction of democratic rule in Europe and the Far East.

Although there have been coalitions which have failed, such as the Central Powers did eventually in World War I, the concept has stood the test of time due to its viability, resilience and flexibility. Thus, as this concept has developed, coalition warfare is no longer just a matter of fighting a common enemy – it is a coalescence of nations that of necessity transcends national core values and beliefs to facilitate positive outcomes of a common cause.

As nations continue to rely upon one another for the provision of security as well as for trade and technological progress, the importance of establishing coalitions becomes more critical by the day. Today, coalitions form up to combat transnational crime, or terror, and it is hard to say which motives will guide tomorrow's coalitions. However, security, whether territorial or indirect, will remain part of the equation, and the financial crisis of the 21st century's first two decades makes the need to coalesce all the more obvious.

Danish officers mentoring Afghan troops questioning local civilians (photo by Major K.V.F. Ahlefeldt-Laurvig who was Force Protection mentor to Garrison Support Unit of 3rd Brigade, 215th Corps).

Biographies

Abdur Rahman Khan, amir of Afghanistan (ca. 1840-1901). The third son of Mohammad Afzal Khan, and grandson of Dost Mohammad Khan, Abdur Rahman Khan is considered a strong ruler establishing central Afghan government in Kabul after the disarray in the wake of the Second Anglo-Afghan War. In May 1881, Abdur Rahman Khan, a nephew of Sher Ali and for some time an Afghan expatriate in Russia, was accepted by the British as new amir in Kabul. He successfully tried re-unite the country, which he ruled with an iron fist for the rest of his life. He died in 1901. At that moment he had not only united the country anew, he had also re-established friendly relations with the British Empire.

Ahmad Shah Massoud, Mujahidin chief (1953-2001) was a Kabul University engineering student turned military leader. He played a leading role in driving the Soviet army out of Afghanistan, earning him the name "Lion of Panjshir." He strongly rejected the interpretations of Islam followed by the Taliban, Al Qaeda or the Saudi establishment. His followers not only saw him as a military commander but also as a spiritual leader. After the collapse of the communist Soviet-backed government of Mohammad Najibullah in 1992, Massoud became the Minister for Defence under the government of Burhanuddin Rabbani. Following the rise of the Taliban in 1996, Massoud returned to the rôle of armed opposition leader, serving as military commander and political leader of the United Islamic Front (also known in the West as Northern Alliance). On 9 September 2001, two days before the 11 September atrocities in the United States, Massoud was assassinated by two suspected al-Qaeda suicide bombers posing as journalists. The following year, Afghan President Hamid Karzai made him a 'national hero.'

Alanbrooke, British field-marshal, Lord, Chief of the Imperial General Staff (CIGS). See Brooke.

Alexander I, Tsar of Russia (1777-1825) served as emperor of Russia from 23 March 1801 to 1 December 1825 and the first Russian king of Poland from 1815 to 1825. He was also the first Russian grand duke of Finland and Lithuania.

Amanulla, amir of Afghanistan (1892-1960) was the Sovereign of Afghanistan from 1919 to 1929, the first Afghan ruler to style himself king. He led Afghanistan to independence over its foreign affairs from the United Kingdom, and his rule was marked by dramatic political and social change. He attempted with some initial success to modernise Afghanistan on western designs. However, eventually he did not succeed and on 14 January 1929, he abdicated and fled to neighbouring British India. From thence he went to Europe where he died in Zürich, Switzerland, in 1960.

Ambrosio, Vittorio (1879-1958), Italian general, WW II chief of the army's general staff. He was an Italian general who served in the Italo-Turkish War, World War I, and World War II. During the latter conflict, Ambrosio was instrumental in the fall of the *Duce*, Benito Mussolini, and the eventual Italian repudiation of her alliance with Germany.

Auckland (1784-1849), Lord, governor General of India from 1836 to 1842. He was born on 25 August 1784 and educated at Christchurch, Oxford. Although Afghan affairs attracted Auckland's attention, his involvement in the first Anglo-Afghan War did not earn him much credit. He deposed pro-Russian Dost Muhammad, and replaced him by pro-British Shah Suja as the Amir of Afghanistan. The Afghan war led to an unfortunate breach of a recently signed treaty with the amirs of Sind. The war ended with considerable British losses. Auckland was recalled, and left for England on 12 March 1842. He was made an earl in 1839 and died on 1 January 1849.
Ayub Khan, Ghazi Mohammad (1857-1914) was amir of Afghanistan from 12 October 1879 to 31 May 1880 and was also the leader of Afghans in the Second Anglo-Afghan War.

Bastico, Ettore (1876-1972), Italian field-marshal. He held high commands during the Second Italo-Abyssinian War (Ethiopia), the Spanish Civil War, and the North African Campaign. In 1932, Bastico was promoted major-general and in 1936 lieutenant-general (*generale di corpo d'armata*). In 1937, during the later stages of the Spanish Civil War, Bastico replaced Mario Roatta as the Commander-in-Chief of the Italian Corps of Volunteer Troops (*Corpo Truppe Volontarie*) in Spain. When Italy entered World War II, Bastico was governor general of the Italian Dodecanese Islands and he was promoted to general (generale *d'armata*). On 19 July 1941, Bastico was appointed commander-in-chief Axis Forces in North Africa. He was promoted Marshal of Italy (*Maresciallo d'Italia*) on 12 August 1942. However, the loss of Libya left him from 2 February 1943 without a command for the rest of the war.

Bhutto, Benazir (1953-2007), the 11th Prime Minister of Pakistan in two non-consecutive terms from 1988 until 1990 and 1993 until 1996. She was the daughter of Zulfikar Ali Bhutto, a former prime minister of Pakistan and the founder of the Pakistan People's Party, which she led. She was assassinated in 2007.

Bernadotte, Jean Baptiste Jules (1763-1844), French marshal, prince of Ponte Corvo later to become crown prince of Sweden. One of the most controversial of Napoleon Bonaparte's marshals, Bernadotte was a committed republican. He rose through the ranks of the French army from 1780 to 1794 moving from private soldier to major-general. Along the way he fought, mainly along the Rhine, and won a victory at Limburg in 1796. Two years later he was sacked for quitting his command due to dissatisfaction with how the war against the Second Coalition was being fought, but had the good fortune to marry Desiree Clary, one of Napoleon's former lovers. Bernadotte was adopted by the childless King Charles XIII of Sweden. He took his new country to heart and put his considerable military talents to excellent use. Initially amicable with Napoleon, things became strained when France occupied Swedish Pomerania in 1812. By 1813, Bernadotte had joined the Sixth Coalition against his former emperor and beat Marshals Oudinot and Ney at Groß Beeren and Dennewitz. He fought on the winning side at Leipzig and in 1814 he incorporated Norway into Sweden. In 1818, Bernadotte became King Charles XIV of Sweden.

Bloch, Jan Gotlib (Bogumil) (1836-1902/1901), Warsaw banker and war theoretician.

Borden, Sir Robert Laird (1854-1937), Canadian prime minister. Borden was a lawyer by profession and a politician. He served as the eighth Prime Minister of Canada from 1911 to 1920, and was the third Nova Scotian to hold this office. After retiring from public life, he served as the chancellor of Queen's University.

Bracken, Brendan Randell, 1st Viscount Bracken PC (1901-1958) was an Irish businessman and a minister in the British Conservative cabinet during WW II. Primarily, the 1st Viscount Bracken is remembered for opposing the Bank of England's co-operation with Adolf Hitler, and for subsequently supporting Winston Churchill's prosecution of the Second World War against Germany. Bracken was also the founder of the modern version of the Financial Times. He served as Minister of Information from 1941 to 1945. Many literary academics believe that it was Bracken who inspired George Orwell to create the character Big Brother in his novel *Nineteen Eighty-Four*. Orwell himself was a civil servant under Brackens department during the war years.

Bradley, Omar Nelson (1893-1981), American general. He was a senior US Army field commander in North Africa and Europe during World War II. During the last German offensive, *Operation Wacht am Rhein* known to English speakers as the Battle of the Bulge, Bradley's command took the initial brunt. For logistical and command reasons, General Eisenhower decided to place Bradley's 1st and 9th Armies under the temporary command of Field-Marshal Montgomery's 21st Army Group on the northern flank of the Bulge, a decision Bradley strongly opposed but eventually accepted. He was later promoted General of the Army (field-marshal), and the first general to be selected Chairman of the Joint Chiefs of Staff.

Brauchitsch, Heinrich Alfred Hermann Walther von (18811948), German *Generalfeldmarschall* [field-marshal]. von Brauchitsch was a *Oberbefehlshaber des Heeres* (Commander of theArmy) in the early years of World War II. When Germany invaded the Soviet Union on 22 June 1941, the Army's failure to take Moscow infuriated Hitler. Things went further downhill for von Brauchitsch as he endured a serious heart attack, and Hitler relieved him from command on 10 December 1941.

Brooke, Alan Francis (1883-1963), 1st Viscount Alanbrooke, British field-marshal. He was Chief of the Imperial General Staff during most of the Second World War, and was promoted field-marshal in 1944. As chairman of the Chiefs of Staff Committee, Brooke was the foremost military advisor to Prime Minister Winston Churchill, and in the rôle of co-ordinator of the

British military efforts was an important but not always well-known contributor to the Allies' victory in 1945. He was born in France and was educated in Pau, where he lived until the age of 16. Thanks to his upbringing in the country he became a fluent French speaker. During the years as CIGS, Brooke had a stormy relationship with Winston Churchill. He was often frustrated with the Prime Minister's habits and working methods, his abuse of generals and constant meddling into strategic matters. At the same time, though, he greatly admired Churchill for the way he inspired the Allied cause and for the way he bore the heavy burden of war leadership.

Broz, Josip "Tito" (1892-1980), Yugoslav revolutionary, marshal and statesman. While his presidency has been criticised as authoritarian, Tito was a popular public figure both in Yugoslavia and abroad, viewed as a unifying symbol for the nations of the Yugoslav federation. He gained international attention as the chief leader of the Non-Aligned Movement, working with Jawaharlal Nehru of India and Gamal Abdel Nasser of Egypt. He was born as the seventh child of Franjo and Marija Broz in the village of Kumrovec within Austria-Hungary (modern-day Croatia). Drafted into the army, he distinguished himself, becoming the youngest sergeant-major in the Austro-Hungarian Army. After being seriously wounded and captured by the Russians, Josip was sent to a work camp in the Ural Mountains. He participated in the October Revolution, and later joined a Red Guard unit in Omsk. Upon his return home, Broz found himself in the newly-established Kingdom of Yugoslavia, where he joined the Communist Party.

Burns, Captain Sir Alexander (1805-1841), British envoy to Afghanistan. He was born in Montrose, Scotland, to the son of the local provost, who was first cousin to the poet Robert Burns. At the age of sixteen, Alexander joined the army of the East India Company and while serving in India, he learned Hindustani and Persian, and obtained an appointment as interpreter at Surat in 1822. Transferred to Kutch in 1826 as assistant to the political agent, he took an interest in the history and geography of north-western India and the adjacent countries, which had not yet been thoroughly explored by the British. Living, from 1837, in Kabul as a British political officer, he advised the governor general of India, Lord Auckland, to support Dost Mohammad on the throne of Kabul, but the Auckland preferred to follow the opinion of Sir William Hay Macnaghten and reinstated Shah Shuja, thus leading to the disasters of the First Afghan War. Burnes remained in Kabul until his assassination in 1841, during the heat of an insurrection. He continued at his post long after the imminence of danger was apparent, and fought ferociously after the killing of his political assistant Major William Broadfoot, killing six assailants in the process.

Bush, George (1946-), 43rd president of the United States of America. Bush, initially a somewhat isolationist president, became known primarily for his engagement of the United States in the wars in Afghanistan (2001-) and Iraq (2003-12) and for his hard-line and activist foreign policy.

Canrobert, François Certain de (1809-95), *Maréchal de France*. The future Marshal was educated at Saint-Cyr; he received a commission as a sub-lieutenant in 1828. Summoned to Paris, he was made *aide-de-camp* to the president, Louis-Napoléon Bonaparte, and took part in the coup d'état in 1851. In the Crimean War he commanded a division at the Battle of Alma, where he was twice wounded. He held a dormant commission entitling him to command in case of General Saint-Arnaud's death, and he thus succeeded to supreme command of the French army a few days after the battle. He was slightly wounded and had a horse killed under him at Inkerman, when leading a charge of Zouaves. Disagreements with the British commander-in-chief, Lord Raglan, and in general, the disappointments due to the prolongation of the siege of Sevastopol led to his resignation of the command, but he did not return to France, preferring to serve as chief of his old division almost up to the fall of Sevastopol. After his return to France in August 1855 he was sent on diplomatic missions to Denmark and Sweden, and created a Marshal of France and a Senator for Life. He was also made grand cross *Légion d'Honneur*, and honorary Grand Cross in the Order of the Bath. He commanded the III Army Corps in Lombardy in 1859 during the Second Italian War of Independence, distinguishing himself at Magenta and Solferino. During the Franco-Prussian War he commanded the VI army corps, which won great distinction at Mars-la-Tour and at the Gravelotte, where Canrobert commanded on the Saint-Privat position.

Carnot, Lazare Nicolas Marguerite, Comte (1753-1823), French secretary for war. Lazare Carnot, was a French politician, engineer, and mathematician. The creation of the French Revolutionary Army was largely due to his powers of organisation and enforcing discipline. In order to raise more troops for the war Carnot introduced conscription by the *levée en masse* ordinance approved by the National Convention in 1793. Once the problem of troop numbers had been solved Carnot turned his administrative skills to the supplies that this massive army would need. In autumn 1793, he took charge of French columns on the Northern Front, and contributed to Jean-Baptiste Jourdan's victory in the Battle of Wattignies. After

Napoleon crowned himself emperor on 2 December 1804, Carnot's republican convictions precluded his acceptance of high office under the First French Empire, and he resigned from public life – although he was later made a Count of the Empire by Napoleon as Lazare Nicolas Marguerite, comte Carnot.

Carter, Nicholas Patrick 'Nick', British major-general Major General. Upon promotion to major-general he became General Officer Commanding 6th Division, which has been deployed to Afghanistan with Carter appointed as Commander ISAF Regional Command South. In September 2009, referring to the efforts of UK and NATO forces, Carter said that 'time was not on our side.' Upon his return to the UK in 2010, Carter gave an interview in which he warned that 'the insurgency is resilient, and alive and well.' Since then he has been Director-General Land Warfare.

Cavagnari, Sir Pierre Louis Napoleon (1841–1879), British envoy to Afghanistan. In September 1878 he was attached to the staff of a British mission to Kabul, which the Afghans refused to allow proceeding through the Khyber Pass. In May 1879, after the British-Indian forces had invaded Afghanistan, and the death of Afghan amir Sher Ali Khan, Cavagnari negotiated and signed the Treaty of Gandamak with Sher Ali Khan's son and successor, Mohammad Yakub Khan allowing a British legation to Kabul. Cavagnari then took up Kabul residence in July 1879. However, on 3 September 1879 Cavagnari and the other European members of the mission, along with their guards were murdered by mutinous Afghan troops. Cavagnari was survived by his wife, Lady Cavagnari (née Mercy Ellen Graves), whom he had married in 1871.

Cavallero, Ugo (1880-1943), *Maresciallo d'Italia*, Chief of the Italian Supreme Command (*Comando Supremo*). After Italy entered World War II in June 1940, Cavallero was made the Commander-in-Chief of the Italian Army Group in Albania. In October, when Italy invaded Greece, he was the commander of the invading Italian forces. In December, he replaced Pietro Badoglio as the Chief of the Comando Supremo. As Chief of the Supreme Command, Cavallero worked closely with German Field-Marshal Albert Kesselring and often asked for Kesselring's advice on military matters. Under Cavallero's leadership, Italy's military forces performed poorly during the war. Nonetheless, he was promoted to Marshal of Italy (*Maresciallo d'Italia*) in 1942.

Chiswell, James R., Brigadier, Commander of Task Force Helmand 2010-11. James Chiswell was commissioned into the (UK) Parachute Regiment in 1983. He commanded the 2nd Battalion The Parachute Regiment in 2004 serving with them in Iraq. He was posted subsequently to the Permanent Joint Headquarters with responsibility for operational-level oversight of the UK deployments in Iraq, and later in Afghanistan. He followed this in 2008 with nine months in Washington DC as the Chief of the Defence Staff's Liaison Officer to the Chairman of the Joint Chiefs of Staff providing military strategic liaison for the current campaigns. During his career he has served in Northern Ireland, Cyprus, Bosnia, Kosovo, Sierra Leone, South America and Iraq. He was awarded the Military Cross for his actions in the Sierra Leone Operation. He completed the Higher Command and Staff Course in 2007 taking over command of 16 Air Assault Brigade in December 2008. He became Commander, Task Force Helmand, in October 2010 when the Brigade deployed to Afghanistan having also under his command the Danish ISAF Team 10.

Christian V(1646-1699) was king of Denmark and Norway 1670-99, the son of Frederick III of Denmark and Sophie Amalie of Braunsweig-Lüneburg. On 14 May 1667, he married Charlotte Amalie of Hessen-Kassel (Hesse-Cassel). It is sometimes argued that Christian V's personal courage and affability made him popular amongst the common people, but this image is stained by his unsuccessful attempt to regain the lost province of Scania from Sweden in the Scanian War. He hired out a Danish expeditionary corps to fight for William III and Mary of Great Britain in the Nine Years War.

Churchill, John Spencer, first Duke of Marlborough, see Marlborough.

Churchill, Winston Spencer (1874-1965), British prime minister. A British Conservative politician and statesman he is primarily remembered for his leadership of the British Empire during World War II. Widely regarded as one of the greatest wartime leaders of the century, he served as prime minister twice (1940–45 and 1951–55). Churchill had resigned his king's commission in the British Army, but carried on as a yeomanry officer and also served in the trenches of the Western Front during parts of World War I. Moreocer, he was a historian, a writer and an artist. He is the only British prime minister to have received the Nobel Prize in Literature, and was the first person to be made an Honorary Citizen of the United States. Churchill was born into the aristocratic family of the Dukes of Marlborough. His father, Lord Randolph Churchill, was a charismatic politician who served as viceroy of Ireland and chancellor of the exchequer; his mother, Jenny Jerome, was an American by birth. As a young army officer, he saw action in British India, the Sudan, and as a journalist in Cuba and the Second Boer

War. Being out of office during the 1930s, Churchill took upon himself to warn about Nazi Germany and in campaigning for British re-armament. On the outbreak of the Second World War, he was again appointed First Lord of the Admiralty. Following the resignation of Neville Chamberlain on 10 May 1940, Churchill became prime minister. His steadfast refusal to consider defeat, surrender, or a compromise peace helped hardening British resolve, especially during the difficult early days of the war when Britain stood alone in its active opposition to Hitler. Churchill was particularly noted for his speeches and radio broadcasts, which helped inspire the British people. He led Britain as prime minister until victory over Nazi Germany had been secured. After the Conservative Party lost the 1945 election, he became leader of the opposition. In 1951, he again became Prime Minister, before retiring in 1955. Upon his death, Queen Elizabeth II granted him the honour of a state funeral, which saw one of the largest assemblies of world statesmen in history. Named the Greatest Briton of all time in a 2002 poll, Churchill is widely regarded as being among the most influential persons in British history.

Ciano, Count Gian Galeazzo, 2nd Count of Cortellazzo and Buccari (1903-1944, Italian foreign secretary and Benito Mussolini's son-in-law. In early 1944 Count Ciano was shot by firing squad at the behest of his father-in-law, Mussolini, under pressure from Nazi Germany.

Clemenceau, George (1841-1929), French prime minister twice, in 1906-09 and from November 1917-20. Clemenceau's staunch republicanism brought him into early conflict with Emperor Napoleon III's government. Although trained as a doctor he travelled to the USA where he remained for several years as a teacher and journalist, returning to France in 1869. Following the 1870 overthrow of Napoleon III, Clemenceau became mayor of Montmartre in Paris. In 1902 Clemenceau was elected senator, and in 1906 became minister of the interior and then premier. In 1909, Clemenceau's government fell and Aristide Briand succeeded him as prime minister. In the following years Clemenceau vigorously attacked Germany and argued for greater military preparedness in the event of war. Clemenceau succeeded Paul Painleve as premier in November 1917, having been appointed by President Raymond Poincare, and remained on the post until 1920. Clemenceau persuaded the Allies to agree to a unified military command under Ferdinand Foch; he energetically pursued the war until its conclusion in November 1918. At the Paris Peace Conference Clemenceau insisted upon the complete humiliation of Germany, requiring German disarmament and severe reparations; France also won back Alsace-Lorraine

Cotton, Sir Willoughby (1783–1860), British-Indian major-general. During his career, Cotton played major roles in the First Anglo–Burmese War, the 1831-32 slave revolt in Jamaica and the First Anglo-Afghan War. Cotton was the Commander-in-chief of Bombay from April 1847 to December 1850.

Crerar, Harry (1888-1965), Canadian general. One of Canada's greatest wartime commanders, General Crerar was born and educated in Hamilton, Ontario. Graduating from the Royal Military College in 1909, he took a position with the Ontario Hydro-Electric Commission, Toronto. At the outbreak of the Great War he immediately joined Canada's First Division, going overseas with the First Contingent. He served in France, initially with the 3rd Field Artillery Brigade, later as brigade major of the 5th Canadian Divisional Artillery. A recognised leader in the development of modern artillery, he designed the largest, most intricate and successful creeping barrage in the later days of the war. After the war he remained in the army and was appointed to the General Staff, Ottawa. Immediately following the declaration of war in 1939, he was promoted and dispatched to Britain to prepare for the arrival of the Canadians. In July the following year, he returned to Ottawa a major-general and as Chief of the General Staff. In 1941, he was promoted lieutenant-general. Late in 1941 he returned to England and, in order to command the 2nd Canadian Division, and in April 1942, he was given permanent command of the 1st Canadian Corps. Crerar assumed command of the First Canadian Army on 20 March 1944, less than three months before the allied assault on Normandy. General Crerar retired in 1946 after serving Canada for more than 35 years. His career spanned two world wars and he was decorated by France, Belgium, the USA, Poland and Holland.

Currie, Sir Arthur William (1875-1933), Canadian lieutenant-general. Born on 5 December 1875 at Napperton in Ontario, Currie was an insurance broker, an estate agent and a militia officer before war broke out - and liable to be prosecuted for embezzlement until a group of friends mounted his financial rescue in 1914. With his name made following his conduct as General Officer Commanding 2nd (Canadian) Brigade during 1914-15, notably during the first German gas attack at Second Ypres, he was handed charge of 1st (Canadian) Division during 1915-16. Again impressing with his sure-footed command and meticulous attention to detail, Currie was promoted GOC Canadian Corps with the elevation of Sir Julian Byng to command of Third Army in June 1917. Largely responsible for the planning and execution of the success assault against Vimy Ridge, Currie remained vocal (and successful) in arguing for the retention of the Canadians as a single coherent fighting force. Notably popular with Sir Douglas Haig, the

BEF Commander in Chief, Currie nonetheless suffered from a reputation as a foul-mouthed, overbearing officer. His preference for managing his troops from far behind the front line further alienated his own troops, although in fact he was a frequent visitor to the front line. Following the war Currie served as inspector general of the Canadian militia and, from 1920 as Principal and Vice Chancellor of McGill University until his death on 30 November 1933.

Davout, Louis-Nicolas (1770-1823), *Prince d'Eckmühl,* French marshal. One of the most distinguished of the Napoleonic field commanders. Despite his noble origin and his training in the best military tradition of the ancien régime, Davout led his regiment in a pro-Revolutionary revolt (1790). He performed with merit in the Belgian campaign of 1792–93 and gained fame for his attempt to stop the treason of General Charles Dumouriez (April 1793). He accompanied Napoleon to Egypt (1798–99) and was promoted to major-general. As a corps commander, he had a significant impact on the victories at Austerlitz (1805), Auerstedt (1806), Jena (1806), Eylau (1807), Eckmühl (1809), and Wagram (1809). Created a duke (1808) and prince (1809) by Napoleon, Davout served as minister of war during the Hundred Days.

Digby, Pamela (Mrs Randolph Churchill, later to become Mrs Averell Harriman, 1920-97). English-born socialite who was married and linked to important and powerful men. In later life, she became a political activist for the United States Democratic Party and American ambassador to Paris. Her only child, Winston Churchill, was named after his famous grandfather.

Dill, Sir John Greer (1881-1944), British field-marshal. From May 1940 to December 1941 Dill was the Chief of the Imperial General Staff, the professional head of the British Army, and subsequently head of the British liaison mission in Washington, as Chief of the Joint Staff Mission and then Senior British Representative on the Combined Chiefs of Staff. Thus, during World War II he was instrumental in the formation of the 'special relationship' between the United Kingdom and the United States.

Disraeli, Benjamin (later 1st Earl of Beaconsfield, 1804-81), British prime minister, Conservative statesman and literary figure. Starting from comparatively humble origins, he served in government for three decades, twice as prime minister of the United Kingdom. He played an instrumental role in the creation of the modern Conservative Party and in the creation, in 1877, of Queen Victoria as Empress of India. From 1852 onwards, Disraeli's career would also be marked by his often intense rivalry with

William Ewart Gladstone, who eventually rose to become leader of the Liberal Party. In this feud, Disraeli was aided by his warm friendship with Queen Victoria, who came to detest Gladstone during the latter's first premiership in the 1870s.

Dost Muhammad Khan (1793-1863), amir of Afghanistan between 1826 and 1863. He first ruled from 1826 to 1839 and then from 1843 to 1863. He was the 11th son of Sardar Pāyendah Khan (chief of the Barakzai tribe) who was killed by Zaman Shah Durrani in 1799. Grandson of Hajji Jamal Khan, who founded the Barakzai dynasty in Afghanistan, Dost Muhammad belonged to the Pashtun ethnic group. Rejecting overtures from Russia, he endeavoured to form an alliance with Great Britain, and welcomed Alexander Burnes to Kabul in 1837. Burnes, however, was unable to prevail on the governor-general, Lord Auckland, to respond to the amir's advances. Dost Mohammad was enjoined to abandon the attempt to recover Peshawar, and to place his foreign policy under British guidance. He replied by renewing his relations with Russia, and in 1838 Lord Auckland set the British troops in motion against him. Subsequent to the First Anglo-Afghan War, Dost Mohammad was received in triumph at Kabul, and set himself to re-establish his authority on a firm basis. In 1855, he concluded an offensive and defensive alliance with the British government. In 1857, he declared war on Persia in conjunction with the British, and in July a treaty was concluded by which the province of Herat was placed under a Barakzai prince. During the Indian Mutiny, Dost Mohammad refrained from assisting the insurgents. In 1862, a Persian army, acting in concert with Ahmad Khan, advanced against Kandahar. The old amir called the British to his aid, and, putting himself at the head of his warriors, drove the enemy from his frontiers. On 26 May 1863, he captured Herat, but on the 9th of June he died suddenly in the midst of victory, after playing a great rôle in the history of Central Asia for forty years. He had named as his successor his son, Sher Ali Khan.

Durand, Sir Mortimer (1850-1924), British-Indian diplomat and civil servant. Durand entered the Indian Civil Service in 1873. During the Second Anglo-Afghan War (1878–1880) he was political secretary at Kabul. From 1884 to 1894, he was Foreign Secretary of India. Durand was appointed minister plenipotentiary at Tehran in 1894 although despite being a Persian scholar and speaking the language fluently he made little impression either in Tehran or on his superiors in London. From 1900 to 1903 he served as British Ambassador to Spain, and from 1903-1906 as Ambassador to the United States of America. In 1893, Mortimer Durand negotiated with Abdur Rahman Khan, the Amir of Afghanistan, the frontier between British India and Afghanistan. This line, the Durand Line, is named after him and remains the international boundary between Afghanistan and modern-day Pakistan, officially recognised by most nations but an ongoing point of contention between the two countries.

Dyer, Reginald (1864-1927), British Indian Army officer who as an acting brigadier-general was made responsible for the Amritsar massacre on 13 April 1919 in the Punjab. In 1919 the British government of India enacted the Rowlatt Acts, extending its World War I emergency powers to combat subversive activities. At Amritsar in Punjab, about 10,000 demonstrators unlawfully protesting these measures confronted troops commanded by Brig. Gen. Reginald E.H. Dyer in an open space known as the Jallianwalla Bagh, which had only one exit. Dyer was subsequently removed from duty but he became a celebrated hero in Britain among people with connections to the British Raj. It left a permanent scar on Indo-British relations and was the prelude to Mahatma Gandhi's non-cooperation movement of 1920–22.

Eden, Robert Anthony (1897-1977), latere 1st Earl of Avon. British Conservative politician, who was prime minister from 1955 to 1957. He was also Foreign Secretary for three periods between 1935 and 1955, including during the Second World War. He is best known for his outspoken opposition to appeasement in the 1930s, his diplomatic leadership in the 1940s and 1950s, and the failure of his Middle East policy during the Suez Crisis in 1956 that ended his premiership.
Eisenhower, Dwight D (1890-1969), American WW II general and 34th President of the United States from 1953 until 1961. 1944-45, he served as Supreme Commander of the Allied Forces in Europe; he had responsibility for the American planning of the invasion of North Africa in Operation Torch in 1942–43 and the successful invasion of France and Germany in 1944–45. In 1951, he became the first supreme commander of NATO.

Ellenborough, Edward Law, 1st Earl of Ellenborough (1790-1871), British Tory politician. He was four times President of the Board of Control (of the Honourable East India Company) and also served as governor-general of India between 1842 and 1844.

Elphinstone, William George Keith (1782-1842), British major-general. Elphinstone entered the British Army in 1804; he saw service throughout the Napoleonic Wars, rising to the rank of lieutenant-colonel by 1813. Elphinstone was promoted major-general in 1837, and, in 1841, during the First Anglo-Afghan War, placed in command of the British garrison in Kabul numbering around 4500 European and Indian troops. The garrison also included 12,000 civilians, including soldiers' families and camp followers. By then, he was elderly, indecisive and unwell, and was responsible for the disastrous retreat from Kabul during January 1842, which saw his command all but wiped out in a massacre. Elphinstone died as a captive in Afghanistan some months later.

Enver Pasha, Ismail (1881-1922), general, Ottoman minister for war. Ismail Enver was an Ottoman military officer and a leader of the Young Turk revolution. He was the main military leader of the Ottoman Empire in both Balkan Wars and World War I. Enver Pasha engineered the Ottoman-German WW I alliance, and expected a quick victory in the war that would benefit the Ottoman Empire. In the early stages of the war he allowed the German men-of-war Goeben and Breslau, to enter the Strait of the Dardanelles to escape British pursuit; and the subsequent transfer of these ships to the neutral Ottoman Fleet worked greatly in Germany's favour. In October 1914, an Ottoman naval squadron including the Goeben and the Breslau entered the Black Sea and raided the Russian ports of Odessa, Sevastopol and Theodosia. Russia duly declared war on Ottoman Empire in November and Britain followed suit later in the same month. Enver proved ineffective as War Minister, and frequently over the next four years the Germans had to support the Ottoman government with generals such as Otto Liman von Sanders, Erich von Falkenhayn, Colmar Freiherr von der Goltz, and Friedrich Freiherr Kress von Kressenstein. The Germans also gave the Ottoman government military supplies, soldiers, and fuel.

Eugene, Prince of Savoy (1663-1736), was one of the most successful military commanders in modern European history, rising to the highest offices of state at the Imperial court in Vienna. Born in Paris to aristocratic Italian parents, Eugene grew up around the French court of King Louis XIV. At the age of 19 he was determined to make a military career. However, as King Louis found him unfit for service in the French army, he moved to Austria to serve the Habsburg Monarchy. He first saw active service against the Ottoman Turks at the Siege of Vienna in 1683 and the subsequent War of the Holy League, before serving in the Nine Years' War, fighting alongside his cousin, the Duke of Savoy. However, the Eugene's fame was secured with his decisive victory against the Ottomans at the Battle of Zenta in 1697, and he further added to his fame during the War of the Spanish Succession where his partnership with John Churchill, 1st Duke of Marlborough, secured the victories at Blenheim (1704), Oudenarde (1708), and Malplaquet (1709). In Austria Eugene's reputation remains unrivalled even today. The city of Vienna was allowed to include his crest in its coat of arms, and the two palaces of Lower and Upper belvedere (today arts museums) still reminds visitors of his extraordinary achievements.

Ezmerai, 2010-11 local police chief, Nahr-e Saraj North, Afghanistan

Foch, Ferdinand (1851-1929), *Maréchal de France*. French soldier, military theorist, and World War I hero credited with possessing 'the most original and subtle mind in the French army' in the early 20th century. Commanding the French Ninth Army Foch was widely credited for the 1914 victory on the Marne. He was subsequently promoted Marshal of France and, in 1918, made Supreme Commander of the Allied Armies, at which time he played a decisive rôle in halting a renewed German advance on Paris in the Second Battle of the Marne. On 11 November 1918, Foch accepted the German request for an armistice. He advocated peace terms that would make Germany unable to pose a threat to France ever again. Prophetically, on what the signature of the Treaty of Versailles would prove, he opined 'This is not a peace. It is an armistice for twenty years.'

Frederick VI (1768-1839), Danish king. He reigned as King of Denmark 1808-39 and as king of Norway 1808-1814. The Royal Frederick University in Norway was named in his honour. His parents were King Christian VII and Caroline Matilda of Wales. His father suffered from serious psychological problems, including suspected schizophrenia expressed by catatonic periods that resulted in his standing down from power as absolute monarch for most of his reign. Thus, during Frederick's childhood the country was governed by various relatives until, on 14 April 1784, he was declared of legal majority, and took over the regency himself. He continued as regent until his father's death in 1808. During this regency, Frederick instituted widespread liberal reforms including the abolition of serfdom in 1788. Crises encountered during his reign include disagreement with the British over neutral shipping, two British attacks on Copenhagen, the First Battle of Copenhagen (1801) and the Second Battle of Copenhagen (1807). The second gave rise to the Gunboat War between Denmark-Norway and the United Kingdom, which lasted until the Treaty of Kiel in 1814 in which Norway was ceded to Sweden. Frederick's wife was his first cousin Marie Sophie of Hesse-Kassel, a member of a German family with close marriage links with the Royal families of both Denmark and Great Britain.

Fredskov, Lennie, Colonel (Royal Life Guards of Denmark), Commanding Officer of Danish ISAF 10

Gaulle, Charles de (1890-1970), French general and statesman. He led the

Free French Forces during World War II. In 1958, he founded the French Fifth Republic and served as its first President from 1959 to 1969. A veteran of World War I, in the 1920s and 1930s de Gaulle came to the fore as a proponent of mobile armoured divisions, which he considered would become central in modern warfare. During World War II, he held the rank of *général de brigade*. De Gaulle was the most senior French military officer to reject the June 1940 armistice with Nazi Germany right from the beginning and being sent as an emissary to the United Kingdom he rallied what was to become the Free French Forces. As the war progressed de Gaulle gradually gained control of all French colonies except Indochina most of which had at first been controlled by the pro-German Vichy regime. Despite earning a reputation for being a difficult man to do business with, by the time of the Allied invasion of France in 1944 he was heading what amounted to a French government in exile. De Gaulle became prime minister in the French Provisional Government, resigning in 1946 due to political conflicts. After the war he founded his own political party, the Rally of the French People (RPF). Although he retired from politics in the early 1950s after the RPF's failure to win power, he was voted back to power as prime minister by the French National Assembly in May 1958. De Gaulle was the creator of a new constitution founding the Fifth Republic, and was elected President of France, an office which now held much greater power than in the Third and Fourth Republics. As President, Charles de Gaulle ended the political chaos that preceded his return to power. Immensely patriotic, de Gaulle and his supporters held the view, known as Gaullism that France should continue to see herself as a major power and should not rely on other nations – particularly not the US – for its national security and prosperity. Often criticised for his Politics of *grandeur*, de Gaulle oversaw the development of French atomic weapons and promoted a foreign policy independent of British and American influence. He withdrew France from NATO military command structure – although remaining a member of the western Alliance – and twice vetoed the UK's entry into the European Community. Despite having been re-elected as President in 1965 by direct popular ballot, in May 1968 he appeared likely to lose power amidst widespread protests by students and workers, but survived the crisis with an increased majority in the Assembly. However, de Gaulle resigned after losing a referendum in 1969. He is considered by many to be the most influential leader in modern French history and a separate and excellent new museum – *l'Historial de Gaulle* – on the lower floor of the *Musée de l'Armée* in *les Invalides* in Paris is dedicated to the memory of his achievements.

Giraud, Henri Honoré (1879-1949), French general. Giraud fought in World

War I and World War II. Captured in both wars, he escaped each time. After his second escape in 1942, he joined the Free French Forces, and participated, to various degrees, with the Allied invasion of North Africa and its subsequent events. He survived the war to become a politician.

Graves, Robert (1895-1985), British poet, translator and novelist. During his long life he produced more than 140 works. Graves' poems, his translations and innovative interpretations of the Greek myths, and his memoir of his early life, including his rôle in the First World War, *Goodbye to All That,* have never been out of print. He earned his living from writing, particularly popular historical novels such as *I, Claudius.* He also was a prominent translator of Classical Latin and Ancient Greek texts; his versions of *The Twelve Caesars* and *The Golden Ass* remain popular today for their clarity and entertaining style.

Habibulla (1872-1919), amir of Afghanistan 1901-19. Born in Samarkand, Uzbekistan, he was the eldest son of the Amir Abdur Rahman Khan, whom he succeeded in October 1901. Habibullah was a relatively secular, reform-minded ruler who attempted to modernise his country. During his reign he worked to bring western medicine and modern technology to Afghanistan. In 1904, he founded the Habibia school as well as a military academy. He also instituted various legal reforms and repealed many of the harshest criminal penalties. Other reforms included dismantling of the repressive internal intelligence organisation that had been put in place by his father. He managed to maintain his country's neutrality in World War I, despite determined endeavours of Sultan of the Ottoman Empire and a German military mission to enlist Afghanistan on its side. He also greatly reduced tensions with British India, signing a treaty of friendship in 1905 and paying an official state visit in 1907. Habibullah was assassinated while on a hunting trip in Laghman Province in 1919. His brother Nasrullah Khan briefly succeeded him as amir and held power for a week between 21 and 28 February 1919, before being ousted and imprisoned by Amanullah Khan, Habibullah's third son.

Hafizullah Amin (1929-1979), Afghan prime minister during the Cold War. Amin was born in Paghman and educated at Kabul University, after which he started his career as a teacher. After working a few years as a teacher, Amin went to the United States to study. He would visit the United States a second time before permanently moving to Afghanistan, and starting his career in radical politics. He ran as a candidate in the 1965 parliamentary election but failed to secure himself a seat, which, however, he obtained in the 1969 parliamentary election. Amin was one of the leading organisers of the Saur Revolution which overthrew the government of Mohammad Daoud Khan. The presidency of Amin, albeit short-lived, was marked by controversies from beginning to end. Amin came to power by assassinating his predecessor Nur Muhammad Taraki. The revolt, which had begun under Taraki, against communist rule worsened under Amin. His government was unable to solve the problem. The Soviet Union, which alleged that Amin was an agent of the Central Intelligence Agency, intervened in Afghanistan on behalf of the Twenty-Year Treaty of Friendship between Afghanistan and the Soviet Union. Amin, after ruling for a bit longer than 3 months, was assassinated by the Soviets in Operation Storm-333 in December 1979.

Haig, Douglas (1861-1928), British Field Marshall. He was commander-in-chief of the BEF during the Somme battle and took much criticism for the sheer loss of life in this battle. Commissioned into the cavalry in 1885, he served both in the Sudan and in the Second Boer War. Serving at the War Office Haig helped to implement the Haldane military reforms. In August 1914, when the WW I started, Haig was the general officer commanding the First British Corps. He fought in the Battle of Mons and the first Battle of Ypres. In December 1915, he succeeded Sir John French as commander-in-chief of the BEF. In 1916, Haig put his belief in one final mighty push against the Germans to be executed on the Somme. The Somme Battle led to the loss of 600,000 men on the Allied side; 400,000 of whom were British or Commonwealth troops. When the battle ended, the Allies had gained ca. 15km of land. Although tanks were employed *en masse* at the Somme, Haig perceived them with scant enthusiasm. Haig soldiered on as C-in-C until the end of the war. He was created an earl for his leadership in 1919. He spent the last few years of his life working for ex-servicemen, and especially those who had been disabled in the war. Haig was a leading light in the "Poppy Day Appeal" and the British Legion movement.

Hamayun, Afghan lieutenant-colonel, commanding officer of 3rd Kandak (2010-11).

Harriman, Averell (1891-1986), American businessman and diplomat. Beginning in the spring of 1941, he served as President Franklin D. Roosevelt's special envoy to the UK and helped co-ordinate the Lend-Lease Programme. He was present at the meeting between Churchill and Roosevelt at Placentia Bay in August 1941, when the Atlantic Charter on Anglo-American wartime co-operation was conceived. He was subsequently dispatched to Moscow to negotiate the terms of the Lend-Lease agreement with the Soviet Union. In the summer of 1942, Harriman accompanied Churchill to Moscow for a second meeting with Stalin. His able assistance in explaining why the western allies were opening a second front in North Africa instead of France earned him the post of US ambassador to the Soviet Union in 1943. At the Tehran Conference in late 1943 Harriman was tasked with placating a suspicious Churchill while Roosevelt attempted to gain the confidence of Stalin. Harriman also attended the Yalta Conference and the final 'Big Three' conference at Potsdam. Subsequently, he served as Secretary of Commerce under President Harry S. Truman and later as the 48th Governor of New York. He served in numerous U.S. diplomatic assignments in the Kennedy and Johnson administrations. In 1971 he married Pamela Digby, former daughter-in-law to Sir Winston Churchill.

Hikmatyar, Gulbuddin (1947-), Mujahidin chief, founder and leader of the Hezb-e Islami political party and paramilitary group. Hikmatyar was a rebel military commander during the Soviet war in Afghanistan during the 1980s, and he was one of the key figures in the civil war that followed the Soviet withdrawal. He was Prime Minister of Afghanistan from 1993 to 1994 and again briefly in 1996. In interviews in the 2000s he has demanded that foreign troops must leave unconditionally. Offers by President Hamid Karzai to open talks with 'opponents of the government' and hints that they would be offered official posts might be seen as bait for Hikmatyar. He is now believed to shuttle between hideouts in Pakistan,s mountainous tribal areas and in northeast Afghanistan.

Hindenburg, Paul von (1847-1934), German field-marshal, WWI Chief of the Great General Staff. In 1866, he was commissioned into a guards regiment and fought against Austria in 1866 and in the Franco-Prussian War, 1870-71. In 1911, he retired from active military service, but was re-called to military service, in August 1914, to lead the Eighth Army in East-Prussia. Mid-September 1914, he achieved national fame by defeating the Russian Army at Tannenberg and on the Masurian Lakes. He was promoted field-marshal and given the supreme command of the Eastern Front in November 1914. In August 1916, Hindenburg was appointed Chief of the Great General Staff. This position gave him vast power even in the civil sphere. The March 1918 German offensive on the Western Front failed miserably, and Hindenburg saw no alternative but to use his considerable with the Kaiser to persuade him to abdicate and to go into exile in the Netherlands. Hindenburg remained in control of the German army until July 1919 when he once again retired from the military. In 1925, Hindenburg was persuaded to stand in the presidential elections – the result of Ebert's death. He won the election and was re-elected in 1931. Although he disliked Hitler, he was persuaded by his son and by Franz von Papen to appoint him chancellor in January 1933. Hindenburg died at his Prussian estate in August 1934.

Hitler, Adolf (1889-1945), Austrian painter and Nazi party leader. During WWI he was a lance corporal (dispatch runner) in a Bavarian regiment. He achieved world notoriety as the *Reichskanzler* (chancellor) and *Führer* of Germany and main responsible for the 3rd *Reich*'s crimes against humanity as well as for the Second World War.

Hopkins, Harry (1890-1946) was one of US President F.D. Roosevelt's closest confidents. In World War II he was Roosevelt's chief diplomatic advisor and trouble-shooter and was a key policy maker in the $50 billion Lend Lease Programme agreed with the UK in 1941. During the war years, Hopkins acted as the president's unofficial emissary to Winston Churchill. Roosevelt wished Hopkins to assess Britain's determination and situation. Churchill chaperoned Hopkins all across the United Kingdom, and converted him to the British cause. Hopkins went to Moscow in July 1941 to make personal contact with Joseph Stalin subsequently recommending the inclusion of the Soviet Union in Lend Lease Programme, which the president duly accepted. Roosevelt brought him along as advisor to his meetings with Churchill at Cairo, Tehran, Casablanca in 1942-43, and in Yalta in 1945.

Hötzendorf, Franz Conrad Graf von (1852-1925), Austro-Hungarian field-marshal, chief of the Austro-Hungarian imperial general staff. In 1914, Conrad was one of the main proponents of war with Serbia in response to the assassination of the designated successor to the Austro-Hungarian thrones, Archduke Franz Ferdinand. Conrad often proposed unrealistically grandiose plans, disregarding the realities of terrain and climate. The plans that he drew up frequently underestimated the power of the enemy. During the Russian 'Brusilov Offensive' Austro-Hungarian forces lost nearly 1.5 million men. However, some historians claim that Conrad was probably the best strategist of the war and that his plans were brilliant in conception, and that the German generals in the East based most of their successful offensive operations on Conrad's plans. Following the accession of Emperor Karl to the throne in November 1916, Conrad was elevated to the rank of field-marshal. Yet under the new emperor, Conrad's powers were gradually eroded. On 1 March 1917, Karl dismissed Conrad, who then requested retirement. The emperor, though, asked him to remain on active duty, and as Conrad accepted he was placed in command of the South Tyrolean Army Group. In the late spring of 1918, the Austro-Hungarian offensives against the Italians, failed dismally, and Conrad was dismissed on July 15.

James II (1633-1701) was, from 6 February 1685, king of England and of Ireland as James II and king of Scotland as James VII. He was the last Catholic monarch to reign over the Kingdoms of England, Scotland, and Ireland. Members of Britain's political and religious elite increasingly opposed him for his designs on a re-Catholisation of the state and for his ambition of becoming an absolute monarch. When he produced a Catholic heir, the tension exploded, and leading nobles called on William III of Orange (his son-in-law and nephew) to land an invasion army from the Netherlands, which he did. In the Glorious Revolution of 1688 James fled Britain. James made one serious attempt to recover his crowns, when he landed in Ireland in 1689 but, after the defeat of the Jacobite forces by the Williamite coalition at the Battles of the Boyne (1690) and (in particular) of Aughrim (1691) as well as the sieges of Limerick (1690 and 1691), the cause was lost. He lived out the rest of his life as a pretender at a court sponsored by his cousin and ally, King Louis XIV of France.

Jodl, Alfred (1890-1946), German major-general, head of the *Wehrmachtführungsstab* (Operations Staff) at the tri-service Oberkommando der Wehrmacht [the Armed Forces High Command] during World War II, acting as deputy to Wilhelm Keitel. At the end of World War II in Europe, Jodl signed the instruments of unconditional surrender on 7 May 1945 in Reims as the representative of Karl Dönitz. At Nuremberg he was tried as a war criminal, sentenced to death and hanged.

Jomini. Antoine Henri Baron de (1779-1869). Born a Swiss, he rose to the rank of brigadier-general in the Napoleonic army, but was later to become a Russian lieutenant-general. He is one of the most celebrated writers on the Napoleonic art of war and on war theory *per se*. His theories of war are based on positivist thinking and a geometric approach. Until WWII his ideas permeated the curricula of military academies around the world – and notably those of Russia and the United States.

Karl (1887-1922), Austro-Hungarian Emperor. Karl was the last ruler of the Austro-Hungarian Empire. He was the last monarch of the House of Habsburg-Lothringen reigning as Karl I as Emperor of the Austro-Hungarian Empire and Károly IV as King of Hungary from 1916 until 1918, when, initially, he stood down as absolute monarch leaving state affairs to an elected government, though he did not abdicate. Fearing for his life, on 24 March 19119 the commanding officer of the British guards detachment, Lt-Col Edward Lisle Strutt, dispatched him into exile in

Switzerland, and he spent the remaining years of his life in vain attempts to restore the monarchies. He has been beatified by John Paul II, known now as the Blessed Charles of Austria. His earthly remains still repose in Madeira, where he died, although a place in the *Kaisergruft* (or *Kapuzinergruft*) in Vienna next to those of his Empress Zita is reserved for him.

Karmal, Babrak, (1929-1996), leader of the Parcham faction of the Afghan Communist Party. Karmal was born in Kamari and educated at Kabul University, after which he started his career as a bureaucrat. Under Karmal's leadership, the Parchamite PDPA were instrumental in Mohammad Daoud Khan's rise to power, and his subsequent regime. The PDPA took power in the 1978 Saur Revolution. Karmal was made Chairman of the Revolutionary Council and Chairman of the Council of Ministers on 27 December 1979, and he retained the latter chairmanship until 1981 when he was succeeded by Sultan Ali Keshtmand. Under Mikhail Gorbachev, the Soviet Union being highly critical of Karmal replaced him by Mohammad Najibullah. Karmal was subsequently exiled to Moscow, whence he returned in 1991. Back in Afghanistan he helped topple the Najibullah government, and he became an associate of Abdul Rashid Dostum, one of the men who brought down the communist government. In 1996, Karmal died from liver cancer.

Karzai, Hamid (1957-), president of Afghanistan. He became a dominant political figure after the removal of the Taliban regime in late 2001. During the December 2001 International Conference on Afghanistan, Karzai was selected by prominent Afghan political figures to serve a six month term as Chairman of the Interim Administration. He was then chosen for a two year term as Interim President during the 2002 *loya jirga* (grand assembly) that was held in Kabul. After the 2004 presidential election, Karzai was declared the winner and became President of the Islamic Republic of Afghanistan. He won a second five-year-term in the 2009 presidential election.

Keitel, Wilhelm (1882-1946), German field-marshal. As head of the *Oberkommando der Wehrmacht* (OKW, Armed Forces High Command) and de facto minister for war under Hitler, he was one of the 3rd *Reich*'s most senior military leaders. At the Allied court at Nuremberg he was tried as a war criminal, sentenced to death and hanged.

Kesselring, Albert (1885-1960), German air-marshal (Generalfeldmarschall der Luftwaffe) during World War II. In a military career that spanned both world wars, Kesselring became one of Germany's most skilful commanders. Kesselring was discharged in 1933 to become head of the Department of Administration (*Luftwaffenverwaltungsamts im Reichsluftfahrtministerium*) on the staff of the Reichs Commissioner for Aviation (*Reichskommissars für Luftfahrt*), Herman Göring, where he was involved in the re-establishment of the aviation industry and the laying of the foundations for the Luftwaffe, serving as its Chief of Staff from 1936 to 1938. He commanded air forces in the invasions of Poland and France, the Battle of Britain and Operation Barbarossa. As Commander-in-Chief South, he was overall German commander in the Mediterranean theatre, which included the operations in North Africa. Kesselring conducted an uncompromising defensive campaign against the Allied forces in Italy until he was injured in an accident in October 1944. In the final campaign of the war, he commanded German forces on the Western Front. He won the respect of his Allied opponents for his military accomplishments, but his record was marred by massacres committed by troops under his command in Italy. After the war, Kesselring was tried for war crimes and sentenced to death. The death verdict unleashed a storm of protest in the United Kingdom. Former British Prime Minister Winston Churchill immediately intervened in favour of Kesselring. Field-Marshal Alexander, now Governor General of Canada, sent a telegram to Prime Minister Clement Attlee in which he expressed his hope that the sentence would be commuted. 'As his old opponent on the battlefield", he started, "I have no complaints against him. Kesselring and his soldiers fought against us hard but clean.' The sentence was duly commuted to life imprisonment. A political and media campaign resulted in his release in 1952.

Khaled, bin Sultan bin Abdul Aziz Al Saud (1949), prince, Saudi commander during the Gulf War of 1991.

King, Ernest Joseph (1878-1956), WW II American fleet admiral and Anglophobia personified. King was Commander-in-Chief, United States Fleet and Chief of Naval Operations. At the start of the US involvement in World War II, blackouts on the US eastern seaboard were not in effect, and commercial ships were not travelling under convoy. His numerous critics attribute the delay to implement these measures to his Anglophobia, as the convoys and seaboard blackouts were British proposals and King was adamant not to have his navy adopt any ideas from the Royal Navy. He also refused, until March 1942, the loan of British convoy escorts. He was, however, aggressive in driving his destroyer captains to attack U-boats in defence of convoys and in planning counter-measures against German surface raiders, even before the formal declaration of war by Germany.

King, William Loyn Mackenzie (1874-1950), Canadian prime minister. He was the dominant Canadian political leader from the 1920s through the 1940s. He served as the tenth Prime Minister of Canada from December 1921 to June 1926; from 1926 to August 1930; and from October 1935 to November 1948. A Liberal with 22 years in office, he was longest-serving prime minister in Canadian history. Trained in law and social work, he was keenly interested in the human condition, and played a major role in laying the foundations of the Canadian welfare state.

Kitchener, Horatio Herbert, 1st Earl of Khartoum (1850-1916), Irish-born British field-marshal. In 1898, Kitchener won the Battle of Omdurman securing control of the Sudan. He played a key rôle in the conquest of the Boer Republics, and succeeded Lord Roberts as commander-in-chief at a time when the Boer forces had taken to guerrilla fighting. His term as Commander-in-Chief (1902–09) of the Army in India saw him quarrel with the Viceroy Lord Curzon, who eventually resigned. Kitchener then returned to Egypt as British Agent and Consul-General (*de facto* administrator). In 1914, at the start of the First World War, Lord Kitchener became Secretary of State for War. One of the few men to foresee a long war, one in which Britain's victory was far from secure, he organised the largest volunteer army that Britain, and indeed the Empire, had seen and a significant expansion of materials production to fight Germany on the Western Front. He was blamed for the shortage of shells in the spring of 1915 – one of the events leading to the formation of a coalition government – and stripped of his control over munitions and strategy. He died in 1916 near the Orkney Islands when the warship taking him to negotiations in Russia was torpedoed by a German submarine.

Laurier, Sir Wilfrid (1841-1919), Canadian prime minister from July 1896 to October 1911. As Canada's first francophone prime minister, Laurier is often considered one of the country's greatest statesmen. He is well known for his policies of conciliation, expanding Confederation, and compromise between French and English Canada. Laurier is also well regarded for his efforts to establish Canada as an autonomous country within the British Empire.

Lawrence, John (1811-1879), Viceroy of India from 1864 to 1869. During the First Sikh War, 1845 to 1846, Lawrence organised the supply of the British army in the Punjab and became Commissioner of the Jullundur district, serving under his brother, the Governor of the province. In 1849, following the Second Sikh War, he became a member of the Punjab Board of Administration under his brother, and was responsible for numerous reforms of the province, including the abolition of internal duties, establishment of a common currency and postal system, and encouraged the development of Punjabi infrastructure. Lawrence was partly responsible for 'preventing the spread' of the Indian Rebellion of 1857 to Punjab, and negotiated a treaty with the Afghan ruler Dost Mohammed Khan, and later led the troops which recaptured Delhi from the rebellious sepoys. For this, he was created a baronet and received an annual pension from the East India Company of £2,000. He returned to Britain in 1859, but was sent back to India in 1863 to become viceroy to succeed Lord Elgin, who had unexpectedly died. As viceroy, Lawrence pursued a cautious policy, avoiding entanglements in Afghanistan and the Persian Gulf. In domestic affairs, he increased educational opportunities for Indians, but at the same time limited the use of native Indians in high civil service posts. Upon his return to England in 1869, he was raised to the peerage as Baron Lawrence.

Louis XIV (1638-1715), king of France, known as Louis the Great or the Sun King (French: le Roi-Soleil), was a Bourbon monarch who ruled as King of France and Navarre. Reigning for 72 years and 101 days he holds the distinction of being the longest-reigning king in European history. Louis began his rule of France in 1661 after the death of his chief minister, the Italian Cardinal Mazarin. An adherent of the theory of the divine right of kings, which advocates the divine origin and lack of temporal restraint of monarchical rule, Louis continued his predecessors' work of creating a centralised state governed from the capital. He sought to eliminate the remnants of feudalism persisting in parts of France and, by compelling the noble elite to inhabit his lavish Palace of Versailles, succeeded in pacifying the aristocracy, many members of which had participated in the Fronde rebellion during Louis' minority. By these means he consolidated a system of absolute monarchical rule in France that endured until the French Revolution. During Louis's reign France was the leading European power and fought three major wars: the Franco-Dutch War, the War of the League of Augsburg, and the War of the Spanish Succession.

Ludendorff, Erich (1865-1937), German WWI *Generalquartiermeister* (quartermaster-general). From August 1916 his appointment as quartermaster-general made him joint head (with Paul von Hindenburg), and chief engineer behind the management of Germany's effort in World War I until his resignation in October 1918. After the war, Ludendorff became a prominent nationalist leader, and a promoter of the stab-in-the-back legend alleging that the German field army had been betrayed by Marxists and Republicans at home. He took part in the unsuccessful *coups d'état* of Wolfgang Kapp in 1920 and the Beer Hall Putsch of Adolf Hitler in 1923, and in 1925 he ran for president against his former colleague, Paul von Hindenburg. From 1924 to 1928 he represented the German Popular Freedom Party in the German Parliament.

Lytton, Edward Robert Lytton Bulwer-Lytton, 1st Earl of Lytton (1831-1891), viceroy of India between 1876 and 1880, British statesman and poet. An extremely accomplished diplomat, who made friends wherever he served, Lytton was afforded the extraordinarily rare tribute – especially for an Englishman – of a state funeral in Paris. While some have questioned his handling of the Indian famine, his diplomatic career was otherwise highly praised and his son, Victor Bulwer-Lytton, 2nd Earl of Lytton, followed him to India as Governor of Bengal and, for a time, as acting viceroy.

Machiavelli, Niccolò di Bernardo dei (1469-1527), Italian Renaissance author, philosopher and writer based in Florence. A founder of modern political science, he was a diplomat, political philosopher, playwright, and a civil servant of the Florentine Republic. He also wrote comedies, carnival songs and poetry. His personal correspondence is renowned in the Italian language. He was Secretary to the Second Chancery of the Republic of Florence from 1498 to 1512, when the Medici were out of power. He wrote his masterpiece, The Prince, after the Medici had recovered power and he no longer held a position of responsibility in Florence. Moreover, his thoughts on the theory of war are explained at length in his work from 1519 *Del Arte della Guerra* [The Art of War].

Mack, Karl *Freiherr* [Baron] von Leiberich (1752-1828), Imperial Roman *Feldmarschall-Leutnant* [Major-General]. He is best remembered as the commander of the Austrian forces that was outmanœuvred and capitulated to Napoleon in the Battle of Ulm in 1805. While most of the Austrian high command was captured at Ulm along with the 100,000 man army, some of his officers, including Prince Schwarzenberg, broke through the French defences in a massed cavalry charge and escaped. General officers received a parole requiring them to abstain from further military action against France thus removing the upper tier of Imperial Roman commanders from the battlefields of the Upper Danube. After Austerlitz, Mack was court-martialled for cowardice and deprived of his rank and honours. As the ultimate allied victory had obliterated the memory of earlier disasters, Mack was, at the request of Prince Schwarzenberg, reinstated in his army rank of *Feldmarschall-Leutnant* and as a member of the Order of Maria Theresa.

Macnaghten, William Hay (1793-1841), British-Indian political officer, who played a major part in the First Anglo-Afghan War. In 1837, he became a trusted adviser of the governor-general, Lord Auckland, with whose policy of supporting Shah Shuja against Dost Mahommed Khan, the reigning amir of Kabul, Macnaghten he agreed. As a political agent at Kabul he strongly disagreed with his subordinate Sir Alexander Burnes, who adviced against a prposed cut in susudies to Afghan local chieftains. As riots followed, Burnes was murdered on 2 November 1841. Macnaghten tried to save the situation by negotiating with the Afghan chiefs and, independently of them, with Dost Mohammad's son, Akbar Khan, by whom he was captured and, on 23 December 1841, assassinated.

Mahmud Shah Durrani (1769-1829), amir of Afghaistan between 1801 and 1803 and again between 1809 and 1818. An ethnic Pashtun, he was the son of Timur Shah Durrani and grandson of Ahmad Shah Durrani.

Mangal. Afghan provincial governor (2010-11)

Marlborough, John Spencer Churchill (1650-1722), first Duke of), English courtier, soldier and statesman whose career spanned the reigns of five monarchs through the late 17th and early 18th centuries. Rising from a lowly page at the court of the House of Stuart, he loyally served James, Duke of York, through the 1670s and early 1680s, earning military and political advancement through his courage and diplomatic skill. His rôle in defeating the Monmouth Rebellion in 1685 helped secure James on the throne, yet just three years later he took an active part in the Glorious Revolution which ousted James. Marlborough served with distinction in the early years of the Nine Years' War, but persistent charges of Jacobitism brought about his fall from office and temporary imprisonment in the Tower. On the accession of Queen Anne in 1702, Marlborough brought back to serve his country in the War of the Spanish Succession. His marriage to Sarah Jennings – Queen Anne's close friend – helped Marlborough on his path to greatness. Becoming de facto leader of Allied forces, his victories on the battlefields of Blenheim (1704), Ramillies (1706), Oudenarde (1708), and Malplaquet (1709), secured him place in the Pantheon of Europe's greatest captains. As after the war he fell out of grace with the Queen, he went into self-imposed exile. He returned in 1714, however, at the accession of George I (of Hanover) to the British throne, but died on 16 June 1722.

Marshall, George Catlett (1880-1959), American general, Chief of Staff of the Army, Secretary of State, and the third Secretary for Defence. Marshall served as the United States Army Chief of Staff during the Second World War and as the chief military adviser to President Franklin D. Roosevelt. As Secretary of State, his name was given to the Marshall Plan, for which he was awarded the Nobel Peace Prize in 1953. He was generally appreciated by Winston Churchill and occasionally by Lord Alanbrooke, the British WW II Chief of the Imperial General Staff.

McNaughton, Andrew George Latta (1887-1966), Canadian scientist, general, cabinet minister, and diplomat. McNaughton went into World War II commanding First Canadian Infantry Division (part of VII Corps). He commanded VII Corps itself from July to December 1940 when it was renamed the Canadian Corps. Then, under his leadership the Corps was reorganised as an army in 1942. McNaughton's contribution to the development of new techniques was outstanding, especially in the field of detection and weaponry, including the discarding sabot projectile. He was unduly blamed for the disastrous Dieppe Raid in 1942. His support for voluntary enlistment rather than conscription led to conflict with James Ralston, the then Minister of National Defence. Due to pressure by critics and weakened by health problems, McNaughton resigned his command in December 1943. McNaughton became Minister of National Defence when Ralston was forced to resign after the Conscription Crisis of 1944, as Canadian Prime Minister King did all he could to avoid introducing conscription. McNaughton was soon pressured into calling for conscription despite King's wishes. He resigned as Defence minister in August 1945. After the war he served on the United Nations Atomic Energy Commission, which he headed between 1946 and 1948, as Canada's Ambassador to the United Nations, during the years of 1948 and 1949, and between 1950 and 1959 he was the President of the Canadian section of the International Joint Commission.

Menshikov, Prince Aleksandr Sergeyevich (1787-1869), Russian general-admiral, Finnish-Russian nobleman and statesman. He was made adjutant general in 1817 and admiral in 1833. He became close with Tsar Alexander I and accompanied him throughout his campaigns against Napoleon. In 1817, Menshikov was appointed acting quartermaster-general. In 1853, Menshikov was sent on a special mission to Constantinople, and when the Crimean War broke out he was appointed commander-in-chief on land and sea. He commanded the Russian army at Alma and Inkerman and showed incompetence and lack of military talent. On 15 February 1855, Menshikov was removed from command, and replaced by Prince Mikhail Dmitrievich Gorchakov.

Metternich, Klemens Wenzel Prince von (1773-1859), German-born Austrian politician and statesman and one of the most important diplomats of his era. He served as the Foreign Minister of the Holy Roman Empire and its successor state, the Austrian Empire, until the liberal revolutions of 1848. He engineered détente with France facilitating the marriage of Napoleon to the Austrian Arch-Duchess Marie Louise. On behalf of the Emperor he signed the Treaty of Fontainebleau that sent Napoleon into exile, and he led the Austrian delegation at the Congress of Vienna which carved up post-Napoleonic Europe between the major powers. In October 1813, in recognition of his service to the Austrian Empire he was raised to the title of Prince.

Milošević, Slobodan (1941-2006), served as the President of Serbia from 1989 until 1997 and as President of the Federal Republic of Yugoslavia from 1997 to 2000. He also led the Socialist Party of Serbia from its foundation in 1990. His presidency was marked by the breakup of Yugoslavia and the subsequent Yugoslav wars. In the midst of the 1999 NATO bombing of Yugoslavia, Milošević was charged with war crimes and crimes against humanity in connection with the wars in Bosnia, Croatia and Kosovo by the International Criminal Tribunal for the former Yugoslavia. Serbian Prime Minister Zoran Đinđić to send him to The Hague to stand trial for charges of war crimes, during which Milošević endeavoured to conduct his own defence. The five-year long trial ended without a verdict as he died in his prison cell in The Hague on 11 March 2006.

Mohammed Daoud Khan (1909-1978), Afghan dictator. He was Prime Minister of Afghanistan from 1953 to 1963. In 1973, he overthrew the monarchy of his first cousin King Mohammed Zahir Shah and declared himself as the first President of Afghanistan serving as such from 1973 until his assassination in 1978.

Mohammed Zahir Shah (1915 – 2007), king of Afghanistan reigning for four decades, from 1933 until he was ousted by a coup in 1973. Following his return from exile, in 2002 he was given the title 'Father of the Nation' which he held until his death.

Molotov, Vyacheslav Mikhailovich (1890-1986), Soviet Foreign Minister, diplomat and a leading figure in Soviet governments from the 1920s onwards. In 1957, he was sacked from the Politburo of the Central Committee by Nikita Khrushchev. He served as Chairman of the Council of People's Commissars from 1930 to 1941, and as Minister of Foreign Affairs from 1939 to 1949 and from 1953 to 1956. Molotov served for several years as First Deputy Premier in Joseph Stalin's cabinet. Molotov was the principal Soviet signatory of the Germano-Soviet non-aggression pact of 1939 (also known as the Molotov-Ribbentrop Pact. After Stalin's death in 1953, Molotov staunchly opposed Khrushchev's de-Stalinisation. He defended his policies and the legacy of Stalin until his death in 1986, and harshly criticised Stalin's successors.

Moltke, Helmuth Karl Bernhard Graf von, (1800-1891), Prusso-German field-marshal. The founder, and for thirty years the chief, of the Prussian general staff and later the Great General Staff of Germany, he is regarded as one of the great strategists of the latter 19th century, and the creator of a new, more modern method of directing armies in the field. He is often referred to as Moltke the Elder to distinguish him from his nephew Helmuth Johann Ludwig von Moltke, who commanded the German Army at the outbreak of World War I.

Moltke, Helmuth Johann Ludwig von (1848-1916), German WWI field-marshal. Also known as Moltke the Younger, he was a nephew of Field-Marshal Count Moltke and served as the Chief of the German Great General Staff from 1906 to 1914. Moltke the Younger's rôle in the development of German war plans and the instigation of the First World War is extremely controversial.

Monclar, Raoul Charles Magrin-Vernerey (1892-1964), *général de corps d'armée* [lieutenant-general], chief-of-staff to the French Korea contingent

Montgomery, Bernard Law, 1st Viscount Montgomery of Alamein (1887-1976), nicknamed 'Monty' and the 'Spartan General', British field-marshal Montgomery. He saw action in World War I, when he was seriously wounded. During World War II he commanded the 8th Army from August 1942 in the Western Desert and in Sicily and mainland Italy before being given responsibility for planning the D-Day invasion in Normandy. He was Allied land component commander during Operation Overlord from the initial landings until after the Battle of Normandy. He then continued in command of the 21st Army Group for the rest of the campaign in north-western Europe. As such he was the principal field commander for Operation Market Garden and for the Allied Rhine crossing. On 4 May 1945 he took the German surrender at Lüneburger Heide in northern Germany. After the War he became Commander-in-Chief of the British Army of the Rhine (BAOR) in Germany and then Chief of the Imperial General Staff.

Murat, Joachim (1767-1815), French marshal, king of Naples. During the Revolutionary and Napoleonic Wars, he rose to *Maréchal de France*, Grand Admiral, 1st Prince Murat, Grand Duke of Berg and King of Naples from

1808 to 1815. He received his titles in part by being the brother-in-law of Napoleon Bonaparte, through marriage to Napoleon's youngest sister, Caroline Bonaparte. He was noted as a flamboyant, daring and charismatic cavalry general.

Musharraf, Perves (1943-), general, president of Pakistan 1999-2007. After years of military service, he rose to prominence when Prime Minister Nawaz Sharif appointed him as the Chief of Army Staff in October 1998. Musharraf was the mastermind behind the controversial and internationally condemned Kargil infiltration, which derailed peace negotiations with Pakistan's long standing enemy India. He previously also played a vital role in the Afghanistan civil war (1996-2001) both in peace negotiations and in trying to end the bloodshed. After months of contentious relations with Sharif, Musharraf was brought to power through a bloodless military coup. As Pakistan's head of state, he was a U.S. ally in the War on Terror. He was credited with the development of Pakistan's economy during the early years of his rule. The last two years of his rule were marred by controversies, including the suspension of the Supreme Court Chief Justice and the Lal Masjid siege.

Mussolini, Benito Amilcare Andrea (1883-1945), Italian politician who led the National Fascist Party and is credited with being one of the key figures in the creation of fascism. Mussolini became Prime Minister of Italy in 1922, though from 1925 styling himself *Il Duce*. Mussolini remained in power until he was replaced in 1943; for a short period after this until his death, he was the leader of the Italian Social Republic.

Napoléon I, Emperor of the French (1769-1821). Napoleon was born in Corsica to parents of noble Genoese ancestry, and trained as an artillery officer in mainland France. He rose to prominence under the French First Republic and led successful campaigns against the First and Second Coalitions arrayed against France. In 1799, he staged a coup d'état and installed himself as First Consul; five years later the French Senate proclaimed him emperor. As Napoleon I, he was Emperor of the French from 1804 to 1815. His legal reform, the *Code Napoléon* or *Code Civil*, has been a major influence on many civil law jurisdictions worldwide. Mainly remembered for his rôle in the wars led against France by a series of coalitions, the so-called Napoleonic Wars, he established hegemony over most of continental Europe and sought to spread the ideals of the French Revolution, while consolidating an imperial monarchy at home, which restored certain aspects of the deposed *ancien régime*. Due to his success in the wars, often against numerically superior enemies, he is generally regarded as one of the greatest military commanders of all time. After final defeat at Waterloo in 1815, Napoleon was deposed and exiled. Napoleon spent the last six years of his life in confinement by the British on the island of Saint Helena. An autopsy concluded he died of stomach cancer, although this claim has sparked significant debate, as some scholars have held that he was a victim of arsenic poisoning.

Napoléon III, Emperor of the French (1808-1873). While Louis-Napoléon Bonaparte was president of the French Second Republic, it was as Napoleon III that he established himself as ruler of the Second French Empire. He was the nephew and heir of Emperor Napoleon I. He ruled as *Emperor of the French* until 4 September 1870, when, following defeat in the Franco-Prussian War he went into captivity in the fledgling German Empire.

Nicolls, Sir Jasper (1778-1849), British East India Company lieutenant-general. Born at East Farleigh in Kent and educated at a private school in Ballygall and at Trinity College, Dublin, Nicolls was commissioned into the 45th Regiment of Foot in 1793. In 1812, he was appointed Quartermaster-General of the Honourable East India Company's armies. In 1815, during the Gurkha War he captured Almora and reduced Kumaon. In 1825, he was made General Officer Commanding a Division of the Madras Presidency and in 1829 he transferred to become GOC 7th (Meerut) Division; in 1838 he was appointed Commander-in-Chief, Madras and in 1839 he took office as Commander-in-Chief, India; he returned to Britain in 1843

Nott, Sir William (1782-1845), British major-general. In 1838, on the outbreak of the First Afghan war, Nott commanded a brigade and from April to October 1839 he was in command of all troops left behind in Quetta. In November 1840, he captured Khelat, and in the following year compelled Akbar Khan and other tribal chiefs to submit to the British. On receiving the news of the Afghan's riot in Kabul in November 1841, Nott took energetic measures. On 23 December the British envoy, Sir William Hay Macnaghten, was murdered; and in February 1842 the aging commander-in-chief, General Elphinstone decided to abandon the country. In March 1842, Nott inflicted a severe defeat on the enemy near Kandahar, and in May drove them with heavy loss out of the Baba Wali Pass. In July he received orders from Lord Ellenborough, the Governor-General of India, to evacuate Afghanistan, with permission to retire by Kabul. Nott arranged with Sir George Pollock, now commander-in-chief, to join him at Kabul. On 30 August he routed the Afghans at Ghazni, and on 6 September occupied the fortress. In 1843 he returned to Britain, where the directors of the East India Company voted him a pension of £1,000 per annum.

Nur Muhammed Taraki (1913-1979), Khalg leader, Afghan politician and statesman during the Cold War. Taraki was born near Kabul and educated at Kabul University, after which he started his political career as a journalist. He later became one of the founding members of the Popular Democratic Party of Afghanistan (PDPA). He ran as a candidate in the 1965 Afghan parliamentary election but failed to secure himself a seat. In 1966 he published the first issue of Khalq, a party newspaper, but it was closed down shortly afterwards by the Afghan Government. The assassination of Mir Akbar Khyber led Taraki, along with Hafizullah Amin (the organiser of the revolution) and Babrak Karmal, to initiate the Saur Revolution and establish the communist Democratic Republic of Afghanistan. His relationship with Amin turned sour and, on 14 September 1979, he was assassinated upon Amin's orders.

Palmerston, Lord (Henry John Temple, 3rd Viscount Palmerston (1784-1865)), British prime minister. He was in government office almost continuously from 1807 until his death in 1865, beginning his parliamentary career as a Tory and concluding it as a Liberal. He is best remembered for his direction of British foreign policy through a period when Britain was at the height of its power, serving terms as both Foreign Secretary and Prime Minister.

Patton, George Smith (1885-1945), American general. Patton is best known for his leadership while commanding corps and armies during World War II. He was also well known for his eccentricity and controversial outspokenness. He was commissioned into the US Army in 1909. In 1916-17, he participated in the unsuccessful Pancho Villa Expedition, a US operation that attempted to capture the Mexican revolutionary Pancho Villa. In World War I, he saw action in France as an officer of the newly raised United States Tank Corps. In World War II, he commanded corps and armies in North African and European theatres-of-war. Patton is notorious for the so-called slapping incident happening in a hospital near Nicosia in 1943, when he encountered a private soldier suffering from shell shock. When Patton asked him where he was hurt, the soldier replied that he was nervous rather than wounded. In response, Patton slapped him across the chin with his gloves, then grabbed him by the collar and dragged him to the tent entrance, shoving him out of the tent with a final kick to his backside. A group of news reporters filed a report on the slapping incident with Bedell Smith, Eisenhower's chief of staff. When General Eisenhower learned of the matter, he ordered Patton to make amends.

Pétain, Henri Philippe Benoni Omer Joseph (1856-1951), Maréchal de

France. WW I war hero later to achieve notoriety as Head of State of Vichy France (*Chef de l'État Français*) from 1940 to 1944. With the imminent French defeat in June 1940, Pétain was appointed Premier of France by President Lebrun, then at Bordeaux, and the Cabinet resolved to make peace with Germany. The entire government subsequently moved briefly to Clermont-Ferrand, then to the spa town of Vichy in central France. His government voted to transform the discredited French Third Republic into the French State, an authoritarian regime. In 1942, the Germans finally occupied the whole of metropolitan France because of the threat from North Africa. Pétain's actions during World War II resulted in his conviction and death sentence for treason, which was commuted to life imprisonment by his former protégé Charles de Gaulle.

Petraeus, David Howell (1952-), American general, NATO Commander-in-Chief for Afghanistan, currently the Director of the Central Intelligence Agency.

Pitt, William the Younger (1759-1806), British prime minister. He served as prime minister from 1783 at the age of 24 relinquishing office in 1801, but was prime minister again from 1804 until his death in 1806. Moreover, he was Chancellor of the Exchequer throughout his premierships. He is known as 'the Younger' to distinguish him from his father, William Pitt the Elder, who previously served as prime minister of Great Britain. The younger Pitt's prime ministerial tenure was dominated by the French Revolution and the Napoleonic Wars.

Pollock, Sir Georg, 1st Baronet (1786-1872), British field-marshal. In 1838, Lord Auckland, the Governor-General of India decided to invade Afghanistan. The initial campaign was a success but at the end of 1841, faced with ever-increasing hostility from the Afghans, the military and political leaders decided to withdraw the 5,000 British and Indian troops and 12,000 camp-followers, wives and children from Kabul and to return to India. The retreat was a disaster and eventually led to a massacre because of inefficient leadership, the cold and the ferocious tribes. There was now almost nothing between the Afghanistan forces and India except for the small British garrison at Jallalabad. Pollock was appointed Commander of the Force sent to relieve Jallabad. He advanced through the Khyber Pass to Jellalabad, whose garrison he relieved in April after defeating an enemy force of 10,000 for the loss of 135. Pollock reached Kabul on September 15 after fighting the battles of Jugdulluck Pass and Tezeen, and Nott arrived on September 17, after fighting the battle of Ghuzmee. Pollock withdrew to India in October after destroying the great Bazaar. Once again he had to fight his way through the Khyber Pass. George Pollock retired in 1870 with the rank of field marshal and was made Constable of the Tower in 1871.

Raeder, Erich Johann Albert (1876-1960), German Großadmiral (grand-admiral), WW II supreme naval commander. Raeder led the *Kriegsmarine* (German Navy) for the first half of the war, but resigned in 1943 being replaced by Karl Dönitz. He was sentenced to life in prison at the Nuremberg Trials, but was later released.

Raglan, Fitzroy James Henry Somerset, 1st Baron (1788-1855), British field-marshal. In 1854 he was promoted general and appointed commander of the British expeditionary force in the Crimea, where he was supposed to co-operate with a strong French army corps under Marshal St Arnaud, later to be succeeded by Marshal Canrobert. Here his diplomatic experience stood him in good stead in dealing with the generals and admirals, British, French and Turkish, who were associated with him during the Crimean War. During the unhealthy winter conditions in the Crimea, the British contingent had 23,000 men sick or wounded and only 9,000 who were fit to fight. Raglan has been rightly blamed for tragically unsuccessful 'Charge of the Light Brigade' as well as the unnecessary losses at Balaklava. However, as the British and the French coalition forces won a decisive victory in the Battle of Inkerman, he was promoted field-marshal. Lord Raglan died in harness due to complications brought on by a bout of dysentery.

Roberts, Sir Frederick Sleigh (1832-1914), British major-general later to become Field-Marshal Lord Roberts of Kandahar, 1832-1914. He was one of the most successful British commanders of the 19th century. In December 1878, in the Second Anglo-Afghan War, he was given command of the Kuram Field Force distinguishing himself enough to receive the thanks of Parliament and be elevated into the peerage. After this success he was appointed commander of the Kabul and Kandahar Field Force, directing his 10,000 troops through Afghanistan to the relief of the latter city. He managed to capture Kabul and defeated Muhammad Yakub Khan, the Afghan emir. Promoted to the substantive rank of lieutenant-general in 1883, he became Commander-in-Chief, India in 1885. This was followed subsequently by his promotion to a supernumerary general on in 1890 and to the substantive rank of general in 1891. On 23 February 1892 he was created Baron Roberts of Kandahar in Afghanistan and of the City of Waterford. He also fought with distinction in South Africa during the Second Boer War after which he was created Earl of Pretoria and Viscount St Pierre.

Rommel, Erwin Johannes Eugen (1891-1944), German World War II field-marshal. He was a highly decorated officer in World War I, and was awarded the *Pour le Mérite* for his exploits on the Italian front. In World War II, he further distinguished himself as the commander of the 7th Panzer Division during the invasion of France in May 1940. However, it was his leadership of German and Italian forces in the North African campaign that established the legend of the Desert Fox. He is considered to have been one of the most skilled commanders of desert warfare in the conflict. He later commanded the German forces opposing the Allied cross-channel invasion in Normandy. In 1944, Rommel was linked to the 20 July conspiracy to kill Adolf Hitler. Because Rommel was widely renowned, Hitler chose to eliminate him quietly; in trade for assurances that his family would be spared, Rommel agreed to commit suicide.

Roosevelt, Franklin D. (1882-1945), American president and a central figure in world events during the mid-20th century, leading the United States during a time of worldwide economic crisis and world war. He worked successfully with Winston Churchill and tolerably well with Joseph Stalin in the Alliance against Germany and Japan in World War II, but died just as victory was in sight. Prior to WW II, Roosevelt spearheaded major legislation and issued a profusion of executive orders that instituted the so-called 'New Deal' – a programme designed to produce government jobs for the unemployed, economic growth and reform through regulation of Wall Street, banks and transportation. As World War II loomed on the horizon, with the Japanese invasion of China and the aggressions of Nazi Germany, FDR gave strong diplomatic and financial support to China and the UK, while his country remained neutral. In March 1941, Roosevelt signed the Lend-Lease Act on assisting countries fighting with Britain against Nazi-Germany. With very strong national support he made war on Japan and Germany after the Japanese attack on Pearl Harbor on 7 December 1941.

Roos-Keppel, Sir George Olaf (1866-1921), British-Indian lieutenant-colonel, governor of the North-West Frontier Province. He was a military officer who served in the capacities of political agent to the governor-general in Kurram and Khyber, and from 1908 till 1919 as chief commissioner in the North West Frontier Province. He is also known for his role in 3rd Afghan War. In 1908, he was knighted and was promoted lieutenant-colonel in 1912. In 1913, Roos-Keppel, along with Nawab Sir Sahibzada Abdul Qayyum established 'Islamia College' in Peshawar. Roos-Keppel was also president of Central Committee of Examiners in Pashto

Rundstedt, Karl Rudolf Gerd von (1875-1953), German

Generalfeldmarschall (field-marshal) during World War II. He held some of the highest field commands in all phases of the war. Born into an aristocratic Prussian family, he entered the *Reichswehr* and rose steadily through the ranks in World War I. In the inter-war years, he continued his career, but ultimately retired. He was recalled for service at the beginning of World War II, when assumed command of Army Group South in the campaign against Poland. He maintained army group command during *Fall Gelb,* the onslaught on France. In *Unternehmen Barbarossa,* the Russian Campaign, he commanded Army Group South, responsible for the largest encirclement in history, the Battle of Kiev. Due to the eventual failure of Barbarossa, he was relieved of command, but was recalled in 1942 as *Oberbefehlshaber West,* i.e. theatre commander western Europe. He retained this command (with several interruptions) until his dismissal by Hitler in March 1945, before he was captured by the Allies. He was charged with war crimes, but never faced any trial due to his poor health. He was released from captivity in 1948.

Saddam, Hussein Abd al-Majid al-Tikriti (1937-2006), fifth president of Iraq, serving in this capacity from 1979 until 2003. A leading member of the Socialist Baath Party, Saddam played a key role in the 1968 coup, later referred to as the 17 July Revolution, that brought the party to long-term power of Iraq. As vice president Saddam created an Iraqi security service through which he tightly controlled conflict between the government and the armed forces. In the early 1970s, Saddam nationalised oil and other industries. Through the 1970s, as oil money helped Iraq's economy to grow rapidly, Saddam buttressed his grip on governmental institutions, and positions of power were filled with Sunni Muslims, a minority group amongst the Iraqi population. Saddam managed to cling on to power during the Iran–Iraq War of 1980 through 1988. In 1990 he invaded and looted Kuwait. An international coalition liberated Kuwait in the Gulf War of 1991, but did not change the Iraqi regime. That happened when, in March 2003, a coalition-of-the-willing led by the U.S. and U.K. invaded Iraq. Saddam's Baath party as well as its armed forces were disbanded and the nation made a transition to a democratic system. Following his capture on 13 December 2003, the trial of Saddam took place under the Iraqi interim government. On 5 November 2006, Saddam was convicted of charges related to the 1982 killings and was sentenced to death. His execution was carried out on 30 December 2006.

Saint-Arnaud, Armand Jacques Achille Leroy de (1798-1854), *Maréchal de France.* A sworn monarchist, he joined the *garde nationale à cheval de Paris* (national guard, mounted division) in 1914. In 1821 he went to

Greece with a regiment of volunteers, but returned to France where he led a poor existence until, in 1827, he re-joined the army. During the colonisation of Algiers he served with *la Légion Étrangère*. In 1852, Napoleon III made him a *Maréchal de France*, Imperial Court Equerry and a senator. He commanded the French expeditionary corps in the Crimea at the start of the war and acquitted himself brilliantly in the Battle of Alma on 20 September 1854. Subsequently he went down with cholera and relinquished command to General Canrobert. Nine days later he died on board a ship bound for Constantinople.

Sale, Sir Robert Henry (1782-1845), British major-general, commandin the Jalalabad garrison during the First Afghan War. In 1838, at the outbreak of the First Anglo-Afghan War, Brevet-Colonel Sale asssumed command of 1st Bengal Brigade of the army poised to move into Afghanistan. His column arrived at Kandahar in April 1839, and in May it occupied the Herat Province. During the onward march on Kabul, the strongly fortified city of Ghazni was stormed. Sale led the assault column personally and distinguishing himself in single combat. For his services Sale was made a Knight Commander of the Order of the Bath (KCB) and received the local rank of brigadier-general. As hostilities re-commenced, Sir Robert Sale was ordered to clear the line of communication to Peshawar. After severe fighting he entered Jalalabad on 12 November 1841 and set about making the old and half-ruined fortress fit to stand a siege. Eventually, General Pollock's relieving army appeared, realising that the garrison had already relieved itself by a brilliant and completely successful attack on Akbar Khan's lines. Sir Robert Sale was promoted within the Order of the Bath to Knight Grand Cross (GCB); a medal was struck for all ranks of defenders, and salutes fired at every large cantonment in India. Pollock and Sale after a time took the offensive, and after the victory of Haft Kotal, Sir Robert's division encamped at Kabul again. His wife, who had shared with him the dangers and hardships of the Afghan war, was amongst the captives. Lady Sale and her daughter were rescued by Sir Robert in person, advancing into hostile territory at the head of a detachment of cavalry. Amongst the few possessions she was able to keep from Afghan plunderers was her diary (Journal of the Disasters in Afghanistan, London, 1843). At the end of the war Sale received the thanks of Parliament. In 1845, as quartermaster-general to Sir Hugh Gough's army, Sale again took the field. At Moodkee (Mudki) he was mortally wounded, and he died on 21 December 1845.

Sanders, Otto Liman von (1855-1929), German general. Like many other Prussian aristocrats, he joined the military and rose through the ranks to lieutenant-general. Like several generals before him (e.g., von Moltke and

Baron von der Goltz), he was, in 1913, appointed head of a German military mission to the Ottoman Empire, which was trying to modernise their army along European lines. Liman von Sanders would be the last German to attempt this task.

Schmidt, Helle Thorning (1966-), Danish leader of the parliamentary opposition, Social Democrat Party. From October 2011 onwards she is Danish prime minister, and she held the presidency of the EU January-June 2012.

Schwarzkopf, Herbert Norman (1934-), General, American commander during First Gulf War, 1991. In 1990, Schwarzkopf deviced a detailed plan for the defence of the oil fields of the Persian Gulf against a hypothetical invasion by Iraq; a plan which later served as the basis of a wargame. As Iraq invaded Kuwait, Schwarzkopf's plan had an immediate practical application as a basis for Operation Desert Shield, the defence of Saudi Arabia. A few months later, the offensive operational plan, called Operation Desert Storm constituted the 'left hook" that brought coalition troops into Iraq behind the Iraqi forces and occupying Kuwait.

Sher Ali Khan (1825-79), amir of Afghanistan from 1863 to 1866 and from 1868 until his death in 1879, third son of Dost Mohammed Khan. Sher Ali Khan initially seized power when his father died, but was quickly ousted by his older brother, Mohammad Afzal Khan. Internecine warfare followed until Sher Ali defeated his brother and regained the title of amir. While his rule was hampered by British and Russian pressure, Sher Ali attempted to keep Afghanistan neutral in their conflict. In 1878, the neutrality fell apart and the Second Anglo-Afghan War erupted. As British forces marched on Kabul, Sher Ali Khan decided to leave Kabul to seek political asylum in Russia. He died in Mazar-e Sharif, leaving the throne to his son Mohammad Yaqub Khan.

Shuja Shah Durrani (1785-1842), governor of Herat and Peshawar from 1798 to 1801, amir of Afghanistan from 1801. Shuja allied Afghanistan with the United Kingdom in 1809, as a means of defending against a combined invasion of India by Napoleon and Russia. On 3 May 1809, he was overthrown by his predecessor Mahmud Shah and went into exile in India. In 1838 he had gained the support of the British and the Sikh Maharaja Ranjit Singh for wresting power from Dost Mohammad Khan Barakzai. This triggered the First Anglo-Afghan War (1838–42). Shuja was restored to the throne by the British in 1839, almost 30 years after his deposition, but did not remain in power when the British left. He was assassinated by Shuja ud-Daula, on April 5, 1842.

Smuts, Jan Christiaan (1870-1950), British field-marshal and prominent South African and British Commonwealth statesman, military leader and philosopher. In addition to holding various cabinet posts, he served as prime minister of the Union of South Africa from 1919 until 1924 and from 1939 until 1948. He led the Transvaal commandos in the Second Boer War. He served in the First World War and as a British field-marshal in the Second World War. During the First World War, he led the armies of South Africa against Germany, capturing German South-West Africa (Namibia) and he commanded the British Army in East Africa. From 1917 to 1919, he was also one of five members of the British War Cabinet, helping to create the Royal Air Force. He became a field-marshal in the British Army in 1941, and served in the Imperial War Cabinet under Winston Churchill. He was the only person to sign the peace treaties ending both the First and Second World Wars. Amongst his greatest international accomplishments were his contributions to the establishment of the League of Nations, to the writing of the preamble to the United Nations Charter, and to setting up the British Commonwealth.

Stalin, Iosif Vissarionovich Dzhugashvili (1878-1953), premier of the Soviet Union 1941-53. He was amongst those Bolsheviks who brought about the third Russian Revolution in October 1917 (the previous two being in 1905 and February 1917). He held the position of CPSU secretary-general 1922-53. After the death of Vladimir Lenin in 1924, Stalin managed to use this position to consolidate more and more power in his hands and gradually put down all opposition groups within the Communist Party. This included Leon Trotsky, a socialist theorist and the principal critic of Stalin among the early Soviet leaders, who was exiled from the Soviet Union in 1929. Whereas Trotsky was an exponent of world revolution, it was Stalin's concept of socialism in one country that eventually became official Soviet doctrine. In August 1939, after the failure to establish an Anglo-Franco-Soviet Alliance, the USSR entered into the 'Molotov-Ribbentrop' non-aggression pact with Germany that divided their spheres of influence in Eastern Europe. After Germany violated the pact by invading the Soviet Union on 22 June 1941 and opening an Eastern Front, the Soviet Union joined the Allies. Stalin headed the Soviet delegations at the Yalta and Potsdam Conferences, which drew the map of post-war Europe.

Stark, Harold Rainsford (1880-1972), American admiral. Stark served as an officer in the United States Navy during World Wars I and II. Stark, From 1939 to 1942 he was Chief of Naval Operations. As such, Stark oversaw combat operations against Japan and the European Axis Powers that officially began subsequent to the Japanese onslaught on Pearl Harbour on 7 December 1941. In March 1942, Stark was relieved on this post and went to the UK to become Commander of US Naval Forces in Europe. From his London headquarters, Admiral Stark directed the naval part of the great build-up in England as well as US naval operations and training activities in the Europe. He built and maintained close relations with British civilian and naval leaders and also with the leaders of other Allied powers. Stark was particularly important in dealing with General Charles de Gaulle. He left active service in April 1946, and since there has been some controversy surrounding his person due to allegations of failures in connection with the Japanese attack on Pearl Harbour - this has only become fiercer since.

Steward, Sir Donald (1824-1900), British major-general, later to become field-marshal, 1st Baronet. for five years Commander-in-Chief, India, and afterwards a member of the Council of the Secretary of State for India.

Stilwell, Joseph Warren (1883-1946), American WW II general. Best known for his service in the China Burma India Theatre-of-War. His generally caustic attitude is reflected in the sobriquet 'Vinegar Joe.' While distrustful of his allies Stilwell – especially the British – he was a reliable and daring tactician in the field. He frequently disagreed as to strategy, ground troops versus air power, with his subordinate, Claire Chennault, who had the ear of Generalissimo Chiang Kai-shek.

Thucydides (460 BC-c. 395 BC), Greek historian and author from Alimos. His generally believed to be one of the first known writers of military history, and his history of the Peloponnesian Wars takes us back to the wars of 431-404 BC between Sparta, heading the Peloponnesian coaltion, and the coalitions dominated by Athens. Thucydides has been dubbed the father of scholarly history because of his strict adherence to the standards of evidence-gathering.

Tito, see Broz.

Victoria (1819-1901), queen of the United Kingdom, empress of India. She was the monarch of the United Kingdom of Great Britain and Ireland from 20 June 1837 until her death. From 1 May 1876 onwards she was also the Empress of India. She inherited the throne when she was 18 since her father's three elder brothers had already died without any legitimate issue. The United Kingdom was already an established constitutional monarchy, but she did, to a certain extent, influence government policy and ministerial appointments. In 1840, she married her first cousin, Prince Albert of Saxe-Coburg and Gotha. Her reign of 63 years and 7 months, which is longer than that of any other British monarch and the longest of any female monarch in history, was a period of industrial, cultural, political, scientific and military development within the United Kingdom, and was marked by a great expansion of the British Empire.

Wavell, Lord Archibald Percival (1883-1950), British field-marshal and the commander of British Army forces in the Middle East during the early phases of the Second World War. He led British forces to victory over the Italians, only to be defeated by the German army. He was the penultimate viceroy of India from 1943 to 1947.

Wellesley, Arthur (1759-1852) later to become 1st Duke of Wellington), British general and statesman, a native of Ireland, from the Anglo-Irish Ascendancy, and one of the leading military and political figures of the 19th century. He is often referred to as 'the IronDuke.' Serving in Ireland as aide-de-camp to two successive lords-lieutenant of Ireland he was also elected as a member of Parliament in the Irish House of Commons. A colonel by 1796, he saw action in the Netherlands and later in India, where he fought in the Fourth Anglo-Mysore War at the Battle of Seringapatam. He won a decisive victory over the Maratha Confederacy in 1803. Wellesley rose to prominence as a general during the Peninsular campaign of the Napoleonic Wars, and was promoted to the rank of field-marshal after leading the allied forces to victory against the French at the Battle of Vitoria in 1813. In 1814, he he was appointed ambassador to France and was granted a dukedom. During the Hundred Days in 1815, he commanded the allied army which, with a Prussian army under Blücher, defeated Napoleon in the Battle of Waterloo. He was twice prime minister and oversaw the passage of the Catholic Relief Act 1829. He remained Commander-in-Chief of the British Army until his death.

William III (1650-1702), King of Great Britain. By birth William was a Prince of the House of Orange-Nassau. From 1672, as Willem III van Oranje, he was the Stadtholder of the United Provinces of the Dutch Republic, viz. Holland, Zeeland, Utrecht, Guelders, and Overijssel. In what became known as the 'Glorious Revolution', on 5 November 1688 William invaded England in an action that ultimately deposed King James II & VII and won him and his spouse Mary the crowns of England, Scotland and Ireland. In the British Isles, William ruled jointly with his wife, Mary II, until her death on 28 December 1694. A Protestant, William participated in several wars against the powerful Catholic king of France, Louis XIV, in coalition with Protestant and Catholic powers in Europe. William's victory over James in the Battle of the Boyne in 1690 is commemorated by the Orange Institution in Northern Ireland and parts of Scotland to this day. His reign marked the transition from the personal rule of the Stuarts to the more Parliament-centred (constitutional) rule of the House of Hanover.

Württemberg-Neustadt, Carl Rudolf duke of (1667-1742), Danish lieutenant-general, contingent commander (under Marlborough) during the War of the Spanish Succession. Fought with distinction at Blenheim 1704.

Württemberg-Neustadt, Ferdinand Wilhelm duke of (1659-1701), Danish lieutenant-general, contingent commander during the Nine Years War. An experienced soldier, he had fought the Turks outside Vienna in 1683 (Kahlenberg) and again in Hungary the year after, where he had received a sabre cut to his forehead, a wound from which he would eventually die.

Yakub Khan, Mohammad (1849-1923), amir of Afghanistan from February 21 to October 1879. He was the son of the previous ruler, Sher Ali Khan. Yaqub Khan was the governor of the Herat Province and, in 1870, he decided to rebel against his father. He was imprisoned in 1874. In the wake of the Second Anglo-Afghan War Yaqub signed the Treaty of Gandamak with the British in May 1879, relinquishing control of Afghanistan foreign affairs to the British Empire. An uprising against this agreement led to his abdication. He was succeeded by Amir Ayub Khan.

Zia ul Haq, Mohammad (1924-88), general, general, Pakistani president from July 1977 to his death in August 1988. He was appointed Chief of Army Staff in 1976 subsequent to Prime Minister Zulfikar Ali Bhutto's sacking of seven senior generals who were tainted with their rôle in the East-Pakistan war. After widespread civil disorder, in 1977 Zia overthrew Bhutto in a bloodless coup d'état. As President, Zia forcefully crushed the secular-communist and socialist democratic struggle led by the eldest daughter of Zulfikar Ali Bhutto, Benazir Bhutto. Zia abandoned the previous economic policies of Bhutto, and replaced them with capitalism and privatisation of the major industries of Pakistan that had been nationalised by Bhutto in 1970s. The Pakistan economy became one the fastest growing economies in South Asia. However, during this period of economic and social change, Zia dealt with the political rivals in a decidedly harsh manner, and his reign is often regarded as a period of mass military repression. He is most remembered for his foreign policy; the subsidising of the Mujahideen movement fighting the Soviet troops in Afghanistan, which led to the Soviet-Russian withdrawal from that country.

In August 1988, Zia died along with top generals and admirals and the United States Ambassador to Pakistan in a suspicious air crash near Bahawalpur.

Bibliography

Interviews

Mr Edward Ferguson, UK Ministry of Defence

Brigadier Tim Radford, UK Ministry of Defence

Brigadier James Chiswell, Commander of Task Force Helmand October 2010 to March 2011 Colonel Lennie Fredskov, commanding officer of the Danish ISAF Team 10

Lieutenant-Colonel Thomas Funch Pedersen, deputy commander, the Danish ISAF Team 10

Major Christian Bach Byrholt, chief operations the Danish ISAF Team 10

Major Michael C. Toft, officer commanding B Company, the Danish ISAF Team 10

Major Kaj Vincent Frederick Ahlefeldt-Laurvig

Major Kim Kristensen

Captain Thomas Larsen, chief intelligence, the Danish ISAF Team 10

Monographs

Barthorp, Michael. *Afghan Wars and the North-West Frontier 1839-1947*. London: Cassel & Co. 1982.

Bourne, Kenneth. *Palmerston: The Early Years 1784–1841*. London: Alien Lane, 1982.

Bruce, George. *Retreat from Kabul*. London: Howard baker, 1967.

Chandler, David. *The Art of Warfare in the Age of Marlborough*. London: B.T. Batsford Limited, 1976.

Childs, John. *The Williamite Wars in Ireland 1688–1691*. London: Hambledon Continuum, 2007.

Clausewitz, Carl von. *On War* (edited and translated by Sir Michael Howard and Peter Paret). Chichester: Princeton University Press, 1976.

Conelly, Owen. *The Wars of the French Revolution and Napoleon, 1792-1815*. London and New York: Routledge, 2006.

Dallaire, Roméo A. *J'ai serré la main du diable*: *La Faillite de L'Humanite au Rwanda*. Montréal : Libre Expression Canada, 2003.

Danchev, Alex and Daniel Todman (Eds). *War Diaries 1939-1945: Field Marshal Lord Alanbrooke*. London: Phoenix Press, 2001.

Dunbabin, T.J. *The Western Greeks. The History of Sicily and South Italy from the Foundation of the Greek Colonies to 480 B.C.* Oxford: Oxford University Press, 1948.

Fuller, J.C.F., *The Decisive Battles of the Western World 480BC-1757*. London: Paladin Books, 1954.

Fuller, J.C.F., *The Decisive Battles of the Western World 1792-1944*. London: Paladin Books, 1954.

Galster, Kjeld Hald. *Danish Troops with the Williamite Army in Ireland,*

1689-91: For King and Coffers. Dublin: Four Courts Press, 2012.

Galster, Kjeld Hald. *Face of the Foe: Pitfalls and Perspectives of Military Intelligence.* Kingston, Legacy Books Press, 2010.

Gaulle, Charles de. *Mémoires de Guerre: le Salut 1944-1946.* No place: Plon, 1959.

Granatstein, J.L. *Empire to Umpire: Canada and the World into the Twenty-First Century.* Toronto: Nelson, 2008.

Hamilton, Nigel. Monty: *The Making of a General 1887-1942.* Toronto: Fleet Books, 1982.

Heathcote, T.A. *The Afghan Wars 1839-1919.* London: Osprey, 1980.

Hillmer, Norman and J.L. Granatstein. *Empire to Umpire: Canada and the World into the Twenty-First Century.* Toronto: Nelson, 2008.

Holmes, Richard (Ed). *The Oxford Companion to Military History.* Oxford: Oxford University Press, 2001.

Holmes, Richard. *Dusty Warriors: Modern Soldiers at War.* London: Harper Press, 2006.

Holmes, Richard. *In the Footsteps of Churchill.* London: BBC Books, 2005.

Holmes, Richard. *Marlborough: England's Fragile Genious.* London: Harper Press, 2008.

Holmes, Richard. *Soldiers: Army Lives and Loyalties from Redcoats to Dusty Warriors.* London: Harper Press, 2011.

Holmes, Richard. *Wellington: The Iron Duke.* London: Harper Collins Publishers, 2002.

Jahn, J.H.F. *Det danske Auxilliaircorps i engelsk Tjeneste fra 1689 til 1697* [The Danish Auxilliary Corps on English Service, 1689–1697]. Kjøbenhavn: Udgiverens Forlag, 1840.

Jenkins, Roy. *Churchill.* Basingstoke and Oxford: Pan Macmillan, 2002.

Keegan, John. *A History of Warfare*. New York: Vintage Books, 1993.

Knox, MacGregor. *Hitler's Italian Allies: Royal Armed Forces, Fascist Regime, and the War of 1949-1943*. Cambridge: Cambridge University Press, 2000.

König, Malte. *Kooperation als Machtkampf: Das fascistische Achsenbündnis Berlin-Rom im Krieg 1940/41*. Cologne: sh-Verlag, 2007.

Lenihan, Pádraig (Ed.). *Conquest and Resistance: War in Seventeenth-Century Ireland*. Leiden, The Netherlands; Brill, 2001.

Lindquist, Herman, trans. Henrik Eriksen, *Napoleon*. Oslo: Schibsted, 2005.

Luft, Gal. *Beef, Bacon and Bullets: Culture in Coalition Warfare from Gallipoli to Iraq*. No place: Gal Luft, 2009.

Lyon, David. *Butcher & Bolt: Two hundred years of Foreign Engagement in Afghanistan*. London: Hutchinson, 1988.

Machiavelli, N. *The Art of War*. Cambridge: Da Capo Press, 2001.

Machiavelli, N. *The Prince, with Related Documents*, Ed. and trans. by W. Connell. Boston: Bedford/St. Martins, 2005.

Maguire, W.A. *Kings in Conflict: The Revolutionary War in Ireland and its Aftermath, 1689–1750*. Belfast: Blackstaff Press, 1990.

McNally, Michael. *The Battle of Aughrim 1691*. Stroud: The History Press, 2008.

McNally, Michael. *The Battle of the Boyne 1690: The Irish Campaign for the English Crown*. Oxford: Osprey Publishing, 2005.

Mitchell, Paul T. *Network Centric Warfare and Coalition Operations: The New Military Operating System*. Oxon: Routledge Global Security Studies, 2009.

Montgomery, Field-Marshal Bernard Law 1st Viscount of Alamein. *The Memoirs of Field-Marshal Montgomery*. London: Fontana Monarchs, 1958.

Morgan, Philip. *Fascism in Europe, 1919-1945*. London : Routledge, 2003.

Murtagh, Harman. *The Battle of the Boyne 1690: A Guide to the Battlefield.* Mell: The Boyne Valley Honey Company, 2006.

Neilson, Keith and Roy A. Prete, Eds. *Coalition Warfare: an Uneasy Accord.* Waterloo, Ontario, Canada: Wilfrid Laurier University Press, 1983.

Nosworthy, Brent. *The Anatomy of Victory: Battle Tactics 1689–1763.* New York: Hippocrene Books, 1992.

Rashid, Ahmed. *Descend into Chaos: Pakistan, Afghanistan and the Threat to Global Security.* London: Penguin Books, 2008.

Robson, Brian. *The Road to Kabul: The Second Afghan War 1878-1881.* London: Arms and Armour Press, 1986.

Roper, A. *The History of King William the Third*, Vol II. London: Black Boy, F. Coggan, 1702.

Roussel, Éric. *Charles de Gaulle.* No place: Gallimard, 2002.

Schneid, Frederick C. *Napoleon's Conquest of Europe: The War of the Third Coalition.* Westport, Connecticut, London: Praeger, 2005.

Smith, E.D. Valour: *A History of the Gurkhas.* Stroud: Spellmount, 1997.

Tinning, Morten and Signe Lund (Eds). *The Distant War: 17 Perspectives on Afghanistan.* Copenhagen: Statens Forsvarshistoriske Museum, 2011 *Thukydides's Historie* [Thucydides' History] translated by M. Cl. Gertz, 1st and 2nd Books. Kjøbenhavn: Karl Schønberg, 1897-98.

Anthologies

Verma, Dinesh. *Network Science for Military Coalition Operations: Information Exchange and Interaction.* Hershey and New York: Information Science Reference, 2010.

Hansen, Birthe and Bertel Heurlin, Eds. *The New World Order.* London: Macmillan Press Ltd.

Articles

Hansen, Birthe "The Unipolar World Order," in Birthe Hansen and Bertel Heurlin, Eds. *The New World Order*. London: Macmillan Press Ltd., 2000. Pp. 112-133.

V. Ilardi, "The Italian league, Francesco Sforza and Charles VII (1454-`461)" in *Studies in the Renaissance* VI, 1959.

Mallett, M. "Diplomacy and War in Later Fifteenth-century Italy" in *Proceedings of the British Academy* LXVII, 1981.

Rubinstein, N. "Das politische System Italiens in der zweiten Hälfte des 15. Jahrhunderts", in '*Bündnissysteme' und Außpolitik im späteren Mittelalter*, ed. P. Moraw (*Zeitschrift der. Historische Forschung. Beihefte*, 5) 1988.

Stone, Norman "Moltke and Conrad: Relations Between the Austro-Hungarian and German General Staffs, 1909-14" in Paul M. Kennedy ed., *The War Plans of the Great Powers, 1880-1914*.

Conference Papers

Cecil, Patrick. Paper presented at the coalition warfare conference in Copenhagen May 2011.

Delaney, Doouglas. Paper presented at the coalition warfare conference in Copenhagen May 2011.

Dover, Paul. Paper presented at the coalition warfare conference in Copenhagen May 2011.

Lambert, Andrew. Paper presented at the coalition warfare conference in Copenhagen May 2011.

Schwartz, Adam and Thomas Heine Nielsen. Paper presented at the coalition warfare conference in Copenhagen May 2011.

Vogel, Thomas. Paper presented at the coalition warfare conference in Copenhagen May 2011.

Newspapers

The Guardian, 3 December 2010. "WikiLeaks cables expose Afghan contempt for British Military."

Berlingske Tidende, 15 Jun 2010.

"I krig med eliten [Waging War with the Èlite]" in *Weekendavisen* 29 October – 4 November 2010.

"Afghanistan – ingen lette løsninger [Afghanistan: no Easy Solutions]" in *Information* 25 August 2010.

Lars Erslev Andersen, "Hvorfor er vi (stadig) i Afghanistan? [Why are we (still) in Afghanistan?]" in *Berlingske Tidende* 25 September 2010.

Anna von Sperling, "Bistandsstrategi lægger op til samarbejde med 'terrorister' [Development Strategy Suggests Co-operation with 'Terrorists'] in *Information* 7 October 2010.

Wikileaks/Afghanistan/DK

Dokumenter fra Wikileaks om danske tropper i Afghanistan.zip

Wikileaks Danskere i Afghanistan 2004-05.doc

Wikileaks Danskere i Afghanistan 2006.doc

Wikileaks Danskere i Afghanistan 2007.doc

Wikileaks Danskere i Afghanistan 2008.doc

Wikileaks Danskere i Afghanistan 2009_1-6.doc

Wikileaks Danskere i Afghanistan 2009_7-12.doc

Internet

Major Willie J. Brown, USMC, *The Keys To Successful Coalition Warfare: 1990 And Beyond*. From internet, URL http://www.globalsecurity.org/military/library/report/1991/BWJ.htm accessed on 11 August 2010.

Robert W. Riscassi's article "Principles for Coalition Warfare." From Internet, URL http://www.dtic.mil/doctrine/jel/jfq_pubs/jfq0901.pdf accessed on 11 August 2010.

Abbreviations

ADV	advisor
ALP	Afghan Local Police
BDA	Battle damage assessment
BDZ	Bridzar
CAT a/b	categories a and b casualties
CIED	Counter IED
CONOP	Concept of Operations
CST	Cimic Support Team
CTN	Clifton
CWIED	Command Wire IMPROVISED EXPLOSIVE DEVICE
DAK	Deh-e Adam Khan (Area I GSK (En))
DCC	District Community Council
DDP	District Delivery Programme
DHG	Danish Home Guard
DKAT	Danish Kandak advisory team
DNK	Denmark
DST	District Stabilisation Team
EXACTOR	Missile-type
FoM	Freedom of manœuvre
FZP	Food Zone Programme
GIRoA	Government of the Islamic Republic of Afghanistan

Abbreviations

HOTO	Hand over/take over
HPC	Helmand High Peace Council
HPTC	Helmand Police Trg Centre
Hw Coy	Heavy Weapons Coy
IEDD IED	disposal
IJC ISAF	Joint Command
Iot	in order to
IRG	Immediate replenishment group (log train)
KIA	Killed in action
KNK	Kah Nikar
LAR	Light armoured recce
LMC	Low Metal Content
LN	Local nationals
MLV	Malvern (patrol base), named after the last battle of the peninsula Campaign of the American Civil War (The Battle of Malvern Hill, also known as the Battle of Poindexter's Farm, took place on 1 July 1862)
MO	Modus Operandi
MOI	Ministry of the Interior
MSST	Military Stabilisation Support Team
NMT	NATO trg Mission
NDS	National Directory of Security
Obj Beefcake	Mid-level commander Haji Ghafar
OCCD	Operational Co-ordination Centre, District
OCC-D	Ops co-ordination cell, district
PB	Patrol base
PCoP	Provincial Chief of Police
PMT	Police mentoring
P-OMLT	Police Operational Mentoring and Liaison Team
PP	Pressure Plate
PRST	Provincial Reintegration Support Team
PRT	Provincial Reconstruction Team
QRF	Quick reaction force
RCIED	Remotely controlled IMPROVISED EXPLOSIVE DEVICE
REVIVOR	Base ISTAR system (Tall pole with a sensor)
RIP	Relieve in place
ROEs	Rules of Engagement
SAF	small arms fire
Silicon	British Police Advisors
SPN	Spondon
SSR	Security Sector Reform
STABAD	Stabilisation Advisor under PRT

STC	Security and transition command
TIC	Troops in Contact
Tiger Team	UK team tasked with find, feel assess
Tolay	Afghan word for "company"
TUFAAN	DA Ops
UAHMMW	Un-armoured HUMV
UK GL	UK Gunline
VICC	Commercial road construction company
Viking	UK Viking Coy
VO	Victim Operated

Index

1st Battalion Irish Guards....... 52, 121
3rd Kandak..... xiv, 78, 80, 82, 96, 109, 110, 116, 121, 261, 291
Aaen, Frank.................... 118
Aallmann, Rev Thomas....... 128, 129
ABC Conference............... 173
Aboukir....................... 17
Abyssinia.............. 175, 178, 276
Advisory bodies working with the Afghan 53
Afghan 3rd Brigade........ 50, 83, 264
Afghan National Army.... 50, 52, 79, 85, 108, 109, 111, 114, 115, 121, 127, 245
Afghan National Army's 3rd Brigade 50
Afghan National Police... 52, 82, 85, 96, 97, 105, 116, 121, 245
Afghan National Security Forces.. 15, 52, 53, 57, 62, 69, 77, 94, 96, 99, 107, 108, 115, 171, 232, 254, 268, 269
Afghanistan. .. viii, x, xii-xiv, 4-6, 8-10, 13, 14, 17, 18, 21-48, 50-62, 64, 66, 68, 69, 71, 73, 76-78, 85, 101, 103, 105-108, 114, 115, 117-119, 122, 124, 127, 132, 134, 140, 167- 169, 225, 230, 232, 240, 244-248, 251, 252, 254, 255, 257-261, 264, 266, 269, 270, 272, 274, 275, 278- 281, 285-288, 290-292, 295, 298, 302, 304-308, 311, 312, 316, 321, 322, 324, 326
aims in the war, see war aims..... 56, 57, 223, 254
aims with the war, see war aims... 51, 57, 162, 223, 248, 251, 254, 255, 271
Air Assault Brigade.. 51, 52, 66, 78, 281
air policing..................... 42
al Qaeda... 8, 9, 45, 46, 49, 50, 106, 240, 254, 274
Alamein.......... 190, 210, 303, 321
Alanbrooke... 16, 19, 180, 184, 190, 191, 193, 195, 199, 201, 205, 208, 212, 213, 275, 277, 301, 319
Alexander, Harold ... 24, 27-30, 34, 152, 154, 199, 210, 217, 247, 253, 275, 278, 285, 296, 300, 301
Allied Expeditionary Forces..... 20, 264
Amanulla......... 38, 39, 41, 275, 290
Ambrosio, Vittorio........... 189, 275
Amin, Hafizullah. xii, 24, 33, 43, 63, 68, 187, 284, 290, 291, 306

Amnesty International. 8
Andersen, Lars Erslev. 87, 119, 324
Anvil (operation). 194
Arcadia Conference. 184, 205
Army of Retribution. 30
Army of the Indus. 29
Army Operational Command. . . viii, xiii, 66, 78, 111
asymmetric warfare. 155
Atlantic Charter. 184, 291
Auckland. . . . 27, 28, 275, 278, 285, 300, 307
Auerstädt. 151
Auftragsbefehl (mission command) . . 264
Austerlitz. 149, 151, 152, 284, 299
Austria. . . 4, 7, 13, 16, 19, 150-155, 162, 163, 168, 202, 223, 237, 239, 240, 243, 251, 262, 269, 278, 287, 292, 293, 295, 299, 301, 302
Axis. . . . . xiii, 29, 30, 172-174, 176-180, 182-189, 197, 198, 201, 253, 254, 256, 259, 263, 264, 276, 314
Bagh. 40, 286
Bala Hissar. 31, 35
Balaclava. 157
Balkans. . xii, 4, 5, 14, 55, 136, 163, 202, 203, 239
Balochistan. 28, 40, 41, 45-47, 197
Bandi Barq Road. . . . 57, 83, 89, 94, 98-100, 104, 105, 107
Basra. 4, 244
Basra Region, see Iraq. 4, 244
Bastion. 94, 112, 113, 179
Battle of the Boyne. 10, 322
Bech, Gitte Lillelund. 127
Beef, Bacon and Bullets. . xiii, 3, 11, 167, 170, 222, 232, 234, 235, 237, 321
Berlin Congress. 33, 34
Bernadotte, Jean Baptiste Jules. 153, 243, 276
Bhutto, Benazir. 47, 276, 316
Bloch, Jan Gotlib. 238, 276
Bocage. 261
Boer War. . 158-161, 163, 170, 218, 252, 272, 282, 291, 308, 313
Bolan Pass. 26-29, 36
Bonaparte, Napoleon, see Napoleon. 149, 150, 152, 154, 229, 279, 304, 305
Borden, Robert. 163, 277
Borodino. 153
Boyne. 10, 322

Brauchitsch, Walther von. 198, 277
Brigade Advisory Group. . . . 83, 99, 110
Britain. . . . 2, 7, 11-14, 16, 17, 22, 25, 27, 30-37, 119, 120, 125, 131, 132, 146-152, 156-158, 162, 164-166, 171-174, 176, 179, 181-187, 189, 199, 207, 216, 218, 231, 233, 237, 241, 243, 246, 248, 253, 257, 262, 264, 269, 272, 281-283, 285-288, 294, 296-298, 305-307, 309, 314, 315
British Expeditionary Force. . . . 256, 308
British Raj. . . . 24, 37, 42, 197, 236, 286
Brooke, Alan, see Alanbrooke. . . . 16, 19, 20, 180-182, 184, 190-193, 195, 199-203, 205-213, 224, 257, 263, 265, 275, 277, 278, 301, 319
Brusilov Offensive. 240
Budwan, Forward Operating Base/Fire Base. 58, 61, 80, 82, 85, 86, 88, 93, 95, 97, 98, 105, 106
Burns, Alexander. 27, 28, 30, 278
Bush, George W.. 7, 46, 97, 123, 241, 279
C4I (command, control, communications, computers and intelligence). . . 18, 227, 228
Caboul, see Kabul. 25
Camberley. 196
Camel's Neck. 40
Cameron, David. 117, 244
campaign plan. . 60, 76-79, 97, 125, 225, 230, 238, 258
Canada. . 2, vii, ix, 10, 12, 158-163, 171, 174, 181, 211, 212, 216-219, 231, 241, 250, 253, 267, 277, 283, 296, 297, 301, 319, 320, 322, 338
Canadian Army Staff College. 217
Carnot, Lazare. 149, 229, 279, 280
Casablanca Conference. 191, 198
Cavagnari, Pierre Louis Napoleon. . . . 35, 232, 280
Central Powers. . 3, 16, 39, 162-165, 167, 239, 240, 250, 259, 264, 272
Chevènement, Jean-Pierre. 235
China. . . 34, 79, 164, 174, 192, 213, 214, 242, 254, 289, 309, 314
Chinese Civil War. 213
Chiswell, James. . . vii, 51, 55-57, 61-65, 70, 76-78, 99, 102, 125, 126, 128, 130, 131, 133, 280, 318
Chitral. 41

Index

Christian norms. 170
Christian V, king of Denmark 7, 146, 242, 253, 281
Churchill, Winston Spencer. . 19, 20, 135, 173, 174, 180-184, 191-194, 201, 204, 205, 207, 209, 210, 212, 213, 223, 250, 256, 263, 265, 277, 278, 281, 282, 284, 287, 291-293, 296, 300, 301, 309, 313, 320
Ciano, Count Gian Galeazzo. . . 189, 282
CIMIC (Civil-military co-operation). . 72, 79, 124, 326
Clausewitz, Carl von. 124, 221, 319
coalitions-of-the-willing. . . xiv, 146, 148, 253
COIN (Counter insurgency). . . 1, 56, 163, 254
Cold War. . . xii, 1, 4, 5, 7, 9, 13, 14, 55, 122, 135, 216, 218-221, 227, 231, 240, 241, 248, 256, 257, 267, 290, 306
Comando Supremo. 188, 280
Confederation of the Rhine. 11, 154, 244, 269
Conrad, Franz Graf von Hötzendorf . 223, 239, 240, 293, 323
conscription. . 15, 16, 149, 212, 267, 279, 301
Constantinople. . . 32, 142, 155, 167, 301, 311
Continental System.. 152
Copenhagen. . . vii, xi-13, 61, 63, 68, 77, 117, 119, 130, 138, 139, 142, 156, 160, 161, 163, 165, 179, 196, 214, 242, 288, 322, 323, 338
Corpo di Spedizione Italiano. 203
Corunna. 153
Cotton, Willoughby. 28, 283
counter insurgency, see COIN. 43, 45, 58, 79, 108, 131, 133, 233, 257, 258, 269
Crerar, General Henry. 20, 197, 283
Crimean War. . . . 32, 155-157, 222, 249, 260, 262, 279, 301, 308
cultural models. 231, 232
Currie, Sir Arthur. . . . 163, 172, 283, 284
Curzon, George. 38, 297
D and F Coys. 52, 101, 123, 128
Dakka. 40
Dallaire, Romeo. 266, 267, 319
DANCON, see Danish Contingent. . . . xiv

Danish Contingent. 147, 148
Danish Kandak Advisory Team. . . 82, 84, 110, 127, 326
Defence Staff. 15
Delaney, Doug. . vii, 160, 161, 163, 197, 323
Denmark. . . . xiii, xiv, 3, 4, 6, 7, 9, 11-13, 15, 21, 50, 53, 55, 62, 63, 65, 66, 68, 69, 104, 111, 118-120, 125, 132, 134, 146, 148, 154, 155, 227, 230, 233, 237, 241, 243, 246, 248, 251, 253, 272, 279, 281, 288, 326, 338
Deutsches Afrikakorps. 178
Digby, Pamela (Mrs Randolph Churchill, later Mrs Harriman). . . . 205, 284, 292
Dill, John. 19, 284
Disraeli, Benjamin. 24, 33, 284, 285
DKAT. 82, 84, 110-112, 121, 326
doctrine. . . 17, 18, 46, 56, 58, 76, 79, 136, 151, 153, 155, 166, 171, 191, 221, 222, 224, 226, 227, 244, 245, 256, 257, 262, 272, 313, 325
Donore. 10
Donore Hill. 10
Dost Muhammad. 26-33, 275, 285
Dover, Paul.. . . . . vii, 141, 142, 323
Dragoon. 57, 174, 203, 255
drugs.. 127
Dunkirk. 21
Durand, Henry. 29
Durand, Sir Mortimer. 37, 285
Dyer, Reginald. 286
Eagle.. . . . . 87, 95
East India Company. . 23-26, 29, 32, 236, 278, 286, 298, 305
École supérieure de guerre. 166
Eden, Anthony. 180, 183, 286
Eid festival. 116
Eisenhower, Dwight D.. . . 20, 194, 199-201, 206, 209, 210, 212, 213, 258, 261, 265, 277, 286, 306
Ellenborough. 31, 286, 305
Elphinstone, William. . 29, 286, 287, 305
Enduring Freedom. 45, 49, 133
Enniskillen Regiment. 10
Entente. 39, 162, 164-167, 223, 239, 242, 250, 255, 256, 264, 272
Entente Cordiale. 256
European Union. 4, 140

EW. 72, 73, 285
Eylau. 152, 284
Ezmerai, Afghan chief of police. . . 78, 92, 105, 288
Falklands. 221
Fascist Italy. 174, 176, 179
FATA, see Federally Administered Tribal Areas. 46, 85, 134, 161, 233, 242
Federally Administered Tribal Areas. 132
Felton, Brigadier. 78
Feltre (conference) 190
Fiin. 123
Foch. . . 12, 165, 166, 247, 264, 282, 288
fodder. 6
Food Zone Programme. . . . 100, 107, 326
force de rupture. 255
France. . . 7, 12-14, 16, 18, 19, 24, 31, 32, 55, 57, 135, 142, 144, 145, 148-159, 163-166, 174, 176, 177, 183, 186, 190-195, 199, 200, 202, 203, 212, 214, 215, 219, 221, 225, 230, 234, 237, 239, 241, 243, 244, 246, 248, 249, 251-253, 255, 257, 262-264, 268, 276, 278, 279, 282, 283, 286, 288, 289, 292, 294, 296, 298, 299, 302-304, 306, 307, 309-311, 315
Francophones. 161
fratricide. 18, 242
Frederick VI. 18, 288
Fredskov, Lennie. viii, 54-61, 65, 66, 68-70, 74, 76-78, 82, 85-87, 89, 95, 96, 99, 109, 116, 128, 130-133, 288, 318
Free French. 205, 289, 290
French Canadians. . . . 160, 161, 218, 253
French Korean Battalion. 214
French North Africa. 191, 206
French Revolution. . . . 50, 148, 243, 279, 298, 304, 307, 319
Friedland. 152
friendly fire. 147, 258
Gallipoli. . 3, 11, 167, 170, 172, 193, 321
Garrison Support Unit. . . . 109, 112-115, 273
Gaulle, Charles de. 288, 320
Gelb. 221, 225, 310
General Service Enlistment Act. 236
Geneva Conventions. 7, 8
Gereshk. . . 50, 52, 54, 57-61, 64, 70, 71, 73, 74, 80-82, 84, 85, 95-97, 99, 102, 103, 107, 109, 110, 116, 121, 125, 130, 131, 133, 261, 267
Ghazni. 29, 34, 305, 311
Ghurkhas. 35, 95, 124
Giraud, Henri Honoré. 191, 206, 207, 289
Goodbye to All That. 128, 290
Grand Alliance. 241, 249
Grandière (frigate). 214
Graves, Robert. 128, 280, 290
Great Game. 28
GSU. 112
Guardian. 131, 132, 233, 324
Guided Multiple Launch Rocket System 88, 106
Gulevich. 238
Gulf War. . . . 55, 226, 234, 235, 256, 296, 310, 312
Habibulla. 38, 39, 290
Habsburg Empire. 32
Hague Conventions. 7
Haig, General Douglas. . . . 164, 166, 167, 255, 283, 291
Hamayun, Mohammad. . 82, 92, 116, 291
Hansen, Birthe. 11, 322, 323
Haq, Zia ul. 43, 316
Haqqani network. 47
Harriman, Averell. 291
Hazrat. . . . 65, 75, 99, 102, 103, 106, 121
Headquarters and Logistics Company 52, 53, 110
Heartbreak Ridge. 215
Heavy Weapons Tolay. . . . 80, 85, 91, 96, 101, 106, 110
Helmand. . vii, xiii-xv, 4-6, 9, 22, 47, 49-59, 61-67, 69-74, 76-79, 81, 83, 85-87, 91, 98, 99, 102-106, 111, 116, 120, 125, 126, 131-133, 225, 227, 228, 230, 233, 248, 251, 254, 256-258, 261, 264, 267, 269, 280, 281, 318, 327
Helmand Plan. 53, 57, 76-78
Herat. . . 26-28, 32, 33, 37, 285, 311, 312, 316
Herkules (Operation Hercules). 179
Herodotos. 137, 138
Herrick. 22, 50, 51, 78, 133, 254
Highway 1. . . . . 57, 81, 85, 86, 102
Hikmatyar, Gulbuddin. 43, 45, 292
Hindenburg, Paul von. 292
Hindu Kush. 29, 33, 34

History of the Peloponnesian War... 137, 138
Hitler, Adolf.... 176-178, 186-188, 190, 198, 203, 204, 246, 277, 282, 292, 293, 295, 298, 309, 310
Hitler's Germany.................. 8
Holmes, Richard. 3, 9, 153-155, 166, 320
Holy Roman Empire.. 144, 145, 149-152, 301
holy warriors..................... 7
Honourable East India Company.. 23, 29, 32, 286
Hopkins, Harry.. 165, 184, 195, 204, 207, 265, 293
human rights..... 2, 4, 8, 49, 51, 99, 252, 254
Human Terrain Mapping.. 72, 75, 83, 96
Humint......................... 73
Hungary... 162, 163, 220, 239, 251, 269, 270, 278, 294, 316
Hussein, Saddam, see Iraq......... 310
IEDs, see improvised explosive devices
.... 58
Ike, see Eisenhower.. 102, 202, 210, 212, 264
improvised explosive devices..... 58, 70, 83, 88, 90, 92, 93, 98-100, 102, 103, 106, 116, 130, 268
India.... 23-40, 42, 44, 46, 47, 167, 174, 181, 192, 208, 232, 235, 236, 253, 254, 275, 278, 280, 281, 283-286, 290, 297-299, 304, 305, 307-309, 311, 312, 314, 315
Indian Mutiny............. 24, 32, 285
indirect security.... 4, 8, 9, 56, 146, 268
Indochina.................. 214, 289
Industrial Revolution................ 6
Influence Operations... 79, 83, 258, 269
Inkerman.......... 157, 279, 301, 308
insurgents.. 30, 45, 47, 52, 57, 59-62, 64, 66, 81-83, 85-91, 93, 94, 96-101, 104, 106-108, 114, 258, 268, 285
intelligence... viii, ix, 35, 44, 48, 53, 63, 66-76, 83, 97, 100, 104, 109, 117, 121, 123, 126, 133, 188, 189, 197, 227, 228, 251, 268, 290, 291, 307, 319, 320
Intelligence collection plan...... 68, 71
Intelligence Cycle................. 76
Intelligence Preparation of the Battlefield
.... 68, 72

Intelligence Processing Application... 73
Inter Service Intelligence, see ISI..... 44
International Joint Staff............ 56
International law...... 2, 7, 8, 51, 54, 252, 269
International Security Assistance Force
.... x, 6, 49, 268
interpretation................ 33, 110
IPA.... x, xiv, 5, 6, 8, 10, 11, 14, 17, 50, 54, 55, 62, 64, 65, 67, 72, 73, 76, 88, 89, 95, 99, 107, 109, 118, 120, 121, 124, 129, 132, 138, 141, 144, 145, 155, 156, 158, 176, 197, 202, 203, 206, 219, 220, 222, 226, 229, 237, 248, 251, 252, 262, 263, 266, 272, 278, 284, 290, 298, 302, 303, 306, 313, 315
Iraq... xii, 3-5, 10, 11, 13, 14, 21, 50, 55, 120, 167, 169, 170, 230, 241, 244, 253, 259, 269, 279, 281, 310, 312, 321
Iraqi Freedom..................... 3
Ireland... 5, 7, 8, 64, 146-148, 205, 241, 251, 259, 264, 272, 281, 294, 314-316, 319, 321
ISAF.... viii, x, xiv, 6, 22, 49-52, 54, 56-59, 61, 62, 65-68, 73-77, 83, 85, 89, 92-94, 98, 99, 103, 107, 108, 114-117, 122, 126, 129, 130, 133, 134, 227, 228, 247, 254, 258, 260, 262, 268, 280, 281, 288, 318, 319, 327
ISAF Team 10.... viii, x, 22, 49, 51, 52, 54, 57-59, 65-68, 73-75, 89, 107, 116, 117, 122, 126, 129, 130, 134, 227, 228, 258, 281, 318, 319
ISAF Team 9................. 77, 85
ISI, see Inter Service Intelligence.. 44-47
ISTAR........ 66, 70, 72, 73, 126, 327
Italian League... 142-144, 249, 252, 323
Jagirdar........................ 26
Jakobsen, Peter Viggo............ 119
James II, king of Great Britain.... 7, 145, 241, 251, 272, 294, 315
Japan... 12, 159, 164, 169, 183, 184, 187, 213, 253, 255, 309, 314
Jena....................... 151, 284
Jenkins, Roy. 20, 174, 181-184, 191, 192, 194, 201, 204, 207, 212, 213, 320
Jensen, Carsten................. 118

Joint Services Command and Staff College 63, 196
Jomini, Baron Antoine. 57, 250, 294
Jutland Dragoons. 52
Kabul. . 26-31, 34-37, 39, 41, 44, 45, 47, 49, 56, 169, 232, 264, 268, 274, 278, 280, 285-287, 290, 295, 300, 305-308, 311, 312, 319, 322
Kamal, Babrak. 43
Kandahar. . 10, 26-33, 35-37, 40, 47, 61, 62, 159, 285, 305, 308, 311
kandaks. 50, 83, 121
Karl, Emperor. . . 137, 139, 150, 293, 294, 299, 303, 308, 309, 322
Karzai, Hamid. 8, 21, 105, 132, 270, 275, 292, 295
Keane, John. 28
Keitel, Wilhelm. 203, 204, 294, 295
Kemal, Mustafa (Atatürk) 41
Kesselring, Albert. . . . 178, 188, 198, 203, 204, 263, 280, 295, 296
kevlar reinforced underwear. 101
Khaled, Prince. 234, 235, 296
Khalg. 43, 306
Khel, Ahmed. 36, 305
Khyber Pass. . . 26-28, 30, 34, 35, 39, 40, 42, 280, 307, 308
King, Admiral Ernest Joseph. . . 207, 296
King, William Lyon Mackenzie. 20, 212
Kitchener, Herbert. 159, 297
Klessheim. 187, 192
König, Malte. 321
Königsberg (Kaliningrad). 152
Kooperation als Machtkampf. . . xiii, 176, 178, 321
Korean War. . . . 214, 215, 217-219, 240, 248, 252
Kosovo. 242, 281, 302
Kurram Valley. 35, 40
Kuwait. 14, 136, 310, 312
Laden, Osama bin. 45, 46
Lakedaimon. 139
Lambert, Andrew. . . . . vii, 156, 323
Land Warfare Centre. 133
Landi Kotal. 40
language. . . . . xiv, 11, 15, 33, 38, 63, 89, 124, 142, 162, 168, 169, 196, 222, 227, 228, 230, 232, 234, 235, 244, 266, 272, 285, 299
Lansdowne. 37
Lashkar Gah. 50, 92, 100
Laurier, Wilfrid. . . 10, 12, 159, 160, 174, 297, 322
lead-nation. 9, 12
League of Nations. 2, 313
Leatherneck, Camp. 50, 112, 113
Leer, G.A. 238
Légion Étrangère. 172, 311
legitimacy. . ix, 3, 8, 9, 15, 103, 215, 220, 241, 244, 248, 252, 269, 271
Leipzig (Battle of the Nations). . 155, 223, 244, 276
levée en masse ordinance. 149, 279
liberal democracies. 3
Light Reconnaissance Company. . . 52, 81
Lisbon, NATO summit, November 2010 . . . . 21, 62, 103
Lisbon summit declaration. 103
Louis XIV, King of France 146, 148, 268, 272, 287, 294, 298, 315
Ludendorff, Erich. 164-166, 298
Luft, Gal. 3, 167, 169, 170, 222, 232, 234, 235, 237, 321
Lytton. 34, 35, 299
MacArthur, General Douglas. . . 208, 218
Machiavelli. 141, 142, 144, 299, 321
Mack, Karl Freiherr [Baron] von Leiberich 150, 299
Macnaghten, William Hay. . 30, 278, 299, 300, 305
Mahan, Alfred Thayer 236
Mahmud Shah. 26, 300, 312
Maiwand. 36, 121
Malta. . 13, 179, 185-188, 190, 211, 256, 259
Malvern East, Patrol Base. 87, 90
Malvern West, Patrol Base. 87, 90
Mangal. 61, 78, 106, 107, 300
Marlborough. 64, 180, 263, 265, 281, 287, 300, 316, 319, 320
Marshall, George 166, 183, 191-193, 195, 207-209, 212, 224, 255, 265, 291, 300
McChrystal, General Stanley. 266
McNaughton, Andrew George. . . 20, 211, 212, 301
Mechanised Brigade. 51, 66
Metternich, Klemens Wenzel. . . 237, 301
military instrument. 7, 146, 271
Military Stabilisation Support Team, see MSST. 72, 327

Milosevic, Slobodan.............. 242
Minimalgruppe Balkan............. 239
Ministry of Defence, Copenhagen..... 15, 77, 130
Ministry of Defence, London..... viii, xii, 62, 116, 130, 131, 197, 270
mission command....... 65, 79, 126, 264
Mission Secret............ 73, 122, 123
Mitchell, Paul T....... 2, 11, 18, 136, 229, 233, 235, 242, 245, 246, 321
Mitterrand, François............... 235
Mohammed, Daoud Khan.. 43, 291, 295, 302
Mohammed, Zahir Shah...... 42, 43, 302
Moltke, Helmuth von...... 124, 167, 239, 259, 264, 303, 311
Monclar, Raoul Charles Magrin-Vernerey 215, 303
Montgomery, General Bernard.... 19-21, 64, 68, 194, 197, 200, 201, 212, 213, 227, 251, 258, 261, 263, 265, 303, 321
Monty, see Montgomery. ... 20, 21, 197, 200-202, 213, 264, 320
MSST, see Military Stabilisation Support Team................ 72, 327
Mujahidin............ 43-45, 274, 292
Murat, Joachim............... 243, 303
Musa Qala..................... 36
Musharraf, Perves.......... 13, 45, 304
musketeer.................... 3, 225
Mussolini, Benito ... 175-178, 189, 190, 198, 203, 204, 256, 275, 282, 304
Napoleon I, Emperor of the French.... 6, 9, 11, 13, 16, 24, 27, 32, 35, 149-156, 158, 229, 232, 243, 244, 247, 249-251, 253, 268, 276, 280, 282, 284, 299, 301, 302, 304, 305, 311, 312, 315, 319, 321
Napoleon III...... 32, 156, 158, 249, 251, 282, 305, 311
Napoleonic era................ 55, 155
Napoleonic France....... 7, 19, 135, 158
Napoleonic Wars... 12, 25, 50, 149, 154, 156, 162, 226, 237, 262, 267, 286, 303, 304, 307, 315
NATO... 3, 4, 6, 8, 13, 14, 17, 21, 47, 55, 56, 61-63, 97, 117, 119, 123, 132, 133, 142, 216, 219, 221, 222, 231, 240-242, 244-246, 248, 254, 256, 264, 266, 272, 280, 286, 289, 302,

307, 327
Nazi Germany..... 16, 172, 174, 176, 179, 180, 182, 189, 203, 248, 256, 268, 282, 289, 293, 309
Network Centric Warfare.. 2, 11, 18, 136, 229, 233, 241-243, 245, 267, 321
Nicolls, Jasper................ 31, 305
Niedermayer, Oskar............... 169
Nielsen, Thomas Heine. .. vii, 118, 138, 323
Nine Years War............ 5, 281, 316
Normandy... 57, 196, 261, 283, 303, 309
North Africa..... 174, 175, 177-180, 183-185, 187, 188, 190, 191, 193, 195, 198, 206, 207, 209, 210, 212, 214, 224, 257, 259, 263, 276, 277, 286, 290, 292, 296, 306, 307, 309
Nott, William............ 30, 305, 307
Oberkommando der Wehrmacht..... 188, 294, 295
Oberste Heeresleitung......... 164, 240
OCCD (Operational Co-ordination Centre District)............. 109, 327
offensive á l'outrance............. 166
Ogdensburg Agreement........... 216
oil..... 14, 180, 190, 208, 209, 254, 310, 312
OKW see Oberkommando der Wehrmacht 188, 295
On War....................... 319
Operation Hercules... 179, 180, 187, 188, 190
Operation Kadesh................ 225
Opération Mousquetaire........... 225
Operation Musketeer............ 3, 225
ordre mixte.................... 149
Ottoman Empire.... 32, 39, 41, 142, 145, 149, 155, 157, 162, 167-169, 172, 239, 259, 287, 290, 312
Overlord (Operation)..... 174, 182, 212, 224, 255, 303
Oxus........................ 33, 34
Parcham................... 43, 295
Patrol Base Line... 57, 58, 61, 69, 73, 80, 81, 83, 85-88, 90, 91, 93, 96, 101, 123, 130
Patton, General George..... 21, 64, 306
PDAT....................... 83
Peacekeeping operations............ 4
Peiwar Kotal................... 35
Peninsular War................. 152

People's Republic of Korea. 213
Personal Role Radios, see PRR. 123
Peshawar. 23, 26-28, 30, 31, 34, 43, 285,
309, 311, 312
Pétain, Philippe. 164, 166, 177, 255,
306, 307
Petraeus, General David. . . 117, 132, 133,
266, 307
Piranha. 93
Pitt, William. 7, 16, 152, 307
Poland. . 14, 16, 151, 173, 182, 183, 194,
225, 256, 275, 283, 296, 310
Police Development and Advisory Team
. . . . 83
Pollock, Georg. . . . 30, 31, 305, 307, 308,
311
poppy planting. 96
Pressburg. 151
Price, Main Operating Base. . . 80, 82, 86,
93, 102, 104, 106, 116, 123
Priority Intelligence Requirements. . . . 70,
71
Protestant (or Evangelical) Union . . . 144,
145, 241, 251, 315, 316
PRR, see Personal Role Radios. 123
Prussia. 4, 7, 13, 124, 135, 149, 151,
152, 155, 163, 168, 234, 237-239,
243, 251, 262, 279, 292, 293, 303,
305, 310, 311, 315
psychological warfare. 222
Punjab. . 23, 26-28, 31, 32, 34, 286, 297,
298
Quetta. 31, 45, 132, 196, 197, 305
Quick Reaction Force. 99, 101, 105,
115, 327
Raglan, Baron Fitzroy. . . . 156, 157, 279,
308
Rahim, Patrol Base. . . . 61, 73, 81, 86, 93,
96, 103, 104, 107, 125
Ramadan. 66, 84, 85, 91, 111, 116
Ranjit Singh. 26, 28, 34, 312
Rasmussen, Lars Løkke. 117, 244
Reformation. 145
Religious and Moral Affairs. 114
Richelieu, Cardinal. 181
Risorgimento. 175
rivalries. 91, 131, 156
Roberts, Lord Frederick. 35, 36, 159,
297, 308
rocket propelled grenades. 87, 92
Rommel, General Erwin. . . 178, 186-188,
198, 204, 263, 309
Roosevelt, F.D. 173, 181, 183, 184,
201, 204, 207, 209, 210, 265, 291-
293, 300, 309
Royal Air Force Regiment. 114
Royal Arsenal Museum. 134
Royal Danish Defence College. . . . . vii, ix-
xi, xv, 63, 338
Royal Life Guards. . . ix, 51, 52, 288, 338
Royal Navy. 152, 168, 179, 185, 197,
204, 256, 296
RPGs, see rocket propelled grenades. . 87
Russia. . . 12-14, 16, 19, 22-28, 31-36, 49,
115, 149-159, 163, 164, 168, 174,
180, 182-184, 187, 189-191, 194,
203, 207, 210, 211, 213, 238-240,
242, 244, 249, 253, 262, 265, 274,
275, 278, 285, 287, 292-294, 297,
301, 310, 312, 313, 316
Ryswick. 147
Sale, Brigadier-General Sir Robert. . . . 30,
311
Sale, Lady Florentia. 31
Sanders, Otto Liman von. . 167, 169, 172,
287, 311, 312
Scanian War. 13, 146, 281
Scapa Flow. 204
School of Combined Operations. . . . 217
Schwartz, Adam. . . . . vii, 138, 323
Schwarzkopf, General Norman. 234, 312
Second Front. . . 174, 182-184, 194, 212,
292
Second Moscow Conference. . . 210, 266
Second Washington Conference. 195,
209, 257, 265
Second World War, see World War II
. . . . . xiii, xiv, 12, 180, 216, 217,
223, 224, 248, 250, 254, 257, 258,
265, 277, 282, 286, 293, 300, 313,
315
Security Council. . . . 2, 3, 8, 49, 213, 216,
242, 252, 268
Sepoy Rebellion, see Indian Mutiny. . 24,
32, 235, 236
Serritzlev, Jeanette. 54
Seven Years War. 24
SHAEF. 264
Sharia law. 48
Sher Ali. 33-36, 274, 280, 285, 312, 316
Shorabak, Camp. 112
Shuja ul Mulk. 26, 28

Sind. 24, 27, 28, 31, 275
Sirdar. 26
Smuts, Jan Christiaan. 250, 313
SOFA, see Status of Forces Agreement
. . . . 147
Soviet Union. . . . 8, 14, 16, 42, 172-174, 178, 186, 188, 189, 203, 215, 245, 246, 253, 254, 277, 291-293, 295, 313
Sparta. 137-140, 272, 314
Spin Baldak. 40
STABAD, see Stabilisation Advisor. . 72, 327
Stabilisation Advisor, see STABAD. . 72, 327
Stahlpakt. 176
Stalin. . 16, 180, 182-184, 210, 291-293, 303, 309, 313
standardisation. 219
State security. 268
Status of Forces Agreement. 147
Stilwell, Joseph Warren. 192, 314
Sublime Porte. 143, 156
Suez Campaign. 3
Suleiman Range. 34, 35
Supreme Headquarters Allied Expeditionary Forces. 264
Swat Valley. 32, 47
Tactical Operations Centre. 121, 124
Taliban. . 9, 17, 24, 29, 34, 41, 45-50, 52, 58, 59, 61, 75, 82, 85-87, 93, 96, 97, 101-104, 106, 107, 114, 116-119, 121, 123, 130, 132, 133, 258, 267, 270, 274, 295
Tannenberg. 164, 240, 292
Taraki, Nur Muhammed. . . . 43, 291, 306
Task Force Helmand. . . . . vii, xiii, 22, 49, 50, 52, 56, 57, 61, 63, 64, 69-74, 76-78, 83, 85, 91, 98, 102, 103, 111, 125, 126, 131-133, 225, 254, 257, 258, 264, 269, 280, 281, 318
technological compatibility. 17
terrorists. 7, 8, 107, 119
Thermopylae. 135
Thin Red Line. 157
Thirty Years War. 5, 6, 50, 145, 251, 272
Thucydides. 136-139, 314
Tilsit. 24, 152, 253
Tobruk. 189, 204, 209, 265
tolays. . . . . xiv, 50, 83, 96, 110, 116, 121

Torch (Operation). . . . 206, 207, 210, 286
Trafalgar. 150
Treaty of Gandamak. 35, 280, 316
Trident Conference. 191
Turkey. 3, 27, 31, 41, 156, 158, 167-169, 182
UK. ix, 10, 18, 21, 50-52, 55, 57, 58, 62-64, 66, 69, 72, 79, 91, 95, 101, 107, 116, 117, 119, 121, 125, 126, 133, 172, 196, 197, 204, 219, 225, 227, 229, 230, 244, 248, 254, 256, 257, 264, 270, 280, 281, 291, 293, 309, 314, 318, 328
Ulm. 149, 151, 299
United Nations. . . . 2, 4, 6, 14, 173, 214-216, 218, 219, 248, 252, 301, 313
United States. . 2, ix, 6, 9, 12-14, 18, 31, 47, 49, 118, 164, 173, 174, 180, 181, 183, 184, 189, 192, 213, 216, 218, 219, 227, 229, 241, 246, 248, 253, 254, 256, 269, 270, 275, 279, 281, 284-286, 290, 294, 296, 300, 306, 309, 313, 317
Upper Gereshk Valley. . . . 58, 61, 64, 70, 71, 74, 80, 81, 125
US Marine Corps. . 85, 86, 114, 123, 217
USMC, see US Marine Corps. . 114, 136, 324
Veritable. 196
Victoria, Queen of Great Britain, Empress of India 24, 284, 285, 314
Vienna (conference) 151, 194, 237, 249, 287, 295, 302, 316, 338
Vietnam War. 240
Viking Group. 53
Villaume, Poul. 117, 118
Vitkevich. 27
Vitoria. 135, 154, 315
Vogel, Thomas. . . . . vii, 179, 323
Wacht am Rhein (operation aka The Battle of the Bulge). 201, 258, 277
wahabi piety. 235
war aims. . . 3, 10, 14, 146, 169, 171, 174, 183, 234, 238, 241, 248, 249, 252, 253, 255, 257, 262, 268, 272
war by proxy. 48
War of Spanish Succession. 50, 146, 180, 225, 249, 263
Warsaw Pact. 9, 220, 267, 270
Warthog Group. 101, 104, 106

Waterloo.. 10, 12, 29, 155, 156, 159, 174, 234, 237, 262, 305, 315, 322
Wavell, Archibald Percival. ... 192, 315
Waziristan. 37, 40-42, 47
weapons of mass destruction. 4, 146
weapons technology.............. 219
Wehrmacht..... 176, 179, 186, 188, 194, 294, 295
Wehrmachtführungsamt........... 185
Weisung für die Kriegführung 1a, Fall Weiß.................. 176
Weitsman, Patricia.............. 9, 10
Wellesley, Arthur, see Wellington. ... 24, 153, 315
Wellington, Arthur, first duke of ... 9, 24, 31, 135, 153-155, 234, 247, 262, 315, 320
Western Front.... 16, 164-166, 191, 199, 239, 240, 242, 264, 281, 292, 296, 297
Westphalia. 6, 145, 252
WikiLeaks. 131, 233, 324
Wilde, Oscar 168
William III, king of Great Britain... 7, 10, 13, 146-148, 241, 264, 281, 294, 315
World War I.... x, 3, 11, 12, 37, 38, 126, 128, 141, 162-166, 168, 169, 171, 193, 238, 239, 242, 250, 255, 260, 272, 275, 281, 286-290, 298, 303, 306, 309, 310
World War II. ... 2, 3, 13, 14, 19, 20, 64, 141, 172-174, 184, 196, 204, 213, 214, 218, 219, 230, 236, 240, 253-255, 272, 275-277, 280, 281, 284, 289, 290, 293-296, 301, 303, 306, 307, 309, 310
Yakub Khan..... 35, 232, 280, 308, 316
Yalta.............. 194, 292, 293, 313
Yugoslavia. . 14, 182, 242, 254, 278, 302
Zumbalay. 60, 61, 82, 95, 97

About the Author

Dr Kjeld Hald Galster was educated at the Royal Military College of Canada, University of Copenhagen, and the Royal Danish Defence College. Commissioned into the Royal Life Guards in 1973, he served for a number of years in various staff and regimental positions in Denmark. As a young officer he served off and on with A Coy of the 5th Battalion (TA), The Queens Regiment (i.e. before, following the 1992 amalgamation with The Royal Hampshire Regiment, it became The Princess of Wales' Royal Regiment). As a lieutenant-colonel, for three years he was the senior military advisor to the Danish Ambassador to the OSCE in Vienna. He has lectured on Military History and the Theory of War at the University of Copenhagen, the Royal Danish Military Academy and the Royal Danish Defence College, and as a guest lecturer at the Royal Military College of Canada and Trinity College, Dublin.

Also available from Legacy Books Press

The Face of the Foe: Pitfalls and Perspectives of Military Intelligence
By Kjeld Hald Galster

ISBN: 978-0-9784652-6-1

Every nation that goes to war has to create images of their enemy. Through intelligence gathering and propaganda, these images are created and used to drive public support and keep soldiers fighting. At the same time, decision-makers must be provided with clear and incisive information on the opposition at hand. Frequently, these aims are mutually conflicting. Carefully balanced and used with circumspection, these images can lead to victory – but they can also drive armies to disaster and entire nations to atrocity.

In this sweeping and fascinating survey, Kjeld Hald Galster explores how intelligence is collected and interpreted. Drawing from examples ranging from the Napoleonic Wars to the 2003 War in Iraq, he examines how military intelligence is used to create the face of the foe – and what makes it a tremendous success...or a disastrous failure.

Bayonets and Blobsticks: The Canadian Experience of Close Combat 1915-1918
By Aaron Miedema

ISBN: 978-0-9784652-9-2

For a long time, it has been accepted that the bayonet was an inadequate weapon in World War I – an anachronism, relied upon by foolish generals eager to relive the glories of the Napoleonic Wars while incapable of coming to terms with the modern battlefield and trench warfare. But was this the reality of the Western Front of the Great War, or a myth perpetuated by historians?

In reality, the soldiers of World War I seemed oblivious to what appears so obvious to critics ninety years removed. They quite liked their bayonets, and they used them – often.

In this fascinating and provocative study, Aaron Taylor Miedema takes a new look at the role of the bayonet and shock tactics on the Western front. Through the experience of the Canadian Corps – the British shock troops of the Western Front – he challenges the conventional view of the bayonet as an obsolete weapon system and rekindles the controversial debate over technologies, old and new, on the field of battle.

www.ingramcontent.com/pod-product-compliance
Lightning Source LLC
Chambersburg PA
CBHW071108160426
43196CB00013B/2509